M-COMMERCE

Experiencing the Phygital Retail

M-COMMERCE

Experiencing the Phygital Retail

Edited by

Punita Duhan, MBA

Meera Bai Institute of Technology

Anurag Singh, PhD

Banaras Hindu University

AAP | APPLE ACADEMIC PRESS

Apple Academic Press Inc.
3333 Mistwell Crescent
Oakville, ON L6L 0A2 Canada

Apple Academic Press Inc.
1265 Goldenrod Circle NE
Palm Bay, Florida 32905 USA

Exclusive worldwide distribution by CRC Press, a member of Taylor & Francis Group
International Standard Book Number-13: 978-1-77188-714-4 (Hardcover)
International Standard Book Number-13: 978-0-42948-773-6 (eBook)

Library and Archives Canada Cataloguing in Publication

M-commerce (Oakville, Ont.)
M-commerce : experiencing the phygital retail / edited by Punita Duhan (MBA, Meera Bai Institute of Technology), Anurag Singh, PhD (Banaras Hindu University).
Includes bibliographical references and index.
Issued in print and electronic formats.
ISBN 978-1-77188-714-4 (hardcover).--ISBN 978-0-42948-773-6 (PDF)
1. Mobile commerce. 2. Retail trade--Technological innovations.
I. Duhan, Punita, 1973-, editor II. Singh, Anurag, 1974-, editor III. Title.
HF5548.34.M46 2018 658.8'72 C2018-905756-4 C2018-905757-2

Library of Congress Cataloging-in-Publication Data

Names: Duhan, Punita, 1973- editor. | Singh, Anurag, 1974- editor.
Title: M-commerce : experiencing the phygital retail / editors, Punita Duhan, MBA, Anurag Singh, PhD.
Description: Toronto; New Jersey : Apple Academic Press, 2019. | Includes bibliographical references and index.
Identifiers: LCCN 2018047350 (print) | LCCN 2018049109 (ebook) | ISBN 9780429487736 (ebook) | ISBN 9781771887144 (hardcover : alk. paper)
Subjects: LCSH: Mobile commerce. | Electronic commerce.
Classification: LCC HF5548.34 (ebook) | LCC HF5548.34 .M23 2019 (print) | DDC 381/.142--dc23
LC record available at https://lccn.loc.gov/2018047350

Apple Academic Press also publishes its books in a variety of electronic formats. Some content that appears in print may not be available in electronic format. For information about Apple Academic Press products, visit our website at **www.appleacademicpress.com** and the CRC Press website at **www.crcpress.com**

ABOUT THE EDITORS

Punita Duhan, MBA

Punita Duhan is a faculty member in Business Administration with the Department of Training and Technical Education, Government of National Capital Territory of Delhi, India, for the last 18 years and is currently pursuing research in social media from the Institute of Management Studies, Banaras Hindu University, Varanasi, India. Her research interests include consumer behavior, behavioral marketing, social and digital technologies, green marketing, digital marketing, e-commerce, and m-commerce. Prior to this, she has worked in the Faculty of Management with University Business School, Punjab University, Chandigarh and Institute of Management and Technology, Faridabad, India. She has attended more than 35 national as well as international seminars/conferences and has presented papers at more than 25 national and international seminars and conferences. She has published more than 12 research papers with prestigious national and international publishers such as Sage, Palgrave, Springer, IGI Global, Bloomsbury, etc. She has published two edited books, titled *Managing Public Relations and Brand Image through Social Media* and *Radical Reorganization of Existing Work Structures through Digitalization* with IGI Global. Another book titled *Green Consumerism: Perspectives, Sustainability, and Behavior* is in press with Apple Academic Press. She is also on editorial boards of several peer-reviewed prestigious journals. In addition, she has been actively involved in curriculum development and e-content development at various universities, National Institute of Open Schooling, Central Board of Secondary Education, and technical education departments of Delhi and Haryana state in the management area. She has attended more than 30 training programs and workshops and has organized around 10 seminars/conferences/workshops.

Anurag Singh, PhD

Dr. Anurag Singh has extensive experience of more than 17 years in teaching, research, and training. He hold as MBA from the Department of Business Administration, Bundelkhand University, Jhansi, India, and is specialized in marketing and human resource management. He obtained his PhD in the marketing area from the Faculty of Management Studies, Gurukul Kangri

University, Haridwar, Uttarakhand, India. Dr. Singh has published more than 50 research papers and chapters in refereed journals and edited books. Apart from organizing the Faculty Development Program on Time Series Analysis, the Workshop on SEM, the National Conference on Emerging Business Practices, and the International Conference on Marketing, Dr. Singh has chaired a number of technical sessions in national and international conferences and seminars. In addition to his active involvement as a trainer to MSME, PVVNL, and Indian Oil, he is regularly invited as a soft skill trainer by various management institutions. Dr. Singh has completed a minor research project funded by Banaras Hindu University and has also edited four books. Dr. Singh has contributed as a National Advisory Board member at several workshops and conferences organized in the field of commerce, management, library science, and technology. He is serving on the editorial boards and as a reviewer of many national and international journals. Prior to taking his present assignment, Dr. Singh has worked with the Faculty of Management Studies, Gurukul Kangri University, Haridwar, Uttarakhand, India, and the Department of Business Administration, Assam University, Silchar, Assam, India. Presently, he is contributing as an Associate Professor at the Institute of Management Studies, Banaras Hindu University, Uttar Pradesh, India, and is guiding doctoral research and master's dissertation students. He has special interests in contemporary issues such as behavioral marketing, digital marketing, and E-HRM.

CONTENTS

To our parents for their support in our endeavors

Ranbir Singh and Sumitra Devi

—Punita Duhan

Shyam Babu Singh and Mahesh Kumari

—Anurag Singh

LIST OF CONTRIBUTORS

Shubhi Agarwal, MBA*
Research Scholar, Department of Business Administration, Faculty of Commerce, University of Lucknow, Lucknow, Uttar Pradesh, India

Anu G. Aggarwal, PhD*
Professor, Department of Operational Research, University of Delhi, Delhi 110007, India

Shivani Arora, PhD
Assistant Professor, Department of Commerce, Shaheed Bhagat Singh College, Delhi University, Delhi, India

Kulvinder Kaur Batth, ICWA, MCom*
Assistant Professor, Department of Commerce, K. C. College, University of Mumbai, Mumbai, India

Arindam Chakrabarty, MBA, MA (Public Administration)*
Assistant Professor, Department of Management, Rajiv Gandhi University (Central University), Rono Hills, Doimukh 791112, Itanagar, Arunachal Pradesh, India

Biplab Datta, PhD
Associate Professor, Vinod Gupta School of Management, Indian Institute of Technology, Kharagpur 721302, India

Punita Duhan, MBA
Lecturer, Business Administration, Department of Training and Technical Education, Meera Bai Institute of Technology, Government of Delhi, Delhi, India

Om Jee Gupta, MBA (reviewer only)
Research Scholar, Institute of Management Studies, Banaras Hindu University, Varanasi, Uttar Pradesh, India

P. Vigneswara Ilavarasan, PhD
Associate Professor, Department of Management Studies, Indian Institute of Technology, Delhi, India

Amandeep Kaur, MBA*
Assistant Professor, Department of Management Studies, Rukmini Devi Institute of Advanced Studies, New Delhi, India

Dilpreet Kaur, MBA
Department of Management Studies, Indian Institute of Technology Delhi, New Delhi, India

Anshu Lochab, PhD*
Dean Academics, Rukmini Devi Institute of Advanced Studies, New Delhi, India

S. Meera, PhD*
Assistant Professor (International Business), Indian Institute of Tourism and Travel Management (An Autonomous Organization Under Ministry of Tourism), Nellore 524321, Andhra Pradesh, India

*Indicates contributors who also acted as reviewers.

Yog Mishra, MBA
Research Scholar, Institute of Management Studies, Banaras Hindu University, Varanasi, Uttar Pradesh, India

Ruhee Mittal, PhD*
Associate Professor, Department of Management Studies, Rukmini Devi Institute of Advanced Studies, New Delhi, India

Madhusudan N. Pandya, PhD
Assistant Professor, Department of Commerce and Business Management, Faculty of Commerce, The Maharaja Sayajirao University of Baroda, Vadodara, Gujarat, India

Devika Prakash*
MBA Student, Department of Management, Waljat College of Applied Sciences, BIT International Centre Muscat, Oman

Vir Ved Ratna, PhD*
Associate Professor, Strategy and General Management Area, Jaipuria Institute of Management, Lucknow, Uttar Pradesh, India

Henrique Schneider, PhD*
Chief Economist, Swiss Federation of Small and Medium Enterprises, Berne 3001, Switzerland
Professor of Economics, Nordakademie Hochschule der Wirtschaft, Elmshorn and Hamburg, Germany

Musarrat Shaheen, PhD
Assistant Professor, HR Department, IBS Hyderabad, a Constituent of ICFAI Foundation for Higher Education, India

Amit Shankar, MBA*
Research Scholar, Vinod Gupta School of Management, Indian Institute of Technology, Kharagpur 721302, India

Himanshu Sharma, MPhil*
Research Scholar, Department of Operational Research, University of Delhi, Delhi 110007, India

Nidhi Sharma, MBA*
Assistant Professor, Department of Management Studies, Rukmini Devi Institute of Advanced Studies, New Delhi, India

Shiv Kumar Sharma, PhD*
Professor, Department of Management, Faculty of Social Sciences, Dayalbagh Educational Institute (Deemed University), Agra, Uttar Pradesh, India

Mithun Shrivastava, MBA*
Assistant Professor, Department of Management, Waljat College of Applied Sciences, BIT International Centre Muscat, Oman

Anuja Shukla, MPhil*
Assistant Professor, Noida International University, Greater Noida, Gautam Budh Nagar, Uttar Pradesh, India

Parag Sunil Shukla, PhD, MCom (Marketing), PGDMM*
Assistant Professor, Department of Commerce and Business Management, Faculty of Commerce, The Maharaja Sayajirao University of Baroda, Vadodara, Gujarat, India

Anurag Singh, PhD
Associate Professor, Institute of Management Studies, Banaras Hindu University, Varanasi, Uttar Pradesh, India

Gursimranjit Singh, MBA*
Junior Research Fellow, Department of Management Studies, Amritsar College of Engineering and Technology, I. K. Gujral Punjab Technical University, Amritsar 143001, Punjab, India

Krishan Pal Singh, MBA (reviewer only)
Research Scholar, Institute of Management Studies, Banaras Hindu University, Varanasi, Uttar Pradesh, India

Maninder Singh, PhD*
Professor and Head, Department of Management Studies, Amritsar College of Engineering and Technology, Amritsar, Punjab, India

Nidhi Singh, MBA, SRF (NET)*
Department of Business Administration, Faculty of Commerce, University of Lucknow, Lucknow, Uttar Pradesh, India

Priyanka Singh, PhD*
Academic Associate (Marketing), Indian Institute of Management Udaipur, Balicha, Udaipur 313001, Rajasthan, India

Gunmala Suri, PhD*
Professor, University Business School, Punjab University, Chandigarh, India

Padmini Tomer, PhD*
Assistant Professor, Commerce Department, University College, Ghanaur, Patiala, Punjab, India

Anushruti Vagrani, MBA*
PhD Research Scholar, Department of Management Studies, Indian Institute of Technology, Delhi, India

Anukriti Verma*
BCom (III Year) Student, Department of Commerce, Shaheed Bhagat Singh College, Delhi University, Delhi, India

A. Vinodan, PhD*
Assistant Professor (Tourism), Indian Institute of Tourism and Travel Management (An Autonomous Organization Under Ministry of Tourism), Nellore 524321, Andhra Pradesh, India

Parimal Hariom Vyas, PhD
Vice-Chancellor, The Maharaja Sayajirao University of Baroda, Vadodara, Gujarat, India
Professor, Department of Commerce and Business Management, Faculty of Commerce, The Maharaja Sayajirao University of Baroda, Vadodara, Gujarat, India
Joint Professor, Faculty of Management Studies, The Maharaja Sayajirao University of Baroda, Vadodara, Gujarat, India

Rambalak Yadav, PhD
Assistant Professor, Department of Marketing, Institute of Management Technology, Hyderabad, India

Farrah Zeba, PhD
Assistant Professor, Marketing and Strategy Department, IBS Hyderabad, a Constituent of ICFAI Foundation for Higher Education, India

LIST OF ABBREVIATIONS

AHP	analytical hierarchy process
ANOVA	analysis of variance
apps	applications
AVE	average variance extracted
BI	behavioral intention
BoP	bottom of the pyramid
CFA	confirmatory factor analysis
CI	consistency index
CoD	cash-on-delivery
CR	consistency ratio
CR	critical ratio
CSI	country-specific issues
D&M model	DeLone and McLean model
e-commerce	electronic commerce
e-WoM	e-word of mouth
EDI	electronic data interchange
EE	effort expectancy
EEI	perceived effort issues
EFA	exploratory factor analysis
EM	expectation maximization method
FAHP	fuzzy AHP
FCI	facilitation issues
FMA	fashion mobile applications
GDP	gross domestic product
GPS	global positioning system
HCE	host card emulation
ICT	information and communication technology
IDT	innovation diffusion theory
IRF	India Retail Forum
ISP	internet service providers
IT	information technology
KMO	Kaiser–Meyer–Olkin
LI	lecturers' influence
m-commerce	mobile commerce

m-CSQ	mobile commerce service quality
m-shopping	mobile shopping
m-WoM	mobile-word of mouth
MC	mobile commerce
MCDM	multicriteria decision-making
MNC	multinational corporation
NASSCOM	National Association of Software and Service Companies
NFC	near-field communication
PC	personal computer
PCI	personal centric factors
PDA	personnel digital assistant
PE	performance expectancy
PEI	performance expectation issues
PI	personal innovativeness
POS	point of sale
PROMETHEE	Preference Ranking Organization METHod for Evaluations Enrichment
QoS	quality of service
RBI	Reserve Bank of India
RFID	radio-frequency identification
SDT	self-determination theory
SEM	structural equation model
SII	social influencing issues
SMS	smart messaging services
TAM	technology acceptance model
TDMA	time division multiple access
TPB	theory of planned behavior
TRA	theory of reasoned action
TSI	technology-specific issues
UTAUT	unified theory of acceptance and use of technology
VIF	variance inflation factor
VSNL	Videsh Sanchar Nigam Limited
WAP	Wireless Application Protocol
WOM	word of mouth

FOREWORD

Mobile is the buzzword for the last few years. It has invaded and touched our personal lives to a great extent and everyone around us is talking about mobile apps and using mobile for carrying out variety of activities. M-commerce, which was initially considered as an extension of e-commerce, is in essence very different from e-commerce. It has opened doors to new technologies, new services, new business models, new applications, and new possibilities in the marketplace. In addition, constraints such as interoperability, usability, security, and privacy need to be looked upon with a renewed focus.

Though the term m-commerce brings forth a picture of buying and selling activities taking place over internet-enabled mobile devices, its scope has transcended these activities. Uses are no longer confined only to purchases via mobile web and apps, to mobile payments, to mobile financial services, and to m-banking, but the possibilities of creating cashless economies and financial inclusion of marginalized sections using mobile technologies are being explored. Customers are hooked into their mobile experiences, and mobile traffic is on rise. Governments are pondering the enactment of legal and policy frameworks related to m-commerce. Seeing the staggering growth of m-commerce worldwide, it is natural for organizations to go mobile in the future.

Accordingly, this volume is an arduous attempt by the editors to investigate the realms of m-commerce in a multifaceted manner. This book enables the reader to comprehend contemporary m-commerce concepts along with the emerging paradigms in a comprehensive manner. The volume systematically puts together the efforts of researchers across the world in the shape of empirical analyses and reviews of the myriad aspects of m-commerce. Overall, the book meets the needs of the researchers, academicians, professionals, and students working in the domain of m-commerce, mobile banking, digital payments, mobile apps, and digital marketing. It will prove to be an excellent reference resource and a handbook for detailed and up-to date knowledge about m-commerce. Editors Punita Duhan and Dr. Anurag Singh deserve to be congratulated for this sincere academic endeavor to cover various facets of m-commerce in a comprehensive manner.

—M. Jagadesh Kumar,
Professor and Vice-Chancellor,
Jawaharlal Nehru University, New Delhi, India

PREFACE

The emergence and evolution of information technologies paved way for the online transactions, termed as "e-commerce." E-commerce was soon rechristened as wired e-commerce due to the rapid rise of mobile phones and other handheld devices facilitating access to the internet while on the go. This led to the coining of the term "m-commerce," the wireless avatar of e-commerce. The term mobile commerce, or simply m-commerce, has been attributed to the collection of location-based commercial services that are delivered by various internet-enabled handheld devices such as mobile phones, tablets, and palmtop devices. Though, essentially, both e-commerce and m-commerce are similar as the transactions are electronic in nature and are facilitated by the internet the element of mobility is one big differentiator between the two. Now, it has become difficult for the people to imagine e-commerce without the mobility dimension appended to it. In today's online business environment, m-commerce is growing as the next stage of e-commerce.

Since its induction in late 2000s, m-commerce has come a long way and is thriving and expanding at an unparalleled rate. Growth of m-commerce is being likened as quickly as weeds, as mobile has generated over 20% of sales in various retail markets in the first quarter of 2015. Such stupendous growth of m-commerce can be majorly attributed to smartphones. Interestingly, though the mobile technologies are largely being developed in the United States and other developed nations, mobile communications access and usage of mobile applications is used almost solely across underdeveloped and developing countries of world, which are characterized by low per capita income and lower computerized internet access.

Further, m-commerce has opened up new vistas for marketing products, for targeting the customers at multiple points, for offering customized services using location-based features, and for enabling enhanced shopping and service experiences. It also offers opportunities for services that are more efficient and more user-friendly. Marketers may also offer new approaches, apps, and in-store solutions for enhanced customer engagement after researching the needs, roles, and use contexts of customers. Marketers have already started talking about the mobile conversion funnel over the desktop conversion funnel. It is quite interesting to note that in mature markets, the shape of the mobile funnel is different and has higher purchase conversion rate.

Applications of mobile now range from money transfer to ticketing, vouchers, coupons and loyalty cards, content purchase and delivery, location-based services, information services, mobile banking, mobile brokerage, auctions, mobile browsing, mobile purchase, in-application mobile phone payments, mobile marketing, and advertising. Everyday, new technologies are being invented and newer applications of mobiles/smartphones are being added to this ever-growing list of applications.

Marketers are posed with new challenges and daunting tasks such as improving mobile websites and apps, better user interface designs, engaging and entertaining browsing experiences, and seamless and safe transactions. As the multiple and diverse devices are being used by customers to browse websites, marketers need to research and understand cross-device behavior for providing better and uninterrupted experiences to customers. Issues such as interoperability, enhanced usability, heightened security and privacy, fit between task and technology and its organizational viability, and ROI also need to be addressed.

The editors' objective, while proposing this book, was to consolidate the contemporary academic and business research. The editors have taken painstaking efforts to present up-to-date research related to m-commerce to various stakeholders in a comprehensive manner and to give them some ideas for future research avenues. Some of the broad areas this book caters to are analysis of preconditions and drivers leading to such stupendous growth of m-commerce, various facets and dimensions of m-commerce, security issues and threats associated with m-commerce, perception of consumers toward m-commerce, m-tourism, mobile learning, m-word-of-mouth, and other ongoing developments in this rapidly expanding area.

The book has been organized in the following four sections:

PART I: M-Commerce: Meaning, Evolution, Trends, and Hindrances
PART II: M-Commerce: Harbinger of "Phygital" Retail
PART III: Facets of M-Commerce and Consumer Perceptions
PART IV: M-Commerce: Miscellany

Each section caters to a distinct dimension of m-commerce. A total of 17 chapters are interspersed in the four sections. A brief description of each of the chapter follows.

The first chapter is an apt opener for the book and takes the readers on a comprehensive journey of the evolution of m-commerce. The authors extensively reviewed the extant literature on m-commerce and brought forth the various definitions propounded by researchers over the last two decades. The chapter further elaborates on the characteristics, features, and applications of m-commerce. Evolution and history of m-commerce have been explained lucidly focusing on the eras, namely, m-portals era, m-internet era, and m-apps era.

The second chapter compels the reader to think whether the mobile commerce is an innovation or just an entrepreneurial alertness. The author stresses that m-commerce is largely the result of recognition of competitive imperfections in market processes, and this has resulted due to entrepreneurial alertness and the customers' alertness. The author concludes that the very fact that mobile commerce is an amalgamation of undetermined, uncoordinated, and unplanned steps of "alertness" is also the prime factor contributing to its growth.

Chapter 3 is a meticulous attempt of the author to explain the evolution, trends, applications, tools and technologies, and SWOT analysis of mobile commerce in India. Building on the unique features of m-commerce carefully gleaned from the literature review, the author has compared and contrasted m-commerce and e-commerce. The chapter puts forward a business model for m-commerce that is affected by mobility, connectivity, and monopolies, and has technology, services, network, regulation, and users as its various stakeholders or participants in model.

Chapter 4 proposes a model based on the various factors, namely, system quality, content quality, use, trust, support, mobility, personalization, and electronic word-of-mouth, to evaluate the success of mobile commerce. Select mobile commerce applications have been ranked using multi criteria decision making (MCDM) techniques and Preference Ranking Order METHod for Enrichments Evaluation (PROMETHEE). Model indicated that customer feedbacks in the form of reviews and customization of products and services are important factors for new age customers.

Chapter 5 focuses on the forces propelling the phenomenon called m-commerce. The author advocates that forces such as simplicity of operating mobile devices, global markets, technological advancements, heightened competition, value addition due to apps and added features on mobiles, enhanced safety and security features, etc., are the major drivers of m-commerce.

Chapter 6 explains the barriers in proliferation of m-commerce in India with specific reference to demographics. The authors bring home the point that m-commerce, despite being such a rage throughout the world, faces various encumbrances that need to be looked into. The authors have taken a novel approach to point out the hindrances in the adoption of m-commerce.

Chapter 7 provides insight into the issues and challenges that have emerged due to the amalgamation of the physical and digital worlds. The result of this amalgamation is a phygital strategy that needs to be evaluated urgently as m-commerce is being widely proclaimed as the "new service frontier" with a capacity to enhance customer experience and to alter the traditional marketing mix in the time to come.

Chapter 8 uses the empirical analysis to bring home the point that m-commerce is the next frontier for shopping and a harbinger of the "phygital" retail in India. The authors conclude that there would be coexistence of physical and digital retail as comparative shopping, Customization, boutique websites, convenience and personalization, and go-digital initiatives are catching up.

Chapter 9 explores the possibilities of m-tourism in India. Researchers studied seven dimensions (along with 35 deterring variables), namely, performance expectation factor, perceived effort factor, socially influencing factors, facilitation factor, technology specific factor, personal centric factor, and country-specific factor to identify what is dissuading travelers from adopting m-tourism in India.

Chapter 10 proposes two sets of factors, namely, personal motivators and technical motivators to identify what affects adoption or usage activities with respect to m-commerce among Indian women so as to help the marketers to have better strategies. Perceived ease of use and perceived purchasing power among women significantly influence the m-commerce adoption, along with technical factors such as internet connectivity, discounts and personalized shopping offers, and security.

In Chapter 11, researchers have explored the possibility of mobile learning in the Indian higher education system. Using the unified theory of acceptance and use of technology (UTAUT), researchers identified performance expectancy, effort expectancy, lecturers' influence, quality of services, and personal innovativeness as significant factors affecting students' adoption of mobile learning. Alignment and modification of the curriculum, attitude and fear of change, excessive usage, and reliance of students on mobile devices have been cited as major impediments.

In Chapter 12, researchers have attempted to investigate the actual experience, satisfaction, and its influence on the buying decisions of m-commerce users in Vadodara, Gujarat, India so as to help marketers realign and reconfigure strategies so as to offer seamless customer experience. The authors conclude that quality of information, comfort, and ease in browsing online catalogues, and usefulness of mobile applications strongly influence user trust and satisfaction.

Behavior of customers toward grocery shopping mobile apps, such as Grofers, PepperTap, Amazon, etc., has been analyzed in Chapter 13. The authors found that perceived ease of use, perceived usefulness, and attitude affect purchase intentions positively, though variations were observed with respect to age, gender, and location. Proximity to market, quality concerns, inconvenience, better service by *kirana* shops (small neighorhood retail store in India), nonavailability of app, etc., are major factors leading to nonusage of apps.

Chapter 14 provides a framework of a mobile commerce service quality (m-CSQ) scale to guide practitioners and scholars through a comprehensive and systematic literature review of various articles related to m-CSQ. The authors identified privacy and security, content, responsiveness, efficiency, reliability, ease of use and usefulness as consistent determinants of m-CSQ measurement scales through critical analysis of scales used in various mobile service areas.

In Chapter 15, the author explores the role that mobile phones and hence m-commerce can play in the economic development of India. The author argues that successive state initiatives such as developing information and communication technology infrastructure, strengthening digital movements, and a developing cashless ecosystem for the inclusion of rural and marginalized populations, can be given greater impetus by making telecom network an integral part of the basic infrastructure ecosystem.

The authors have tried to understand the role of mobile payments and transactions for the bottom-of-pyramid (BoP) in Chapter 16. The authors, while enumerating various initiatives undertaken by various governments and private players for proliferation of mobile transactions, conclude that m-commerce can be a game changer for organizations in tapping BoP markets.

Due to extensive usage of mobiles by consumers, mobility dimension has been appended to the word-of-mouth (WoM) concept. Chapter 17 attempts to throw light on the conceptual issues related to mobile word-of-mouth in relation to consumer behavior on m-commerce platforms. The chapter also studies various determinants of mobile WoM and highlights the change in users' perceived experience and behavior due to emergence of mobile-word-of-mouth.

In nutshell, this book is an attempt to comprehend the contemporary m-commerce concepts along with the emerging paradigms in a pragmatic way. Endeavors have been made to systematically put together the efforts of researchers across the world in the shape of empirical analyses and reviews on the myriad aspects of m-commerce. The book delivers a complete and comprehensive idea of m-commerce to its readers and the editors are sure the book will compel its readers to think, imagine, and innovate beyond the evidence. Overall, the editors have tried to ensure that the book meets the requirements of marketing practitioners, academicians, researchers, and students of e-commerce, m-commerce, and digital marketing.

Punita Duhan
Department of Training & Technical Education, Meera Bai Institute of Technology, New Delhi, India

Anurag Singh
Institute of Management Studies, Banaras Hindu University, Varanasi, Uttar Pradesh, India

INTRODUCTION

The term mobile commerce, or simply m-commerce, has been attributed to the collection of location-based commercial services that are delivered by various internet-enabled handheld devices such as mobile phones, tablets, and palmtop devices. M-commerce, though it emerged as an off-shoot of e-commerce and is considered the next wave of e-commerce, has become a phenomenon and a force to reckon with in itself. This is largely due to proliferation of mobile phones with smart features, such as smartphones; innovations in telephony technologies; emergence of 3G, 4G, and 5G networks; initiatives of governments to spread digital to every nook and corner of their countries; availability of cheaper network plans; and a number of telecom players. In addition, lower infrastructure investments, lower costs, and greater use of mobile services, ease of access of mobile networks in difficult terrain and geographies, scalability of mobile networks, ease of procuring of mobiles, shorter payback periods, ease of sharing of mobile handsets, and the possibility of micropayments have contributed to such astonishing success of m-commerce globally.

Diminishing differences between average order value of mobile and desktop attests the fact that mobile has become the more engaging and preferred internet-browsing medium, with growing desktop being used only during office hours. Mobile share in e-commerce in mature and developed markets, such as Japan, South Korea, and the United Kingdom, is close to or above 50%. Global average growth of mobile transactions was 34% in first quarter of 2015, with a forecast of 40% growth by the end of 2015, thereby reaffirming the status of m-commerce as an unstoppable and reckonable phenomenon. Added features, such as bigger screens are accelerating the trend of making smartphones the dominant mobile channel vis-à-vis other mobile channels, such as tablets and palmtops, and no competing alternative appears to be in the vicinity of this compelling medium.

The rapid rise of m-commerce has led to the emergence of new paradigms in the marketplace. Rampant use of internet-enabled mobile devices has titled power scales in favor of customers. Customers are ever connected to the organizations, other customers, and competitors. Even while shopping offline, they are looking and comparing the products, services, and prices online. The difference between physical and digital retail is diminishing, and a new "phygital retail" is on the rise. Information is power and the customer now carries copious amounts of it in his pocket. Customers' scale and speed of acquiring information has increased manifold. Marketers need to sit back and take a note of this emerging paradigm. New opportunities as well as threats are emerging in the marketplace.

The book is an attempt to present a pragmatic approach for comprehending the contemporary m-commerce concepts along with the emerging paradigms. The book encapsulates the evolution, emergent trends, hindrances, customer perceptions about various facets of m-commerce, and embryonic applications in the arena of m-commerce quite comprehensively. It is a painstaking endeavor to systematically put together the efforts of researchers across the world in the shape of empirical analyses and reviews of the myriad aspects of m-commerce. The book delivers a complete and comprehensive idea of m-commerce to its reader and may compel him/her to think, imagine, and innovate beyond the evident.

Punita Duhan, MBA
Department of Training & Technical Education,
Meera Bai Institute of Technology, New Delhi, India

Anurag Singh, PhD
Institute of Management Studies, Banaras Hindu University,
Varanasi, Uttar Pradesh, India

PART I

M-COMMERCE: MEANING, EVOLUTION, TRENDS, AND HINDRANCES

CHAPTER 1

M-COMMERCE: MEANING, EVOLUTION, AND GROWTH

MITHUN SHRIVASTAVA[1*], DEVIKA PRAKASH[1], and VIR VED RATNA[2]

[1]*Department of Management, Waljat College of Applied Sciences, BIT International Centre Muscat, Oman*

[2]*Strategy and General Management Area, Jaipuria Institute of Management, Lucknow, Uttar Pradesh, India*

[*]*Corresponding author. E-mail: mithunshrivastava@gmail.com*

ABSTRACT

The term "m-commerce" comes from the union of two words "mobile" and "commerce," and refers to any transaction of monetary value occurring over a wireless communication network through the use of wireless handheld devices. M-commerce was initially hampered by issues such as slow Internet speeds, lack of standards in hardware and software, and limited capabilities of handheld devices. However, rapid technological development in various domains such as Internet, wireless communication, handheld devices, mobile payment systems, and mobile applications has ensured growing use and adoption of m-commerce today. Widespread acceptance of m-commerce is attributed to its various features such as ubiquity, convenience, localization, personalization, and identifiability. Nevertheless, m-commerce suffers from disadvantages such as the cost of setting up wireless infrastructure and restricted functionality of handheld devices. Furthermore, major challenges faced by m-commerce also include security risks involved in online mobile payment systems and lack of consumer trust in the same. Today, m-commerce applications used widely include mobile banking, mobile purchasing, location-based services, mobile ticketing, and information services. This chapter provides an introduction to the concept of m-commerce. Its meaning is explored through a comparison of definitions published over the last two decades. The evolution and growth of m-commerce is traced parallel to advances in the domains of hardware, software, and networking. Moreover, the advantages, disadvantages, and challenges of m-commerce are described. The applications of m-commerce today in various countries such as Japan, Finland, the United States, the United Kingdom, and India are also outlined. In today's dynamic world of rapid development, the future of m-commerce is quite difficult to speculate. However, based on the trends related to m-commerce growth as discussed in the chapter, m-commerce can be expected to grow and impact and get impacted by human lives.

1.1 INTRODUCTION

Internet has ushered in a new business era. Though initially developed to share scientific documents, the possibilities for business by communicating latest product information to customers were realized in e-commerce (Herzog and Gottlob, 2001). The rapid developments in telecommunications industry, especially with reference to the proliferation of various handheld devices, such as smartphone, have paved way for a significant increase in the reach of e-commerce, and have crafted an era marked by strong presence of mobile commerce.

The term "m-commerce" comes from the union of two words "mobile" and "commerce." It was first used in 1997 by Kevin Duffey, and refers to any transaction of monetary value taking place over a wireless mobile telecommunications network. It is considered to be an extension of e-commerce, whereby handheld devices enable users to access Internet-related services such as purchasing, selling, and searching for information, without being tethered to a wired desktop. M-commerce is an integral part of contemporary transactions, providing various services such as mobile banking, location-based information (news, weather, traffic reports, etc.), mobile ticketing, and mobile browsing, to name a few. This chapter deals with the meaning of m-commerce, its evolution, and growth over the years.

1.2 PURPOSE

M-commerce has been spearheading developments in multitudinous domains related to business. The most significant trend associated with m-commerce is its rapid growth especially during the past two decades, due to availability of high-speed Internet and affordable smartphones. This chapter seeks to explore the meaning of m-commerce as defined by numerous authors over the years. It then attempts to trace its origin and evolution parallel to that of communication technologies, especially Internet, and developments in handheld devices. The salient aspects, related to historical timeline for m-commerce evolution and growth, are highlighted. The chapter then focuses on the features of m-commerce, followed by disadvantages and challenges faced by it. Finally, thc chapter ends with a discussion on a multitude of applications offered by m-commerce today in different sectors such as banking, education, providing access to information, location-based services, and mobile purchasing.

This chapter seeks to provide readers a review and comparison of definitions of m-commerce published over the past two decades. Readers can understand the evolution of m-commerce in synchronization with that of e-commerce. The paradigm shifts in e-commerce from 1950s to date and synchronization with m-commerce advances over the past two decades are explored. Further, the impact of technology in the domains of hardware, software, and networking catalyzing developments in m-commerce are highlighted. Growth of m-commerce in terms of features and applications available now and expected in the future are also discussed, along with contributions of select countries to m-commerce.

1.3 METHODOLOGY

This chapter is based on a review of existing literature from articles concerning m-commerce, published in reputed journals. A comprehensive set of references available on or related to the domain of m-commerce was obtained by means of a thorough online search of select databases such as ABI/INFORM Global (ProQuest), Academic Search Premier (EBSCO), Emerald, Google Scholar, JSTOR, and SAGE. The keywords used to perform the search include "m-commerce" and "mobile commerce" which were used to obtain a list of references pertaining to literature available on the subject. The search yielded peer-reviewed articles from top-tier journals and articles in a number of conference proceedings published till date. By perusal of both abstract and full text of each article, the list of references was further narrowed by eliminating those articles not directly related to the themes or sub-themes of m-commerce. Furthermore, theses, white papers, and web links on m-commerce were referred. Back referencing of the list of articles thus obtained helped in identifying additional seminal works related to the topic.

1.4 M-COMMERCE: DEFINITIONS, EVOLUTION, AND HISTORY

1.4.1 *M-COMMERCE: A REVIEW OF DEFINITIONS*

The term "m-commerce" has been defined by a number of researchers, with subtle variations in these definitions over the years. Durlacher (2000) defined m-commerce as "any transaction with a monetary value that is conducted via a mobile telecommunications network." M-commerce is considered by some as a subset of e-commerce, and is defined as "a natural extension of e-commerce that allows users to interact with other users or businesses in a wireless mode, anytime/anywhere" (Coursaris, 2003).

The definitions of m-commerce are scattered over a number of books (including edited volumes), journal articles, and other forms of academic publications concerning a range of disciplines such as advertising, computing and information technology, electronics, business, and management (Table 1.1). This section focuses on highlighting trends associated with definitions of m-commerce over the years.

TABLE 1.1 M-Commerce: A Review of Definitions (2000–Till Date).

Authors	Purpose			Relation to e-commerce	Usage of wireless mobile device	Usage of wireless networks
	Communication					
	Business purposes	Other purposes	Transaction			
Friedman, 1999	✓	✓				
Durlacher, 2000			✓			✓
Senn, 2000			✓		✓	
Clarke, 2001			✓		✓	✓
Herzog and Gottlob, 2001				✓	✓	
Ghosh and Swaminatha, 2001	✓	✓	✓			✓
Balasubramanian et al., 2002	✓	✓	✓		✓	✓
Kalakota and Robinson, 2002	✓				✓	
Varshney and Vetter, 2002				✓	✓	

TABLE 1.1 *(Continued)*

Authors	Purpose			Relation to e-commerce	Usage of wireless mobile device	Usage of wireless networks
	Communication		Transaction			
	Business purposes	Other purposes				
Tarasewich et al., 2002			✓		✓	✓
Coursaris, 2003	✓	✓		✓	✓	✓
Sadeh, 2003					✓	✓
Stafford and Gillenson, 2003	✓	✓	✓			
Wang et al., 2005			✓	✓	✓	
Tiwari and Buse, 2007			✓		✓	✓
Wang and Liao, 2007			✓	✓	✓	✓
Wei et al., 2009					✓	
Kourouthanassis and Giaglis, 2012				✓	✓	✓
Chong, 2013			✓		✓	
Narang and Arora, 2016			✓		✓	
Kalinic and Marinkovic, 2016			✓		✓	

Source: Compiled by authors.

1.4.1.1 PHASE 1: BEFORE 2000

Prior to 2000, mobile commerce has not been explicitly defined in journal articles; rather the focus was on e-commerce. The growing applications and potential of Internet meant that e-commerce was on the rise, whereas the integration of Internet technology and mobile handheld devices had not yet progressed enough to warrant a detailed discussion on m-commerce.

1.4.1.2 PHASE 2: 2000–2005

M-commerce is seen as an extension of e-commerce, with focus on the conduct of commercial transactions and communication for private or business purposes. The ability to purchase goods and services irrespective of the user location using handheld devices such as PDAs (Personal Digital Assistants) or cell phones is highlighted. The purpose of m-commerce is split between communication and commercial transactions. This can be further understood by the fact that mobile payment systems were still in the initial development phase and associated security and privacy risks were considered high.

1.4.1.3 PHASE 3: 2005–2010

The definition of m-commerce has remained the same, with subtle variations. The focus on use of Internet-enabled wireless handheld devices increased. These devices included cell phones, palm-sized devices, or interfaces mounted on vehicles. Though previous definitions focused on the communication aspect of m-commerce, the concept gradually shifted toward incorporating all activities facilitating commercial transactions. This can be accounted for by research and development in mobile payment systems as explained in detail in Section 1.5 later in this chapter. Research into Near-field Communication (NFC) technology and its integration with smartphones paved way for mobile payments as seen today.

1.4.1.4 PHASE 4: 2010–TILL DATE

The pervasiveness of smartphones and associated software and mobile apps has changed the public perception of m-commerce. Widespread use of location-based services, mobile banking services, mobile ticketing, payment through mobiles, and mobile purchasing through a number of mobile applications have made m-commerce a predominant part of contemporary transactions. Kourouthanassis and Giaglis (2012) state the difference between m-commerce and traditional e-commerce as "m-commerce services are accessible on the move through devices (such as smartphones and tablets) with fundamentally different presentation, processing, and interaction modalities compared to a desktop computer; such services enable a whole new set of unprecedented service capabilities, including location awareness, context sensing, and push delivery."

It is interesting to note that the back-end systems of Internet underwent major changes from Web 1.0 to Web 2.0 in phase 3 and is now further shifting toward Web 3.0 in phase 4. However, the definition of m-commerce has not varied much with regard to the changing web. Though definition of m-commerce has evolved over the years, the concept did not undergo much significant change. Hence, it can be summarized that m-commerce is, in essence, the buying and selling of products and services over the Internet on an anytime, anywhere basis using wireless handheld devices.

1.4.2 M-COMMERCE: EVOLUTION AND HISTORY

At the start of the millennium, m-commerce faced several problems due to heterogeneity in technological environment marked by the presence of multiple standards concerning hardware, software, and networks. Different mobile operating systems, slower Internet speeds, lack of supporting websites, and incompatibilities in early versions of different standards such as the Wireless Access Protocol (WAP) led to slow development of m-commerce (Herzog and Gottlob, 2001). Early handheld devices also suffered from disadvantages such as smaller screens with lower screen resolution, smaller keypads, limited processing power, limited memory, and lack of user-friendly interface (Wu and Wang, 2005).

The main forces behind the success of m-commerce today are considered to be rapid proliferation of mobile devices, convergence of Internet and other telecommunication networks, advancement to 4G and associated higher data rates, and the development of highly personalized, context aware, and location sensitive applications (Sadeh, 2003). Apart from this, the development of m-commerce can mainly be attributed to development of Internet, progress in wireless communication technology, evolution of handheld devices, growth of Internet companies, development of mobile applications, advancements in mobile payment systems, and development of back-end systems of the Internet. These are explained in detail as follows (Fig. 1.1).

The *development of Internet* can be divided into three phases—Internet innovation phase, Internet institutionalization phase, and Internet commercialization phase (Laudon and Traver, 2006). The Internet innovation phase (1961–1974) is marked by development of the idea of Internet and involved the development of packet switching techniques and Transmission Control Protocols/Internet Protocols. The Internet institutionalization phase (1975–1995) concentrated on developing a robust network by focusing on Internet security and redundancy. Developments of the institutionalization phase played out against the backdrop of heightened tensions between the United States of America and the Union of Soviet Socialist Republics during the cold war. As a result, advancements made on Internet focused on building a robust Internet platform that could outlive a nuclear war, with emphasis on improving security and redundancy. Several websites were launched toward the end of this phase with the aim of marketing products/services. Examples include the online website of HotWired magazine, the first to feature banner ads that are still widely used today in online marketing (Laudon and Traver, 2006). The concept of m-commerce existed in the pre-Internet era via telemarketing, whereby customers dialed a given phone number to purchase a product advertised on television or billboards. The final phase, Internet commercialization phase (1995–present), saw the commercialization of Internet and includes private corporations taking over and developing Internet backbone and services (Laudon and Traver, 2006). It also includes advances in wireless communication technology. According to Evans (2014), the number of people online grew from about 0.4 billion in 2000 to nearly 3 billion in 2014. This number is expected to surpass 4 billion by 2020.

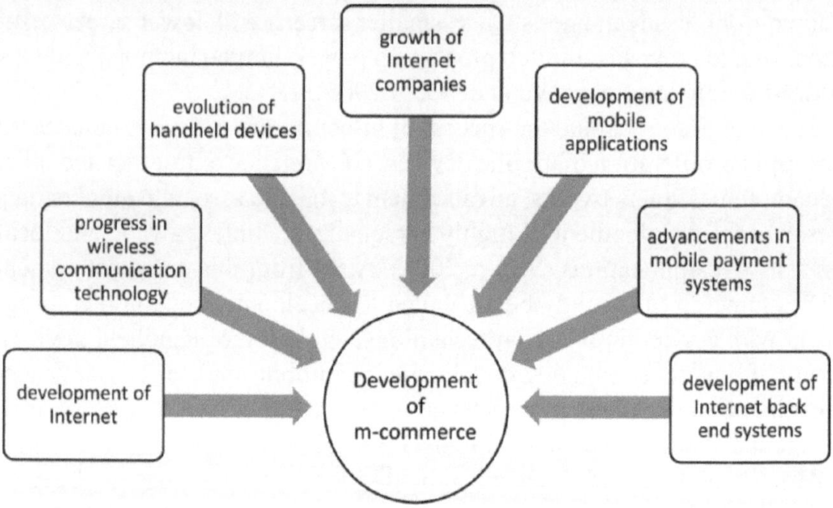

FIGURE 1.1 Factors contributing to the growth of m-commerce.

In the early days of m-commerce, mobile network operators bundled Internet services and m-commerce applications into mobile portals, such as NTT DoCoMo's i-Mode. The concept of m-portals never took off except in a few Asian countries, and this resulted in their exit from the modern m-commerce market (Kourouthanassis and Giaglis, 2012). This period of m-commerce is referred to as "m-portal era" by Kourouthanassis and Giaglis (2012). It is interesting to note that the conception of m-portal era coincides with the end of second phase of Internet development, the Institutionalization Phase, which spanned from 1975 to 1995 (Laudon and Traver, 2006). The start of the m-portal era also saw the birth of Amazon (1994) and eBay

Inc. (1995), both of which have emerged as some of the major players in m-commerce today. One of the first uses of m-commerce in the way it is widely used now was in 1997. Customers in Finland could purchase Coca-Cola drinks from vending machines via short message service (SMS), for which they would be charged by their mobile operators (Balasubramanian et al., 2002). In 1999, NTT DoCoMo's i-Mode was introduced, and became a huge success with 30 million users from a population of 126 million people in 2002 (Sadeh, 2003). The same year financial services group Nordea launched its online banking service WAP Solo portal. In 2001, Webraska set up SmartZone, a platform offering prepackaged location-based services to wireless carriers who in turn customized them for mobile users (Sadeh, 2003). Thus, though m-portals helped to catalyze the development of and use of m-commerce in several countries, the lack of personalization in standard applications led to its downfall.

The speed, ubiquity, and convenience of m-commerce today are the result of swift advancements in supporting infrastructure such as *wireless communication technology*. Wireless communication today has come a long way from wireless telegraphy through Morse code in 1886. Rather than a series of dots and dashes, wireless networks are now capable of transmitting video and audio-rich files instantly. In 1971, the first modern-day mobile phone system was put in place by AT&T, by dividing cities into "cells" (Melton et al., n.d.). In its early days, the development of m-commerce was hampered by lack of uniform standards of the WAP, lack of security for transactions, slow Internet speeds, limited device processing power, small screen size, and relatively impoverished sites (Clarke, 2001; Herzog and Gottlob, 2001). In 2000, the WAP, a communications protocol used for wireless data access through mobile networks had emerged, and most wireless handheld devices had adopted this standard (Senn, 2000). Internet speeds, though varying from country to country, were restricted to 2G including Time Division Multiple Access (TDMA) and Code Division Multiple Access (CDMA) in the early 2000s. TDMA phone technology was created by the Cellular Technology Industry Association which was established in 1988 (Melton et al., n.d.). Limited speed and bandwidth of 2G Internet meant that multimedia transmission was challenging, and most m-commerce applications till this time had extremely simple interfaces or used SMS to send the requests. This was followed by the introduction of General Packet Radio Service (GPRS) and Enhanced Data-rates for Global Evolution (EDGE) technology (Tiwari and Buse, 2007). Higher Internet speeds benefitting m-commerce came about with the launch of Universal Mobile Telecommunications System (UMTS), that is, third-generation mobile cellular system in 2002 and then with 4G which is being used today. Thus, m-commerce has benefited due to development of wireless communications technology starting from heterogeneous WAPs and expanding to uniform standards.

The concept of m-commerce was further revolutionized by the *evolution of handheld devices*, including introduction of the concept of "smartphones" with ability to access Internet. Specifications of smartphones available now easily surpass those of personal computers available a few years ago. Though the notion of m-commerce today brings to mind Internet-enabled smartphones, the invention of mobile handheld devices predates Internet by several years. The first handheld phone was demonstrated by Motorola in 1973 and was called the Motorola DynaTac (Murphy, 2013). In 1975, AT&T formed its own cellular plan in Chicago. Cell phone testing was then permitted in the city by the Federal Communications Commission (FCC) in 1977; in the same year, Bell Telephone Company partnered up with AT&T to receive the license for cellular plan (Melton et al., n.d.). However, by 2002, times had changed. The share of Internet-enabled mobile phones was increasing in Europe, Japan, and Korea (Sadeh, 2003). By 2005, mobiles were capable of several Internet services such as web information search, e-mail, gaming, chat, Global Positioning System (GPS) and Multimedia Messaging Service (MMS), to name a few (Okazaki, 2005). The launches of iPhone in 2007 as well as Apple App Store the succeeding year led to an era of mobile apps catering to every need of users. Massive competition between various phone manufacturers led to development of phones with larger screen, lesser weight, and better processing capability. Today, WAP has mostly disappeared to be replaced by phones supporting full Hyper Text Markup Language (HTML). Smartphones can now connect to Internet at speeds starting from 50 Mbps and support high definition mobile gaming, TV, video conferencing, and even 3D video. The longer battery life, faster processing capabilities, larger screens, and high-speed net capabilities offered by affordable smartphones have led to increased acceptance of m-commerce.

Kourouthanassis and Giaglis (2012) have labeled the period of m-commerce after 2000 as "m-Internet era," where the abovementioned advancements in wireless communications technology and Internet-enabled mobile phones led to several e-commerce websites optimizing their websites for mobile devices to tap the m-commerce market. This era can be linked with the early stages of third phase of Internet development (commercialization phase). The period got ushered amidst the rapid development of mobile phone communication and next-generation mobile technologies which resulted in the growth of Internet services and products. The first commercial microbrowser, Opera Mobile, was introduced in 2000 followed by Microsoft's Mobile Explorer 3.0 in 2001. During the m-Internet era, numerous other m-commerce applications also emerged. In 2003, Helsinki introduced mobile ticketing in its public transport system. In the same year, American technology company Apple also started to offer 99 cent mp3 downloads (Mallat et al., 2004). The largest single use of m-commerce transpired in 2005 when mobile phone users in the United Kingdom donated more than £1 million to the Asian tsunami relief fund (Harris et al., 2005). The m-Internet era thus led to the development of microbrowsers and Internet companies focused on optimizing their websites for the m-commerce market.

Commercialization of the Internet in 1995 led to a large number of *Internet companies* being set up. The dot-com bubble burst of autumn 2001 was the result of speculative investments in large numbers of Internet companies set up in the late 1990s due to perceived value and potential of the Internet (Geier, 2015). This resulted in the shut down or takeover of several e-commerce websites, with few websites such as amazon.com being able to get back on their feet. The dot-com burst also led to a shift in focus from applications with clichéd value propositions to Internet applications that contributed true value to consumers. Major Internet companies of today include Amazon, Alibaba, eBay, and Uber. Personalized services offered by these companies have led to the development of m-commerce and encouraged more people to adopt the same.

The growth of m-commerce can also be attributed to the *development of user-friendly and context aware mobile applications*. The year 2007 was marked by the launch of Apple's iPhone which revolutionized the cell phone industry and thereafter ushered the era of mobile applications. Subsequent competition between various phone manufacturers and supporting software led to Apple (iOS), Google (Android), and Microsoft (Windows Phone) developing hardware and platform innovations and engaging customers through mobile applications (Kourouthanassis and Giaglis, 2012). The huge market success of mobile applications can be attributed to a large customer base that in turn attracts a large number of developers who compete to develop and provide the best features. Mobile apps today are highly personalized, context aware, user friendly, and location sensitive. According to a report by Nielsen from 2014, smartphone users spend 89% of their mobile media time using mobile apps, while only 11% of their time is spent in mobile web (Nielsen, 2014). Today, the number of apps available on Google

Play Store is over 2.8 million, while the number of mobile app downloads worldwide in 2016 was over 149 billion.

It is quite interesting to note the *developments in mobile payment systems* overlapping aforesaid technological advancements. Mobile payment systems are an increasingly popular application of m-commerce today. In 2005, Nokia launched the first NFC-enabled cell phone. This is significant as it led to the development of mobile payments through host card emulation (HCE). Parallel to the development of NFC technology, American Internet payment company PayPal launched PayPal Mobile in 2006 using which users could make payments through SMS (Arrington, 2006). In 2007, Vodafone launched M-Pesa in Africa. M-Pesa is a mobile phone-based money transfer, financing, and microfinancing service, and has become the most successful mobile financial service in the developing world. In 2010, Android launched its first NFC-enabled smartphone, the Samsung Nexus. Subsequently, Google Wallet, a P2P payments service which enables a user to send/receive money to/from a smartphone, debuted in 2011. In 2013, HCE was shipped within Android operating system KitKat. Apple introduced its own digital wallet service, the Apple Pay, in 2014. The year also saw the launch of Starbucks' popular mobile app using which customers can place orders and make payments through mobile devices. The same year, MasterCard and Visa announced its support for HCE and published specifications for using the technology in NFC payment transactions. Further, Google revamped its mobile payment service with the introduction of Android Pay in 2015. Mobile wallets in NFC-enabled smartphones can be used in stores with an NFC-enabled payment terminal. Though hardly commonplace today, NFC-enabled payment terminals are growing in number and are expected to be a significant element in the future of m-commerce. Today, China has the largest adoption of mobile wallets (47%) followed by Norway and the United Kingdom (IAB, 2016). Thus, rapid developments in mobile payment systems over the past one decade have increased the adoption of m-commerce.

Another factor that led to significant change in the nature in which m-commerce is conducted was the *development of back-end systems* of Internet from Web 1.0 to Web 2.0 and subsequently, to Web 3.0. Web 1.0, also known as static web, linked web pages together using hyperlinks. Information updating was sporadic and user interaction with web pages was minimal. In Web 1.0 era, opportunities for m-commerce were limited, typically characterized by advertisements displayed on web pages, which provided toll-free numbers to order a product. The development of Web 2.0, also called the social web, lead to considerable changes in m-commerce, giving it the face we see today. Much of the growth of m-commerce has been due to Web 2.0 tools such as online social networking, social bookmarking, blogs, and wikis. The emphasis on user participation and user-generated content revolutionized the way companies interact with their customers in the aspects of marketing, selling, and customer feedback. Some social media websites such as Facebook and Pinterest have now introduced "buy buttons" which enable users to purchase products without leaving the platform. Web 3.0, successor to Web 2.0 and also called semantic web, uses context-aware applications which collect user data such as their location, preferences, and previous search and order history to display personalized advertisements and suggestions to users. Personalized suggestions and services will reduce the time spent in hopping from one website to another in search of the desired service. Web 3.0, using artificial intelligence and complex algorithms, is set to revolutionize m-commerce in the near future.

The growing trend of m-commerce is further evidenced by select facts and figures provided by We Are Social Ltd. According to them, out of the global population of 7.395 billion, the number of Internet users is 3.419 billion while unique mobile users are 3.790 billion. Smartphone penetration is also on the rise, with South Korea at the forefront with 88% of the population owning a smartphone (Poushter, 2016). Mobile penetration in 2016 can be seen in the graph presented in Figure 1.2.

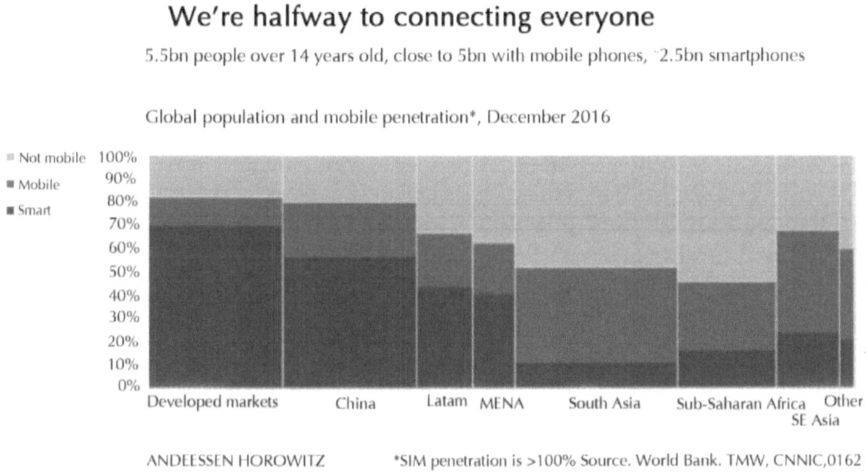

FIGURE 1.2 Comparison chart of trends for global population and smartphone penetration. *Source*: Adapted with permission from Evans (2016).

Further, the mobile's share of web traffic has grown from 0.7% in 2009 to 38.6% in 2016 (We Are Social). M-commerce can thus be expected to grow exponentially in the coming years.

It is evident from above discussion that though definitions of m-commerce have not undergone revolutionary changes since its inception, developments in all areas of m-commerce have certainly seen a marked transformation. The evolution and growth of m-commerce was supported by parallel advancements in Internet, wireless communication technology, handheld devices, Internet companies, mobile applications, mobile payment systems, and Internet back-end systems. Progress in all these domains supported one another and led to the development of m-commerce. Nevertheless, it is yet to be seen whether technological developments in the domains of hardware, software, and networking are able to grow in parallel with advancements of m-commerce and related areas. The acceptance and use of m-commerce by individuals in the light of challenges faced by it, as discussed in detail in Section 1.5.3 later in this chapter, remains to be seen.

1.5 M-COMMERCE: FEATURES, DISADVANTAGES, CHALLENGES, AND APPLICATIONS

1.5.1 M-COMMERCE: FEATURES

The features of m-commerce have been mentioned by several authors and overlap considerably.

Clarke (2001) outlined the advantages of m-commerce over other forms of e-commerce with reference to select features mentioned hereunder:

Ubiquity: Ubiquity of both mobile devices and Internet offers significant advantages. Mobile handheld devices of today are lightweight and portable, and are in close proximity to users during entire day. High-speed affordable Internet access is increasingly common in many regions across the world and instantaneous access to real-time information leads to use of m-commerce in time-critical applications such as stock quotes, and traffic updates, etc. Communication thus takes place irrespective of the user's location, and hence mobile devices offer the capability of sending and receiving information on an anywhere, anytime basis.

Convenience: The flexibility provided by mobile devices offers greater advantages for m-commerce than other forms of e-commerce. Using mobile applications, consumers can avail services at a time and place of their choice. Instantaneous access to a variety of options to choose from adds to this convenience.

Localization: With the advent of GPS technology and the fact that handheld devices are always in close proximity to its user, the location of smartphone users can be determined with high accuracy. Knowledge of user location is a big advantage in m-commerce over wired e-commerce. Location-based services such as information on nearby ATMs, restaurants, hospitals, or turn-by-turn navigation services are possible using location data. Thus, information on a number of applications relevant to the consumer's position can be provided in real time.

Personalization: Mobile devices have emerged as an effective marketing tool using which businesses are able to administer push-based strategies driven by personalization of content/services, complementing pull-based strategies. High smartphone penetration combined with location data and tracking of consumer behavior online makes mobile devices ideal for target marketing. Businesses can use location data to personalize messages to different regions by modifying the advertisement visually or auditorily, thus catering to individual preferences.

Identifiability: M-commerce has the ability to identify user of a service, as mobile devices are typically used by an individual, as opposed to a Personal Computer (PC) where the identity of the user is ambiguous. A mobile phone also has an inbuilt identifier to facilitate secure transactions, whereas a desktop is anonymous. Using GPS technology, it is possible to identify and locate a customer to provide personalized content and/or services.

Stanoevska-Slabeva (2003) described an additional feature:

Immediacy: It is the ubiquitous availability of m-commerce services that facilitate an instant action and reaction to changing demand.

Zeng et al. (2003) put forth the feature of currentness, which is interpreted similar to the feature of Immediacy mentioned above:

Currentness: M-commerce can provide real-time information to clients anywhere, regardless of their location.

Xiaojun et al. (2004) added the following feature:

Accessibility: M-commerce enables one to be contacted anywhere at any time. Moreover, there is an ability to limit accessibility to m-commerce to particular people or times.

The advantages of m-commerce also include its social spillovers due to commercial transactions being conducted online, such as a reduction in unregulated markets and unaccounted money.

1.5.2 M-COMMERCE: DISADVANTAGES

Based on the writings of Schwiderski-Grosche and Knospe (2002) and Niranjanamurthy et al. (2013), the following list summarizes the main disadvantages of m-commerce:

1. Limited functionality and screen size of mobile phones as compared to traditional desktops or laptops.
2. Wireless communication between mobile device and the network makes way for additional security threats such as eavesdropping.
3. Mobile devices are liable to loss by theft or misplacement. According to a study by IDG Research and Lookout, 44% of stolen phones were taken after being forgotten at a public place. High security standards must be in place to protect mobile devices due to the increasing amounts of personal and confidential information stored in it.
4. Additional cost of setting up mobile and wireless broadband infrastructure.
5. Security-related risks concerning data moved across mobile devices and wireless networks.

1.5.3 M-COMMERCE: CHALLENGES

Several challenges faced by m-commerce stated in research articles from the early 2000s include limitations of wireless devices and wireless communication technology. Limitations of wireless devices prevalent at that time include limited screen size, limited memory, absence of user-friendly keyboards, and higher dependency on text-only services with limited interactivity and visual appeal, and user interfaces that were hard to deal with (Varshney and Vetter, 2002). Wireless communication technology faced several challenges such as the absence of common wireless networking standards, and limited bandwidth available for wireless communication. However, the subsequent advancements in communications and Internet technology, as well as the emergence and proliferation of smartphones, have kept up with business expectations and have thus solved many of these issues.

Though m-commerce has come a long way from being hindered by slow Internet speeds and mobile devices of limited capabilities, it still faces challenges such as privacy and security which affect consumer trust. According to the 2016 LexisNexis report, m-commerce companies experience 880 fraud attempts each month of which 66%, that is, 581 attempts are successful (LexisNexis, 2016). Majority of the people are unaware of whether their information is being collected, stored, and used for questionable purposes. As a result, consumers are becoming more wary of divulging sensitive information to online vendors (Kim et al., 2011). This stems from a lack of trust in, and the subsequent perceived risk of using m-commerce services.

The decision to conduct transactions online involves an evaluation of the risks involved in doing so. These risks depend on the nature of purchase and vary from one service to another. M-commerce transactions involve business agreements without the observation of others, without personal interaction and without the touch and feel of products. This leads to customers feeling insecure, especially in a collectivist culture like India, where shopping tends to take place in groups. Unlike the purchase of tangible goods where only minimal personal information is required, the purchase of services online requires a significant amount of personal information (Featherman et al., 2010). This risk associated with m-commerce is due to a lack of trust in m-commerce security. According to Siau and Shen (2003), trust in m-commerce can be divided into two categories—trust in mobile technology and trust in mobile vendors (Siau and Shen, 2003). According to Laudon and Traver (2006), security of online transactions deals with the concepts of:

1. Integrity: ability to ensure that information transmitted via Internet is not altered by unauthorized third party
2. Nonrepudiation: ability to ensure that a party cannot deny an action undertaken online
3. Authenticity: ability to know and identify with certainty the party with whom one is dealing with online
4. Confidentiality: ability to ensure that the information provided by one party is only viewed by authorized person(s)
5. Privacy: ability to control the use of one's information provided online
6. Availability: ensuring that the online service continues its operations as intended

Distrust in availing m-commerce services can be due to several factors. These include unreliability of payment gateways, fraudulent sellers, loss of personal information and potential misuse of the same, phishing attacks, pharming attacks, madware (mobile adware), and cyber laws, to name a few.

However, studies suggest that consumer trust in m-commerce transactions can be increased through aesthetically designed websites (Li and Yeh, 2010) and mobile applications, with greater interactivity and customization. Brand image as well as responsiveness of m-commerce vendors affects consumer inclination to perform m-commerce transactions (Shao Yeh and Li, 2009). Moreover, displaying security statements which provide information to consumers for financial operations and security solutions have been found to increase consumer trust (Kim et al., 2010).

1.5.4 M-COMMERCE: APPLICATIONS

Mobile applications form an integral part of m-commerce today. The ever-increasing consumer base of mobile apps has led to larger number of developers competing to put forward applications with the best functionality and content. Mobile applications today take advantage of improved hardware in terms of mobile device screen size, processing power, screen resolution, and battery life; connectivity in terms of fast and ubiquitous nature of the Internet; personalization through tracking of customer behavior and customer location; and user-friendly software interfaces (Mahatanankoon et al., 2005). The multitude of advantages that m-commerce provides has led to its use in a variety of applications. Countries such as Japan, Finland, Hong Kong, the United States, and the United Kingdom have greater mobile penetration as compared to others and are often at the forefront in exploring new avenues of applications of m-commerce. As of today, they have some of the most developed and fast-growing m-commerce markets. Some of the most prevalent contemporary uses and/or applications of m-commerce are discussed next.

Mobile banking: Convenience is one of the key drivers for growth of mobile banking. Mobile banking services have been around since 1999 with the launch of i-Mode by Japan's NTT DoCoMo, and WAP Solo portal by Nordea, a financial services group in the Nordic region (Sadeh, 2003). In its earliest form, mobile banking involved obtaining information of account balances through SMS (Mallat et al., 2004). Today, most of the leading banks have developed their own mobile banking applications which offer a wide range of services such as checking account history, notifications on account activity, review of card/loan statements, fund transfers between linked accounts, bills payments, real time stock quotes, etc.

Mobile purchasing: Scandinavian countries have been pioneers in m-commerce, with the first application of m-commerce in Finland in 1997, where Coca Cola introduced a small number of vending machines that allowed customers to purchase drinks via SMS (Balasubramanian et al., 2002). The customers were billed through their mobile operator. Today, the nature of shopping has changed with mobile applications of various e-commerce sites such as Amazon, e-Bay, and Flipkart, to name a few, providing customers with access to all services otherwise available through their websites. In fact, many other e-commerce businesses have developed websites optimized for smartphones to facilitate mobile shopping.

Location-based services: Through the development and integration of GPS technology in mobile phones in the late 1990s and early 2000s (Clarke, 2001), location-based services have become very popular. Some of the earliest location-based services include SmartZone offered by an American location-based services company Webraska (Sadeh, 2003). Users could seek directions to nearby restaurants or gas stations using location-sensitive services. Another early example is the use of GPS-enabled smartphones in Tokyo, through which frequently transferred employees find their way to their destinations. Today, these services include location-based real-time weather reports, traffic updates, turn-by-turn navigation, and requesting a nearest service such as restaurant/hospital/ATM, to name a few.

Mobile ticketing: Mobile ticketing refers to the process through which users can order, pay for, receive and/or verify tickets using mobile handsets. Earlier, mobile ticketing was done via SMS, where the customers were billed through the mobile phone operator. In 2001, Helsinki Public Transport launched an SMS system of selling tickets (Mallat et al., 2009). Ticketing can also be done via Internet, such as in India, where many passengers use the IRCTC (Indian Railway Catering and Tourism Corporation) mobile application to book and pay for tickets using their mobile phones (Kapoor et al., 2013). Today mobile ticketing is used in several parts of the world for booking tickets in airlines, tourist attractions, zoos, museums, cinema, and public transport.

Information services: Among the m-commerce applications that first gained attention were Japan's NTT DoCoMo's i-Mode portal offering Internet services such as e-mail, ticketing, banking, and shopping; Nordea's WAP Solo mobile banking service; and Webraska's SmartZone platform

(Sadeh, 2003). Information on stock prices, sports scores, weather updates, emergency alerts, flight information, news, and traffic information can be delivered to handheld devices in real time. This is especially advantageous for highly time-sensitive data such as stock quotes.

Other applications: Further applications of m-commerce include mobile reverse auctions, content purchase and delivery, and mobile marketing. Mobile marketing has led to a shift from customers entering the retail market to retailers entering the customer environment through anytime, anywhere mobile devices (Shankar et al., 2010). Mobile marketing, especially social media advertising, provides numerous advantages to sellers. They gain instant connectivity to their customers who can interact with brands regardless of location or time (Pavel and Vlad, 2016). Social media advertising has low cost and broad reach. Advertising is thus no longer controlled by large businesses with enough financial support to launch advertising campaigns through television and print media. Rather, social media advertising through Facebook, Twitter, Instagram, and other social media networks enables instant dispersion of information. Consumers can "share" an advertisement with their friends or colleagues, creating a snowball effect.

The applications of m-commerce are fast growing and wide ranging. Initially used to order public transport tickets or download ringtones through SMS, m-commerce is now used in mobile banking, mobile purchasing, location-based services such as traffic updates and weather reports, mobile ticketing, and information services, to name a few. M-commerce is used everywhere and these applications will only grow in scope and number. The growth of wearable devices such as smartwatches is expected to aid further development of applications in m-commerce. Social buy buttons which enable users to purchase a product without exiting social media platforms is predicted to grow in popularity. M-commerce applications will further be influenced by the increasing convergence of physical and online worlds. According to Google, 82% of smartphone users turn to their phones inside a store when making purchasing decisions (Mooney and Johnsmeyer, 2015). Amazon's first physical store, the Amazon Go, further blurs lines between physical and online worlds. Customers can walk in, pick up their desired products, and walk out without waiting at a check-out counter. Item are scanned automatically and then billed to the customers' through their online Amazon accounts (Amazon, 2016). Thus, the plethora of services and applications offered through m-commerce platforms is expected to increase exponentially in near future.

1.6 CONCLUSIONS

M-commerce has made use of the changing face of Internet and telecommunication technologies to tap into the huge market space of cell phone users with continuous Internet access. Though the term m-commerce is described differently by several authors, it is generally defined to be an extension of e-commerce, whereby all activities related to a commercial transaction are carried out through mobile handheld devices using wireless telecommunication networks. The evolution of m-commerce was driven by advances in Internet, progress in wireless communication technology, evolution of handheld devices along with user-friendly applications, development of mobile applications, growth of Internet companies, advancements in mobile payment systems, and development of back-end systems of the Internet. The subsequent growth of m-commerce was expedited due to several advantages it offers to the modern user such as ubiquity of service, convenience provided by flexibility of usage, location-based services facilitated by mobile tracking, and personalization of desired services. On the other hand, it suffers from the disadvantages of increased security threats, vulnerability of smartphones to theft, and limited screen size of smartphones as compared to laptops and desktops. M-commerce also faces the challenges of security and privacy, which affect consumers' trust in the notion of m-commerce. From the simple act of purchasing a can of Coca Cola from a vending machine through SMS, m-commerce today has developed such that the process of searching for a product, comparing its price with similar offerings, reading customer reviews, ordering, and online payment can all take place within the daily work commute. M-commerce has a myriad of applications including mobile banking, mobile purchasing, mobile ticketing, and information-based services, to name a few.

Advances such as mobile devices with greater speed and processing power equipped with NFC technology bode well for the future of m-commerce. With the explosion of smartphones and the associated mobile applications, mobile commerce is rapidly approaching the business forefront. Countries such as Japan, Finland, Hong Kong, the United States, and the United Kingdom have greater mobile penetration and more advanced m-commerce markets. However, Asian countries such as China and India, which have a huge potential customer base, have recently started focusing on m-commerce developments. In today's fast-paced dynamic world, it would be quite difficult to speculate the future of m-commerce. Nevertheless, based on trends related to growth of m-commerce discussed in this chapter, it can be concluded that the cutting edge tools, technologies, and applications in and/or related to the domain of m-commerce will definitely impact and get impacted by human lives.

KEYWORDS

- **m-commerce**
- **Internet**
- **mobile device**
- **wireless communication**
- **e-commerce**

REFERENCES

Amazon. Introducing Amazon Go and the World's Most Advanced Shopping Technology, 2016. https://www.youtube.com/watch?v=NrmMk1Myrxc (accessed June 14, 2017).

Arrington, M. PayPal Mobile Launches–And It's Awesome, 2006. https://techcrunch.com/2006/03/22/paypal-mobile-launches-and-its-awesome/ (accessed June 15, 2017).

Balasubramanian, S.; Peterson, R. A.; Jarvenpaa, S. L. Exploring the Implications of M-commerce for Markets and Marketing. *J. Acad. Market. Sci.* **2002,** *30* (4), 348–361. https://doi.org/10.1177/009207002236910 (accessed Jan 2, 2017).

Chong, A. Y. L. Predicting M-commerce Adoption Determinants: A Neural Network Approach. *Exp. Syst. Appl.* **2013,** *40* (2), 523–530. https://doi.org/10.1016/j.eswa.2012.07.068 (accessed Jan 2, 2017).

Clarke, I. Emerging Value Propositions for M-commerce. *J. Bus. Strateg.* **2001,** *18* (2), 133–148.

Coursaris, C.; Hassanein, K.; Head, M. M-Commerce in Canada: An Interaction Framework for Wireless Privacy. *Can. J. Admin. Sci.* **2003,** *20* (1), 54–73. https://doi.org/10.1111/j.1936-4490.2003.tb00305.x (accessed Jan 7, 2017).

Durlacher Research Ltd. Mobile Commerce Report. In *Mobile Networking with WAP*; SCN Education B. V., Ed.; Vieweg+Teubner Verlag, 2000. https://doi.org/10.1007/978-3-322-86790-2_25 (accessed Jan 4, 2017).

Evans, B. Mobile is Eating the World [PowerPoint slides]; 2014. http://ben-evans.com/benedictevans/2014/10/28/presentation-mobile-is-eating-the-world (accessed Feb 26, 2017).

Evans, B. Mobile is Eating the World [PowerPoint slides]; 2016. http://ben-evans.com/benedictevans/2016/12/8/mobile-is-eating-the-world (accessed Feb 26, 2017).

Featherman, M. S.; Miyazaki, A. D.; Sprott, D. E. Reducing Online Privacy Risk to Facilitate E-service Adoption: The Influence of Perceived Ease of Use and Corporate Credibility. *J. Serv. Market.* **2010,** *24* (3), 219–229. https://doi.org/10.1108/08876041011040622 (accessed May 29, 2017).

Friedman, T. L. *The Lexus and the Olive Tree: Understanding Globalization*; Farrar, Straus, and Giroux: New York, 1999; pp 228–230.

Geier, B. *What Did We Learn from the Dotcom Stock Bubble of 2000?* TIME: New York, 2015 (March 12). http://time.com/3741681/2000-dotcom-stock-bust/ (accessed May 17, 2017).

Ghosh, A. K.; Swaminatha, T. M. Software Security and Privacy Risks in Mobile E-Commerce. *Commun. ACM.* **2001,** *44* (2), 51–57.

Harris, P.; Rettie, R.; Kwan, C. C. Adoption and Usage of M-Commerce : A Cross-cultural Comparison of Hong Kong and the United Kingdom. *J. Electron. Commer. Res.* **2005,** *6* (3), 210–224.

Herzog, M.; Gottlob, G. InfoPipes: A Flexible Framework for M-Commerce Applications. In *Technologies for E-Services. TES 2001*; Lecture Notes in Computer Science; Casati, F., Shan, M. C., Georgakopoulos, D., Eds.; Springer: Berlin, Heidelberg, 2001; Vol. 2193, pp 175–186. https://doi.org/10.1007/3-540-44809-8_13 (accessed Dec 29, 2016).

IAB. A Global Perspective of Mobile Commerce, 2016. http://www.iab.com/wp-content/uploads/2016/09/2016-IAB-Global-Mobile-Commerce-Report-FINAL-092216.pdf (accessed June 15, 2017).

Kalakota, R.; Robinson, M. *M-business: The Race to Mobility*; McGraw-Hill: New York, 2002.

Kalinic, Z.; Marinkovic, V. Determinants of users' intention to adopt m-commerce: an empirical analysis. *Inf. Syst. E-Bus. Manag.* **2016,** *14* (2), 367–387.

Kapoor, K.; Dwivedi, Y. K.; Williams, M. D. In *Role of Innovation Attributes in Explaining Intention and Adoption: A Case of the IRCTC Mobile Ticketing Application in the Indian Context*, UK Academy for Information Systems Conference Proceedings, Berlin, Heidelberg, 2013; Springer: Berlin, Heidelberg, 2013; p 19. https://aisel.aisnet.org/cgi/viewcontent.cgi?article=1018&context=ukais2013 (accessed Jan 2, 2017).

Kim, C.; Tao, W.; Shin, N.; Kim, K. S. An Empirical Study of Customers' Perceptions of Security and Trust in E-payment Systems. *Electron. Commer. Res. Appl.* **2010,** *9* (1), 84–95. https://doi.org/10.1016/j.elerap.2009.04.014 (accessed May 25, 2017).

Kim, M. J.; Chung, N.; Lee, C. K. The Effect of Perceived Trust on Electronic Commerce: Shopping Online for Tourism Products and Services in South Korea. *Tour. Manag.* **2011,** *32* (2), 256–265. https://doi.org/10.1016/j.tourman.2010.01.011 (accessed May 25, 2017)

Kourouthanassis, P. E.; Giaglis, G. M. Introduction to the Special Issue Mobile Commerce : The Past, Present, and Future of Mobile Commerce Research. *Int. J. Electron. Commer.* **2012,** *16* (4), 5–18. https://doi.org/10.2753/JEC1086-4415160401 (accessed Jan 6, 2017).

Laudon, K. C.; Traver, C. G. *E-commerce;* Pearson Ed: New Delhi, 2006.

LexisNexis. True Cost of Fraud Study, 2016. https://www.lexisnexis.com/risk/downloads/assets/true-cost-fraud-2016.pdf (accessed June 15, 2017).

Li, Y. M.; Yeh, Y. S. Increasing Trust in Mobile Commerce Through Design Aesthetics. *Comput. Human Behav.* **2010,** *26* (4), 673–684. https://doi.org/10.1016/j.chb.2010.01.004 (accessed May 23, 2017).

Mahatanankoon, P.; Wen, H. J.; Lim, B. Consumer-based M-commerce: Exploring Consumer Perception of Mobile Applications. *Comput. Stand. Inter.* **2005,** *27* (4), 347–357. https://doi.org/10.1016/j.csi.2004.10.003 (accessed Jan 4, 2017).

Mallat, N.; Rossi, M.; Tuunainen, V. K. Mobile Banking Services. *Commun. ACM* **2004,** *47* (5), 42–46.

Mallat, N.; Rossi, M.; Tuunainen, V. K.; Öörni, A. The Impact of Use Context on Mobile Services Acceptance: The Case of Mobile Ticketing. *Inf. Manag.* **2009,** *46* (3), 190–195. https://doi.org/10.1016/j.im.2008.11.008 (accessed Jan 7, 2017).

Melton, J.; Miller, G.; Morris, T. History of Mobile Commerce—Timeline, n.d. https://sites.google.com/site/mobileecommerce/history-of-mobile-commerce---timeline (accessed June 14, 2017).

Mooney, A.; Johnsmeyer, B. I-Want-to-Buy Moments: How Mobile Has Reshaped the Purchase Journey, 2015. https://www.thinkwithgoogle.com/marketing-resources/micro-moments/i-want-to-buy-moments/ (accessed June 14, 2017).

Murphy, T. 40 Years After the First Cell Phone Call: Who is Inventing Tomorrow's Future? *IEEE Consum. Electron. Mag.* **2013,** *2* (4), 44–46. https://doi.org/10.1109/MCE.2013.2273653 (accessed Jan 5, 2017).

Narang, B.; Arora, J. B. Present and Future of Mobile Commerce: Introduction, Comparative Analysis of M Commerce and E Commerce, Advantages, Present and Future. In *Securing Transactions and Payment Systems for M-Commerce*; IGI Global: Hershey, Pennsylvania, 2016; pp 293–308.

Nielsen. *An Era of Growth: The Cross-platform Report Q4, 2013*; March, 2014. http://www.nielsen.com/us/en/insights/reports/2014/an-era-of-growth-the-cross-platform-report.html (accessed June 15, 2017).

Niranjanamurthy, M.; Kavyashree, N.; Jagannath, S.; Chahar, D. Analysis of E-Commerce and M-Commerce: Advantages, Limitations and Security Issues. *Int. J. Adv. Res. Comput. Commun. Eng.* **2013,** *2* (6), 2360–2370.

Okazaki, S. New Perspectives on M-commerce Research. *J. Electron. Commer. Res.* **2005,** *6* (3), 160–164.

Pavel, C.; Vlad, F. Trends in the Evolution of Mobile Commerce. *Quaestus* **2016,** *8*, 381.

Poushter, J. Smartphone Ownership and Internet Usage Continues to Climb in Emerging Economies; February, 2016. http://www.pewglobal.org/2016/02/22/smartphone-ownership-and-internet-usage-continues-to-climb-in-emerging-economies/ (accessed Dec 30, 2016).

Sadeh, N. *M-commerce: Technologies, Services, and Business Models*; Long, C. A., Ed.; John Wiley and Sons: Boston, 2003.

Schwiderski-Grosche, S.; Knospe, H. Secure Mobile Commerce. *Electron. Commun. Eng. J.* **2002,** *14* (5), 228–238. https://doi.org/10.1049/ecej:20020506 (accessed June 14, 2017).

Senn, J. A. The Emergence of M-commerce. *Computer* **2000,** *33* (12), 148–150. https://doi.org/10.1109/2.889097 (accessed Dec 30, 2016).

Shankar, V.; Venkatesh, A.; Hofacker, C.; Naik, P. Mobile Marketing in the Retailing Environment: Current Insights and Future Research Avenues. *J. Interact. Mark.* **2010,** *24* (2), 111–120. https://doi.org/10.1016/j.intmar.2010.02.006 (accessed Jan 3, 2017).

Shao Yeh, Y.; Li, Y. Building Trust in M-commerce: Contributions from Quality and Satisfaction. *Online Inf. Rev.* **2009,** *33* (6), 1066–1086. https://doi.org/10.1108/14684520911011016 (accessed May 28, 2017).

Siau, K.; Shen, Z. Building Customer Trust in Mobile Commerce. *Commun. ACM* **2003,** *46* (4), 91–94.

Stafford, T. F.; Gillenson, M. L. Mobile Commerce: What It Is and What It Could Be. *Commun. ACM* **2003,** *46* (12), 33–34. https://doi.org/10.1145/953460.953485 (accessed Jan 5, 2017).

Stanoevska-Slabeva, K. In *Towards a Reference Model for M-Commerce Applications*, ECIS 2002 Proceedings, 2003; pp 159.

Tarasewich, P.; Nickerson, R. C.; Warkentin, M. Issues in Mobile E-commerce. *Commun. Assoc. Inf. Syst.* **2002,** *8* (3), 41–64.

Tiwari, R.; Buse, S. *The Mobile Commerce Prospects: A Strategic Analysis of Opportunities in the Banking Sector*; Hamburg University Press: Hamburg, 2007.

Varshney, U.; Vetter, R. Mobile Commerce: Framework, Applications and Networking Support. *Mobile Netw. Appl.* **2002,** *7* (3), 185–198. https://doi.org/10.1023/A:1014570512129 (accessed Jan 6, 2017).

Wang, Y. S.; Liao, Y. W. The Conceptualization and Measurement of M-commerce User Satisfaction. *Comput. Human Behav.* **2007,** *23* (1), 381–398. https://doi.org/10.1016/j.chb.2004.10.017 (accessed Jan 6, 2017).

Wang, J. J.; Song, Z.; Lei, P.; Sheriff, R. E. In *Design and Evaluation of M-commerce Applications*, 2005 Asia-Pacific Conference on Communications, Perth, Western Australia, 2005; pp 745–749. https://doi.org/10.1109/APCC.2005.1554161 (accessed Jan 6, 2017).

We Are Social. Digital in 2016, 2016. https://www.slideshare.net/wearesocialsg/digital-in-2016 (accessed June 10, 2017).

Wei, T. T.; Marthandan, G.; Chong, A. Y.-L.; Ooi, K.-B.; Arumugam, S. What Drives Malaysian M-commerce Adoption? An Empirical Analysis. *Ind. Manag. Data Syst.* **2009,** *109* (3), 370–388. https://doi.org/10.1108/02635570910939399 (accessed Jan 3, 2017).

Wu, J.-H.; Wang, S.-C. What Drives Mobile Commerce? An Empirical Evaluation of the Revised Technology Acceptance Model. *Inf. Manag.* **2005,** *42* (5), 719–729. https://doi.org/10.1016/j.im.2004.07.001 (accessed Jan 3, 2017).

Xiaojun, D.; Junichi, I.; Sho, H. Unique Features of Mobile Commerce. *J. Electron. Sci. Technol.* **2004,** *2* (3), 205–210.

Zeng, E. Y.; Yen, D. C.; Hwang, H.-G.; Huang, S.-M. Mobile Commerce: The Convergence of E-commerce and Wireless Technology. *Int. J. Serv. Technol. Manag.* **2003,** *4* (3), 302–322.

CHAPTER 2

WHAT'S NEW WITH MOBILE COMMERCE?

HENRIQUE SCHNEIDER*

Swiss Federation of Small and Medium Enterprises, Berne 3001, Switzerland

Nordakademie Hochschule der Wirtschaft, Elmshorn and Hamburg, Germany

E-mail: hschneider@gmx.ch

ABSTRACT

This chapter applies the idea of alertness to mobile commerce. The term alertness describes how individual agents in market processes perceive imperfections of these processes and build a business model addressing and solving these imperfections. In mobile commerce, most innovation comes out of entrepreneurial alertness. But many novelties also come from the alertness of the customers that use feedback loops and arbitrage to solve what they perceive as imperfections. Since mobile commerce relies heavily on these decentralized, undetermined dynamics of alertness, mobile commerce—despite all of its technical, logistical, and operational novelties—follows the typical economic logic of innovation as spontaneous order.

2.1 INTRODUCTION

Subscribing to the general definition of Mobile Commerce, M Commerce, or MC, as any transaction with monetary value that is conducted via a mobile network (Barnes, 2002), several questions emerge: Is MC boon or bane? What are the advantages of MC? What are its driving forces? How does MC change stationary or traditional commercial practices? Many more questions would follow. While they are all important, they can only be answered after the clarification of an even more fundamental issue: What is the novelty, if any, of mobile commerce?

When analyzing the novelty of MC, at least two very different dimensions can be differentiated. The first is the technological-logistic or operative dimension. It involves such aspects as network-infrastructure, transportation, payment methods, distribution, marketing, and so on. But then, there is a second dimension, the one about the economic theory behind MC. Analyzing this dimension involves discerning if MC changes the logic of commerce as it is, that is, the activity of selling goods and/or services at a price. Hayden and Webster (2014), Turban et al. (2003), or also Keen and Mackintosh (2001) claim it does. If MC does indeed change the logic of commerce, that is, the economic theory behind commerce, then these changes and how these occur must be explained. If it does not, then it has to be explained why it does seem, at a first glance, to do so, or why many observers think it does.

This conceptual chapter addresses this fundamental question of the novelty of MC offering two sets of responses: First, and as far as the first dimension is concerned—technology, logistics, and operation—MC offers a lot of innovations and indeed changes much of the way in which commercial transactions can occur. Second, however, MC does not change the nature of commerce or the economic theory behind it. MC is perfectly comparable to all other forms of commerce and to all other forms of economic development. It does, however, rely heavily—perhaps more heavily than other forms of commerce—on agents taking advantage of imperfections in the market processes, thus searching for more efficient ways in connecting supply and demand. That is, those elements that prima facie look like the innovations of MC is all things considered nothing more than an application of the economic logic of commerce using modern technology. This chapter will emphasize this second dimension of MC.

This taking advantage of imperfections in the market processes, and why at the same time MC seems novel, will be explained by using the idea of "alertness." Entrepreneurs are alert to possibilities that markets offer them. Sometimes, these possibilities involve developing new products; and sometimes, the possibilities are about how to reach consumers, how to market products, and how to use transport and information technology. "Alertness" means taking advantage of a perceived opportunity. Opportunities arise out of imperfections in the market processes. There is, however, one dimension of alertness that has been understudied and might be interesting, especially in the context of MC—consumer alertness.

The chapter will begin by rolling out the question of what the novelty in MC is. The first surveys recent literature and attempt at establishing the innovation of the logistic, technical, and operational dimension explained before. The second section turns to economic theory, introducing and explaining the notion of "alertness," especially entrepreneurial alertness. The third section applies this notion to MC. In the fourth section, the understudied notion of consumer alertness will be discussed; it will be emphasized how MC profits from consumer alertness and how it is in itself a driver of MC. Brief conclusions end this chapter.

The main conclusion will be that MC is not genuinely something new, but it is the amalgamation of undetermined, uncoordinated, and unplanned steps of "alertness." It is because of all these properties that MC has potential for growth. And it is equally because of all these properties that MC has several points of interest for business and management research.

2.2 THE INNOVATION OF MOBILE COMMERCE

Popomaronis (2016) gives an overview and summary about the current state of MC—from the point of view of its most advanced markets, the United States and Canada:

> "Technology has given retail a face lift, spurring the dramatic rise of e-commerce. Millennials are particularly in love with online shopping, with 67% preferring it to in-store buying. With this mobile-loving group leading e-commerce, over half (53%) of consumers use smartphones to meet their shopping needs. That doesn't mean, however, that companies are doing the greatest job providing an exceptional mobile shopping experience. In fact, consumers are due for big improvements. According to Google's 2016 report, "Mobile Has Changed How People Get Things Done," about one out of three (36%) of consumers pick their smartphone for online shopping/research because the device gives them the best experience. That might seem like a decent enough rate, except that roughly two out of three (69%) of consumers pick their smartphone to address their need just because it's the closest device to them at the time. Google's findings support the idea that the increasing use of smartphones in e-commerce has more to do with the prevalence and proximity of the devices than particularly good innovations or perks from developers, and that retailers can do more to make using their apps and mobile sites attractive. It's somewhat

understandable that retailers still aren't quite on the money yet with mobile shopping. E-commerce still makes up just 8.1% of total US sales, although major companies like Facebook are fueling growth with new innovations that merge retail, chat, and social media."

But, continuing with Popomaronis (2016) and his reporting on the study commissioned by Google, there are still several aspects on how to improve MC:

"Retailers and consumers would have more instant communication with personalized recommendations. 42% of consumers would give out a telephone number for text messages (e.g., geofencing), 44% want on-demand communication in stores, and 41% want to be targeted based on purchase history or preference. Consumers would have more visuals. The ability to see product pictures ranks as the biggest priority for 62% of mobile shoppers. Consumers likely would welcome more or 360 degree views. Buyers would know exactly what they're getting and if they're getting the best deal. Having product descriptions, being able to read reviews and having the ability to compare prices all rank strongly for consumers at 44% each. Site navigation would be fast and simple, with retailers understanding the difference between regular and mobile websites. Users want pages to load in four seconds or less (64%), will abandon carts if getting around is too complex (25%) and expect a mobile-friendly website (62%). Navigating the checkout would be easier without the annoyance of entering personal data on a small screen without a mouse. 35% of consumers would like to buy items on smartphones if checkout were easier. Retailers would nix—or at least reduce—shipping fees. 79.5% of online buyers would buy more if initial shipping were free, 72.4% would buy more if return shipping were free and 67.5% always choose the cheapest delivery option. Consumers already are willing to pick up their phones and buy because the device is nearby, but that willingness shouldn't translate to complacency from businesses about improving the mobile buying experience. If companies pay attention and give online buyers what research says they want, investing in mobile development, both businesses and buyers can benefit."

Skeptics might now argue that this picture is typical or even exclusive for the US economy and not even representative of similar economies, let alone of emerging or frontier markets such as India or Vietnam. There are some compelling factors to this reasoning: Many aspects of the hype of MC are a North American peculiarity that cannot be transferred without modification to other markets. For example, MC, needs many adjustments to succeed in India, where more than 65% of the population reside in rural areas with poor or no connectivity to the internet and sometimes even without telephone networks (Khokhar, 2016). On the other hand, the state of MC in the United States as outlined above, including the *desiderata* for future improvements, might provide the blueprint for such countries as India, Vietnam, or even the European Union to catch up. For example, Zinnov (2015) projects estimated revenues of USD 125–160 billion by 2025 in India. According to Zinnov (2015), the driving factors of MC in India are favorable demographic, growth and size of traditional retail, growing smart phone adoption, and growing internet user base. These factors would allow India to leapfrog on the US experience.

It is no wonder, then, that research on MC has been thriving in emerging and frontier markets (Deshmukh et al., 2013; Chowdhury et al., 2010; Solomon, 2014). Those economists that subscribe to the idea of alertness as taking advantage of imperfections in the market processes would even claim that emerging and frontier markets provide the most fruitful grounds for the advancement of MC. After all, in these markets, there are enough imperfections to be dealt with innovatively and imaginatively.

Popomaronis (2016) and the study commissioned by Google that he reports show, especially and abstractly, which measures have to be set in place on the technological and logistic level in order for sellers and buyers to take advantage of MC. They are principally, on the level of infrastructure: roads, mobile networks with good coverage and high quality, and broadband; on the level of supply: code and software, interface, real-time response, product and stock management, just-in-time-techniques, transportation, payment possibilities including online, parabanking, cash management and data protection, as well as quality management; on the level of consumers: connectivity, mobile devices, access to payment options, access to distribution networks, real-time response, willingness to search for information and compare information, as well as responsibility (Madden et al., 2016; Ahuja and Khazanchi 2016; Carey and Helfert 2016).

Wu and Wang (2005) study the outset drivers of MC empirically. They aim to devise the overarching variable that explains and summarizes the diffusion or adoption of MC. Their findings show that:

> "The most important determinant for behavioral intention to use is compatibility. Our findings suggest that MC providers and managers should improve their compatibility with various user requirements, past experience, lifestyle, and beliefs in order to fulfill customer expectations. Similarly, privacy and security problems are less than satisfactory and must be overcome for MC to become an accepted merchandizing practice. Despite

> most consumers being concerned with the various risks, including transaction security, merchant information, products, online privacy, and personal data, these problems are often ignored by online commerce site providers. From a theoretical perspective, it seemed reasonable that a higher perceived risk in MC will lead to a lower rate of intention to use, which will result in lower MC use. Furthermore, perceived risk was believed to be a predictor and barrier to online transactions, and expected to negatively influence consumer's behavioral intent."

According to Wu and Wang (2005), at the end, there are two variables that explain MC's diffusion and use: compatibility and intention. While intention is formed on the demand side, that is, consumers, compatibility is driven by the supply side, that is, by producers of goods through MC. Both, intention and compatibility, affect each other mutually, but both can be influenced separately. All of the abovementioned measures that have to be set in place in order for sellers and buyers to take advantage of MC can be subsumed under either compatibility or intention. The more compatible MC's offerings are with the intentions of the demand side, the easier this intention will lead to action, that is, the demand side engaging in MC. An example of compatibility is usability (Bilgihan et al., 2016). The more easily consumers can navigate a mobile commerce site, the easier it will be for them to buy and therefore, the initial intention that led the consumer to look up the MC's site can translate into the action of buying through MC. Compatibility, at the end, means aligning the whole MC experience to the wishes of the customers, because these wishes are at the root of the customers' intentions to buy. Making the MC experience compatible with the customers' intentions is the most important task of the supply side.

Bhatti (2015), on the other hand, examines the demand side. The question here is how the intentions of potential MC customers are formed. According to Bhatti (2015):

> "The user's intention to adopt mobile commerce is affected significantly by subjective norms, perceived ease of use and perceived behavioral control by that user. Perceived usefulness and personal innovativeness does not influence directly and indirectly on behavioral intention. However, perceived behavioral control has been found to significantly influence behavioral intention to adopt mobile commerce. Furthermore, the results also show that the perceived ease of use and perceived behavioral control directly influence the intention to adopt which is consistent with previous studies. Perceived behavioral control is found to influence directly on intention to use and perceived ease of use. Mobile applications service providers may

> offer free use of service for a period that would enable the users to learn the service, thus increasing their perceived control. Another finding from this study is that the subjective norms directly and indirectly influence the intention to adopt mobile commerce. Subjective norms are often used to explain the rapid adoption of technology and the results of this study reveal a significant and effect of normative pressure to use. This emphasizes the importance of managers to consider the social context in which the mobile commerce is used. Social influences also help to shape an individual's estimation of his or her confidence in or the ability to use a system well."

It becomes clear that the consumers' intentions are affected by intrinsic and extrinsic factors. And this is why compatibility is possible. Extrinsic influences such as social norms or trends play an important role in the customers' adoption of MC and then formulating their own intentions of using MC for a specific transaction. Since extrinsic factors are as important as they are, the supply side in MC can influence norms and trends—for example, by marketing aggressively, by educating the consumers on how to use MC, and even by investing in more comfortable and secure payment options, in bandwidth or even in networks—in order to ease the individual customer's intention building and decision-making. Also, the more the customer feels in control of the MC transactions, the more likely the adoption of MC is. Therefore, increasing transparency, giving customers control over their data and ownership over purchase, and delivery and payment processes are supply side policies that lead to demand side adoption of MC.

This all having been reviewed, the question still stands: What, then, is the novelty of MC? As the brief literature survey in this sections shows, on technical, logistic, and operational level, there are many novelties about MC. Examples of those are the technological innovations, say, on network, hardware, software, or code level. Then, there are innovations in the interaction of the supply and demand sides, say, process ownership, transparency, logistics, or delivery. There are also novelties at the level of marketing, business case, product management, or business engineering. In sum, there are considerable novelties of MC of the technologic-logistic or operational level. But does it revolutionize economic theory, as it is often claimed or hailed (see the examples mentioned above, especially Hayden and Webster, 2014)? The answer is no. The fundamental logic of economics and economic innovation does not change because of MC. In fact, MC is a direct fruit of that logic. The next sections will examine the economic logic of innovation and its relationship to MC.

2.3 THE ECONOMICS OF ALERTNESS

Entrepreneurial alertness is the ability that some people have to recognize competitive imperfections in market processes. Competitive imperfections exist in market processes when information about technology, demand, or other determinants of competition in an industry is not widely understood by those operating in that industry. The existence of competitive imperfections in market processes suggests that it is possible for at least some economic actors in these market processes to earn economic profits, or to capitalize on solving these imperfections. Thus, entrepreneurial alertness can be thought of as the ability of some people to recognize opportunities to earn economic profits out of imperfections in the market processes. However, entrepreneurial alertness does not imply that individuals are systematically and rationally searching their environment for competitive imperfections. Rather, these individuals become aware of these competitive imperfections through their day-to-day activities. Indeed, they are often surprised that these imperfections exist, and that they have not been previously exploited by someone else.

Before examining more in-depth this notion of alertness, the idea of the market process must be clarified. Alertness as an economic theory sees markets as the sum of an unlimited amount of unique transactions by individuals or groups of individuals. The sum of these transactions forms a dynamic, nondetermined process or many equally dynamic nondetermined processes. These processes are called market processes because they are social interactions envisaging the transaction of economic goods involving payments. However, market processes do not have a preestablished outcome or a result. They are spontaneous order, or, in the positive version of the expression, chaotic (Schneider, 2017).

It is from this background that the economist Israel Kirzner (1973) introduces the concepts of "pure entrepreneurship" and "alertness" in his book *Competition and Entrepreneurship*. Alertness is an aspect of consciousness and not an action. It leads to action, but it is primarily when a market agent has a moment of clarity—that of being alert to changing buying and selling possibilities. For the idea of alertness, it is not important if anyone has ownership over means of production or finance. Important is the use of entrepreneurial alertness; that is, its employment in competitive agency envisaging making a profit. In Kirzner's terms, everyone in the market process is open for entrepreneurial alertness. In this way of thinking, the market is not divided between entrepreneurs and consumers, but the market has participants whose perceptions and decisions have an entrepreneurial dimension.

Kirzner (1963, p. 5) conceptualizes all participants in the market process as individuals with intentions and purpose, who are constantly making plans, choosing among many different alternative actions, acting and adjusting their actions or plans in response to what they perceive to be done by others. The actions of every individual constrain the actions by others. Entrepreneurship is when in the amalgamation of all these dynamic relationships, some agents are more alert than others taking advantage of what they consider to be a gap in knowledge or a gap in adaptation.

"At the center of the entrepreneurial market process are the purposive, planning, deciding and acting human beings, imperfectly knowledgeable of all the data of the market and uncertain of the future, but pursuing their chosen goals, and being alert to changing profit opportunities in a necessarily disequilibrium world. The result of actions taken because of entrepreneurial alertness to profit opportunities is a competitive process that enables market participants to more completely realize their goals by increasing their knowledge of market conditions and more closely coordinating their decisions and actions with those of others (Kirzner 1963, p. 17)." For Kirzner (1963 and 1973) the entrepreneur notices—not: creates—opportunities.

Later in 1979, Kirzner adapts this idea incorporating to the alertness of entrepreneurs—suppliers and demanders—not only to each other's needs but also the stock of knowledge produced by market processes. As the market process is a cooperative practice, it becomes more and more standardized. If an agent takes time to observe what happens in the process and how other agents navigate it, the observing agent might derive probabilities of how the others will act. There is still no way of knowing what and how a specific exchange will happen or what any other individual agents might act or react. But the stock of knowledge a market produces can be translated into a series of alternative scenarios and probabilities.

Innovators use this knowledge. This dispersed knowledge and particular intentions of market agents are brought into agreement by entrepreneurialism—of course, as often as entrepreneurs succeed in that task, they also fail in it. It is entrepreneurial alertness that provides the mechanism for agents in the market processes to discover useful information and to revise their plans so as to create more plan coordination in markets. This makes the argument theoretical, rather than empirical, because entrepreneurial alertness is an element of all human action.

How do, then, entrepreneurs access information? It cannot be a calculated process of searching, since that would assume that the entrepreneur had prior knowledge of what one wanted to search. But if one had that knowledge, search would be unnecessary. Rather, Kirzner (1979) sees the discovery of new information as a spontaneous movement out of ignorance into the knowledge of the plans of other market participants. As such, it is the result of entrepreneurial alertness, which differs in quality from one person to another. This leads Kirzner (1979, p. 150) to conclude: "What the market process does is to systematically translate unnoticed opportunities for mutually profitable exchange among individuals into forms that tend to excite the interest and alertness of those most likely to notice what can be spontaneously learned."

Does this entail that the beginning of an entrepreneurial activity is ignorance? Yes, but in a nonnormative way. Ignorance is the general state of agents in the market process because other than their own preferences and cost constrains, individual agents know nothing of the market process. They do not know anything because there is not anything to know. All other individual agents act only according to their own context, preferences, and constrains. Knowing these aspects of another agent is epistemically impossible.

Here is an example using two imagined agents: Fatima can only speculate about why Maria is investing in iron ore. And even Maria cannot know every preference of the ore company in which she is investing. If market processes are just series of spontaneous, undetermined, open-ended exchanges, there is nothing to know about them. Ignorance is the state every agent in the market process is in. Recognizing it is the first step toward entrepreneurialism. But the second step is also important. Action itself, entrepreneurial or otherwise, is always directed toward the future but its effects shape the present. Maybe Maria noticed before everyone else that the economic growth of India depends on iron ore and that alone is a reason driving the price of iron ore. New investments are as innovative as Fatima finding out the cure for mad cow disease.

All these elements of human action led Kirzner (1979) to establish his concept of "alertness" as the motivated propensity of people to formulate an idea for themselves of their future and to be aware of the ways all human agents can be imaginative or determinate. The motivation for a more accurately envisaged future is that of profit seeking. Kirzner even says that "the human agent is at all times spontaneously on the lookout for hitherto unnoticed features of the environment (present or future), which might inspire new activity on his part (Kirzner 2000, p. 124)."

Instead of being restricted to present acts of arbitrage, Kirzner's later work (see 2000, 2009) expands the concept of entrepreneurship to specifically include arbitrage in the present, arbitrage through time or "speculative arbitrage," and innovation—the creation (for a future more or less distant) of an output, method of production, or organization that is not in use yet. Kirzner also added an emphasis on competition—the rivalrous activities of market participants trying to win profits by offering the market better opportunities than are currently available—as a stimulus to the exercise of entrepreneurial alertness.

2.4 ALERTNESS IN MOBILE COMMERCE

How does the concept of alertness help in explaining the innovation of MC? Does alertness also help in explaining why MC is often perceived as something completely new or even revolutionary? It does. MC is an amalgamation of many instantiations of alertness. MC is made up of several small innovations. Each of them has, of course, its own merit, but they happen rather because entrepreneurs seize opportunities than because of a "masterplan of innovation." On the other hand and because of the fast-paced dynamics of alerted innovations, the bulk of those processes that are subsumed under MC seem novel as a whole. The following examples shall clarify how these series of rather one-stepped acts of alertness can lead to an overarching dynamics of innovation.

Take, for example, the beginning of MC: over the phone sales. Phone networks were thought of as systems for voice-to-voice communication. They evolved into a sales channel not because they were "destined" for it or because telecommunication enterprises planned for it, but because some entrepreneur realized that it is possible to sell something over the phone. The possibility was there from the outset, but it was idle until the first person had a moment of alertness to use it. Naturally, entrepreneurship does not end here. Finding the imperfection—in this case, the idle sales channel—is necessary but not sufficient. Making use of the imperfection in order to attain profit is what makes alertness an entrepreneurial endeavor. In order the imperfection to be used, some business concept needs to be developed, such as, for example, advertising for products that can be sold over the phone, educating customers how to order goods over the phone, making sure payment systems work, and delivering the goods timely. Also, as the one alert entrepreneur set out to turn the discovered imperfection into a business model, other entrepreneurs follow. These other entrepreneurs can discover further imperfections or just continue on the path set out by the first, alert, one. The likelihood that these entrepreneurs find another imperfection, however, increases, the more customers they can attract to the new business model.

The same applies to all other elements that make up MC: broadband was originally thought of as a channel for sharing information. Distribution networks were originally conceived as peer-to-peer channels and are becoming, because of MC, part of business-to-customers logistics. The drone, originally a military device, became a toy because of an alert entrepreneur finding an imperfection in the market and using it. The imperfection was that no one realized that the drone could be used as a toy and the business model was to manufacture smaller and cheaper drones for leisurely use. With the onset of MC and its logistical requirements, some other alert entrepreneur found a new imperfection: on-land distribution channels might be troublesome and unreliable, whereas aerial might be used to the advantage of the seller over MC. Therefore, some alert entrepreneur saw the drone as a new possibility to distribute products bought via MC and set out to develop and sell drones fit for MC's logistics.

On software and code level there are a lot of advancements in MC and most of them can be attributed to alertness. These are also the most interesting types of alertness, because they do not exclusively come from the individual entrepreneur but especially, from the growth of MC and its dialogue with customers. On software and code level, it is often the systematic search for imperfections in different systems that drive a series of alerted innovations. For example, customers provide feedback on how their experience was. This feedback is used in developing new software, making interfaces more usable, or individualizing algorithms as well as information. Through the automatization of most contemporary MC platforms (at least, those based on apps), many feedback loops occur instantly and without the need of an extra effort for customer feedback. Even the use of "big data"—datasets that need to be structured, compiled, and analyzed by special programs (Mayer-Schönberger and Cukier, 2013)—goes back to entrepreneurial alertness. It was alertness that made some entrepreneurs realize that the massive amount of data produced by MC transactions itself can be a source for identifying imperfections and even possible solutions—besides other things—in order to further improve MC. Using MC data to improve MC and automatizing this feedback loop itself is an instantiation of alertness.

All these examples serve to make the claim plausible that MC itself is not a new logic to economic theory. While MC might revolutionize commerce, it fits perfectly into actual economic theory. It might even be claimed that MC is a good example of the logic of alertness: entrepreneurs challenge each other in the market processes in a series of taking advantage of perceived imperfections. Therefore, the advancement of MC is not the development by plan of a clear-cut entity with a predetermined end state. It is a series of small and dynamic improvements that occur because of market- and technology imperfections. This series of small and dynamic improvements neither follows a plan nor can the direction of these improvements be predetermined. In short: What some chose to name MC is, in reality, the product of steps of alertness and therefore a spontaneous order.

There is, however, one aspect of alertness that might play a more important role for the future development of MC. This role is more important than what has been mentioned before and certainly more than Kirzner gives credit to it. In most account of alertness, it is the entrepreneurial side that is seen as the driving factor. And it is. But in the context of MC, as well as in the contexts of platforms, transparency, or collaboration, the customers' alertness is important too. This will be explained in the next sections.

2.5 CONSUMER ALERTNESS AND MOBILE COMMERCE

Often, economic theory sees customers as passive consumers of the goods supplied (Solomon, 2014). This passivity is addressed by the discipline of marketing that generally engineers communication and experience channels with the customers in order to directly and indirectly influence not only their decision-making processes but also their very specific decisions. However, this image of the consumer is oversimplified. If the Kirznerian framework is to be applied here, then it is true that the consumer does not build a business model around the perceived imperfection of the market processes. But much like the entrepreneur does, some consumers perceive imperfections and act on it. Many instances of Kirznerian entrepreneurial alertness apply to consumer alertness.

Take the following examples. Without the simplest arbitrage by customers, that is, searching for the lowest price, there would be no price-competition among suppliers. Customers try to optimize prices not just by comparing them between, say, two outlets of stationary commerce, but by comparing different channels for buying the good that they are searching. These channels can be stationary, mobile, catalogues, and the like. Sometimes, it is the customer that finds a completely different channel, not because he set out to do so, but out of a moment of alertness. For example, at the beginning of Amazon's expansion to Europe around the year 2000, customers based in Switzerland would order on Amazon but prefer the ordered good to be shipped to Germany because shipment to there was considerable cheaper than to Switzerland. This led Amazon to offer shipment free of charge to Switzerland if a purchasing threshold was surpassed by the customer. It was the alertness of customers at the root of the free shipment offer. Similarly, systematic feedback, call centers, individualized communication channels between customers and suppliers of MC increase the probability of moments of alertness on both sides.

In a more systematic analysis of the elements of customers' alertness, four of its elements stand out. (1) Customer alertness is an aspect of human action that is akin to the context of supplier choice. (2) The context of choice entails four major supply market alternatives where an actual/potential supplier can be considered in light of other actual/potential suppliers. (3) Hirschman's (1970) theory of exit, voice, and loyalty can be drawn upon concerning how customer alertness relates to prevailing relationships, and when it comes to anchor potential buyer–seller relations. (4) Customer alertness is conditioned by specificity (mentioned above as compatibility) and information.

The first element is not surprising, for it just states that customer and entrepreneurial alertness are similar and both are explained by Kirzner's theory of action as discussed above. The second element is about the specific means-ends calculations of the customers and concerns how they chose suppliers. They momentarily choose the ends (to buy or not to buy in light of prevalent opportunities), whereas suppliers adapt their means offers accordingly, for example, by raising or lowering prices, by optimizing stock and distribution, or by creating new offers. The four broad alternatives of customers are: (1) An actual supplier can be considered in light of other actual suppliers; (2) An actual supplier can be considered in light of potential suppliers; (3) A potential supplier can be considered in light of other potential suppliers; and (4) A potential supplier can be considered in the light of actual suppliers.

In each of these alternatives, the actual decision-making, that is, the acting upon intention can be driven by alertness. In fact, the more suppliers there are in a series of market processes, the more this decision-making is likely to be guided by customer alertness. There are many reasons for this. First, the more participation in market processes, the more information becomes available to the individual agents. This information hints at what works well and which imperfections still exist. Second, the more transparency in the market processes, the more individual agents are reached by this information. Third, the more rapidly a type of market processes develops, that is, MC, the more imperfections are produced. These imperfections can be noticed by suppliers and demanders alike. The constellations of market processes that make alertness more probable are the same or at least similar on the supply and demand side. They can be summarized by the following relationship: The more participants in similar market processes, the more likely it is that suppliers and demanders will have moments of alertness.

Turning to the third above-enumerated element, Hirschman (1970) explicitly refers to customers as alert or inert. His argument revolves around the observation that supplier performance deterioration (reflecting organizational slack) carries with it two major behavioral reactions or consequences on the part of demanders. Either they "exit" (they cease buying) or they address their dissatisfaction by "voice," that is, telling openly about it to anyone who cares to listen. Whatever is undertaken, it impacts the market processes in an undetermined, open-ended way. Customer alertness can be discerned in both exit and voice.

When exit occurs, it means alertness invokes a change of suppliers. This behavioral consequence thus safeguards openness (competition) in market processes, but also provides the incumbent supplier with a valuable feedback mechanism. That is to say, the supplier will alternate its plan setup in order to better attract and keep customers in the future by matching its plans in an improved manner with those of buyers. Voice typically occurs when there are a few, if any, alternative sellers at hand but does not necessarily relies upon such market characteristics. It means to try to make change come by, to articulate one's interests by "alerting [an] organization to its failings" (Hirschman, 1970, p. 33). Any market situation displaying customer alertness is likely to experience a complementary mix of exit and voice expressed by different customers, something thus driving the market process and subject also to prevailing elasticities of demand. Under some circumstances (for instance, when voice cannot constitute a threat of exit despite unhappiness creating uneasiness, or when deterioration is not perceived at all), which Hirschman labels loyalty, customers remain inert and customer behavior does not result in voice. That is to say, customers do not exit and it is not necessary that they resort to voice either. These, of course, are not situations of customer alertness. These situations might, however, provide important information to the supply side. This information can regard, for example, how much imperfection in the processes customers are ready to support without exiting or voicing. This again might lead other entrepreneurs to either use the imperfection or try to solve it by creating a business model around it.

The fourth and last element enumerated above was customer alertness being conditioned by specificity (mentioned in the first section as compatibility) and information. This element seems straightforward and is a conclusion of the third. In market processes, such as MC, where customer feedback, transparency, customer ownership over data, and payment or quality processes are increasingly the rule and play a more important role, these channels make entrepreneurial and customer alertness more compatible.

This section makes the case for consumer alertness playing an important role in MC. While there is no claim that it is as important as entrepreneurial alertness, there is the claim that both complement each other, especially in MC. Both can be synchronized by using the many logistical, technological, and operational innovations happening around MC and other areas of electronic connectivity. At the same time and allowing for alertness to emerge and develop, MC is subscribing to the most established logic in economic theory, that is, spontaneous order.

2.6 CONCLUSION: NOT NEW, BUT MUCH INNOVATION IN MOBILE COMMERCE

This article started by asking what the novelties of MC are. On technical, logistic, and operational level, there are many innovations. However, claims that MC escapes or changes economic theory either misinterpret the logic of economics or misunderstand how MC evolves.

It has been shown how MC is the amalgamation of different, often uncoordinated and always undetermined, open-ended steps of innovation best described by the concept of alertness. Drawing on Kirzner's theory of entrepreneurial alertness, this chapter extended it to the alertness of customers. Both, supply and demand sides contribute to taking advantage of imperfections in the market processes. The more agents in MC, the more transparent they communicate and the more open they are to each other's influences, the better MC will advance and bring profit to both consumers and entrepreneurs. MC works best if it maintains its characteristic as spontaneous order.

KEYWORDS

- **mobile commerce**
- **innovation**
- **alertness**
- **consumers**
- **market processes**

REFERENCES

Ahuja, V.; Khazanchi, D. Creation of a Conceptual Model for Adoption of Mobile Apps for Shopping from E-Commerce Sites–An Indian Context. *Procedia Comput. Sci.* **2016,** *91*, 609–616.

Barnes, S. J. The Mobile Commerce Value Chain: Analysis and Future Developments. *Int. J. Inf. Manag.* **2002,** *22* (2), 91–108.

Bhatti, T. Exploring Factors Influencing the Adoption of Mobile Commerce. *J. Internet Bank. Commer.* **2015,** *20* (07), 56–81.

Bilgihan, A.; Kandampully, J.; Zhang, T. Towards a Unified Customer Experience in Online Shopping Environments: Antecedents and Outcomes. *Int. J. Qual. Serv. Sci.* **2016,** *8* (1), 102–119.

Carey, K.; Helfert, M. Improving the Front End of Innovation: The Case of Mobile Commerce Services. In *International Conference on HCI in Business, Government and Organizations*; Springer International Publishing: Dordrecht, 2016; pp 491–501.

Chowdhury, H. K.; Parvin, N.; Weitenberner, C.; Becker, M. Consumer Attitude Toward Mobile Advertising in an Emerging Market: An Empirical Study. *Marketing* **2010,** *12* (2), 206–216.

Deshmukh, S. P.; Deshmukh, P.; Thampi, G. Transformation from E-commerce to M-commerce in Indian Context. *Int. J. Comput. Sci. Issues* **2013,** *10* (4), 55–60.

Keen, P. G.; Mackintosh, R. *The Freedom Economy: Gaining the M-commerce Edge in the Era of the Wireless Internet*; McGraw-Hill Professional: New York, 2001.

Khokhar, A. S. Digital Literacy: How Prepared Is India to Embrace It? *Int. J. Digit. Liter. Digit. Competence* **2016,** *7* (3), 1–12.

Kirzner, I. Rational Action and Economic Theory: Rejoinder. *J. Political Econ.* **1963,** *71* (1), 84–85.

Kirzner, I. *Competition and Entrepreneurship*; Chicago University Press: Chicago, 1973.

Kirzner, I. *Perception, Opportunity, and Profit: Studies in the Theory of Entrepreneurship;* University of Chicago Press: Chicago, 1979.

Kirzner, I. The Limits of the Market: the Real and the Imagined. In *The Driving Force of the Market–Essays in Austrian Economics;* Routledge: Milton Park, 2000; pp 115–36.

Kirzner, I. The Alert and Creative Entrepreneur: A Clarification. *Small Bus. Econ.* **2009,** *32* (2), 145–52.

Hayden, T.; Webster, T. *The Mobile Commerce Revolution: Business Success in a Wireless World*; Que Publishing: New York, 2014.

Hirschman, A. *Exit, Voice, and Loyalty: Responses to Decline in Firms, Organizations, and States;* Harvard University Press: Cambridge, 1970.

Madden, G.; Banerjee, A.; Rappoport, P.N.; Suenaga, H. E-commerce Transactions, the Installed Base of Credit Cards, and the Potential Mobile E-commerce Adoption. *Appl. Econ.* **2017,** *49* (1), 21–32.

Schneider, H. *Uber: Innovation in Society*; Palgrave Pivot: Hoboken, 2017.

Mayer-Schönberger, V.; Cukier, K. *Big Data: A Revolution That Will Transform How We Live, Work, and Think*; Mifflin Harcourt: Houghton, 2013.

Solomon, M. R. *Consumer Behavior: Buying, Having, and Being*; Prentice Hall: Engelwood Cliffs, 2014.

Popomaronis, T. *Think Great Innovation Drives Mobile Commerce? It's Actually This*, *Forbes*, Sep 26, 2016.

Turban, E; King, D.; Wang, J. *Introduction to E-commerce*; Prentice Hall: Upper Saddle River, 2003.

Wu, J. H.; Wang, S. C. What Drives Mobile Commerce? An Empirical Evaluation of the Revised Technology Acceptance Model. *Inf. Manag.* **2005,** *42* (5), 719–729.

Zinnov. *Indian M-Commerce: Riding High Tides*; Zinnov: Delhi, 2015.

CHAPTER 3

MOBILE COMMERCE: EVOLUTION AND TRENDS

PADMINI TOMER*

Commerce Department, University College, Ghanaur, Patiala, Punjab, India

**E-mail: padmini.tomer@gmail.com*

ABSTRACT

Mobile commerce being an extension to electronic commerce refers to the commerce that is carried out by using wireless devices such as mobile phones, Personal Digital Assistant, or other handheld devices. It is an innovative way to attract customers and provides greater convenience to its subscribers. Different sectors such as telecom, finance, banking, and real estate are some of the sectors using mobile commerce. In recent years, a large number of people in India have adopted m-commerce to perform transactions. Further, with increasing number of applications in mobile technology, it is now more secure and convenient for subscribers to use it. Personalization factor, mobility, immediacy, and easy connectivity are some of the factors contributing positively toward m-commerce's future prospects. Moreover, in the year 2015, out of 7.32 billion people worldwide, 7 billion had mobile phones. However, as per study conducted by Telecom Regulatory Authority of India, India has over 935 million mobile connections. Due to increasing trend in mobile uses, now business world is starting to see m-commerce as a major focus for the future. This chapter's theoretical contribution is to explain the trends in mobile commerce in India.

3.1 INTRODUCTION

Commerce is concerned with the flow of goods and services from producer to consumer. Initially, all business transactions were performed in person only. The reason was businesses were operating at small level and personal interaction was possible all the time. But due to globalization, liberalization, and industrialization, now business activities are performed at large scales. From business point of view, whole world is now considered as business village. Further, with development of new technologies, numerous new business methods of commerce take place. Due to development of both print and electronic media, now customers can get product-related information very easily. They are more aware about latest products and services which increase competition among business houses. Moreover, to cope up with competition, now it is essential for each and every business to adopt such methods and technologies which are convenient and useful for customers. Further, time factor is also a matter of discussion in commerce. Most of the people have lack of time to shop from small markets. Further, lack of choice and quality of products are some other matters of concern. To overcome these problems, m-commerce is one of the innovative ways of marketing introduced recently. It is a sophisticated service adopted, both in India and worldwide, that provides a way to many companies to project themselves as an initiator in adoption of new marketing strategies by way of mobile commerce. It directly helps companies to stand out of clutter of traditional ways of marketing. Mobile commerce may be defined as an ability of buyer and seller to buy and sell product and/or information on internet or other online sources. Further, as a subset of e-commerce, all the aspects involved in e-commerce can be extended and applied to m-commerce also (Kwon and Sadesh, 2004). So any transaction having definite consideration and completed via mobile terminal equipment and communication network can be considered as a part of mobile commerce (Yan, 2005). Moreover, as an extension of e-commerce, mobile commerce managed transaction of products and services through wireless mobile equipment. It is featured by free flow of transactions without any time or place constraints and this directly enhance the profitability and efficiency of business.

3.1.1 REVIEW OF LITERATURE

Panneerselvam (2013) in his paper explained that due to increase in number of mobile subscribers, there is a sea change in the way of doing business in

India. Further, mobile commerce is quite popular in different sectors such as financial, telecom, banking, and real estate, etc. He opined that in spite of m-commerce being a subset of e-commerce, it has many more advantages over e-commerce. Moreover, it is more secure and convenient. He found that it will be next generation mode of business. Camponovo (2002) in his study highlighted suitable business model for mobile business market. He revealed that being a newly emerging industry, it includes several technological, demand, and strategic uncertainties. The paper highlighted that a large number of players and participants are involved in mobile commerce activities and they try to fit themselves in its value system. Moreover, when they perform such activities, they try to find out a suitable business model for the same. He concluded that a best business model for m-commerce is that which includes mobility, network effects, and natural monopolies. Kaur and Singh (2016) in their study evaluated the significant growth of mobile commerce in India. They highlighted that mobile commerce is at emerging level in India and mobility is the main factor behind applications such as mobile marketing, mobile banking, and mobile entertainment. They found out that earlier business strategies were formulated after considering geographical limitations and rigidity in mobility. However, these two factors are now totally missing while formulating strategies in case of mobile commerce. Sharma et al. (2012) in their study highlighted the impact of government policies and regulations on m-commerce in India. They explained that due to uncertainties and complexities attached with mobile commerce business, a proper regulatory framework is required for the same. They revealed that the main reason behind complexities attached is involvement of large number of stakeholders and regulatory agencies. Further, for the solution of mobile commerce complexities, system dynamics simulation models are presented by them. The paper suggested that with the help of these models the policy makers can easily understand the relation between different policy approaches and their impact on mobile business markets. Satinder and Niharika (2015) in their research article highlighted that to attain good and fast business transactions, Indian people are now switching to m-commerce. Further, they opined that easy availability of mobile apps is another reason behind attraction of people toward use of mobile commerce. They highlighted that complexities in mobile application, payment issues, and security problems are some main problems attached with mobile commerce those need to be addressed. So due to these unique features of m-commerce, this chapter has been planned. The main objectives of this chapter are as follows:

1. To study the unique features of m-commerce and comparison between m-commerce and e-commerce.
2. To explain evolution of mobile commerce and technology used for the same.
3. To find out suitable mobile business market model with its major determining factors.
4. To highlight the uses of mobile commerce and Strength, Weakness, Opportunity and Threat analysis (SWOT) of Indian mobile commerce business.
5. To evaluate the recent trends and growth of mobile commerce in India and worldwide.

Further, this chapter has been divided into three sections. First section deals with the evolution of m-commerce and its unique features and technology used for the same. A suitable mobile commerce model with its major determining factors and application of mobile commerce is discussed in Section 2. Further, section three deals with trends in mobile commerce worldwide and in India and a SWOT analysis of Indian mobile commerce market.

3.1.2 MOBILE COMMERCE VERSUS E-COMMERCE

E-commerce is concerned with buying and selling of goods and services with the help of electronic medium having internet connectivity. E-commerce users can send goods and services anywhere and anytime while sitting in their office and home. A large number of benefits such as unlimited shelf space, no geographical boundaries, no time bondage, and lower operating cost are provided by e-commerce to its users. Moreover, easy internet connectivity is another main reason behind increased use of e-commerce.

M-commerce (mobile commerce) is the process of buying and selling of goods and services or transmitting of funds or data via a mobile device. M-commerce also includes many other activities such as business trading, m-banking, m-ticketing, etc. M-commerce is the extended form of e-commerce. Further, the domain of e-commerce is wider than m-commerce in which internet-enabled hand held devices and increased use of smartphones formulate foundation for mobile commerce. M-commerce as an expansion of e-commerce is also known as wireless e-commerce in which a business transaction having consideration or with money value is conduct through mobile phones. (Kaur and Singh, 2016). Some of the main technologies supporting mobile commerce are as under:

- High Speed Circuit Supported Data (HSCD)
- Global System for Mobile Communication (GSM)
- General Packet Radio Services (GPRS)
- Enhanced Data rate for Global Evolution (EDGE)
- Wireless Application Protocol (WAP)
- Bluetooth
- Universal Mobile Television System (UMTS)

3.1.3 FEATURES OF M-COMMERCE

M-commerce have number of unique features which make it an effective way of doing business transactions. Some of these features are discussed as under:

1. **Independent of geographical location:** In mobile commerce, consumer can buy product and services directly by use of their mobiles anywhere. Producers and customers need not be at same place. Further, producers can sell their products anywhere in the world without any regional boundaries. This proves to be a strategic advantage for producers.
2. **Effective time management:** Due to connectivity feature, products and services are provided by producer in time. In the same way, consumers also get benefited by real-time availability of products.
3. **Customized services:** By use of mobile commerce, companies can provide services as per need of the customers. Moreover, more personalization and connectivity is possible with customers.
4. **Keep in touch:** Due to full time connectivity with customers, companies using mobile commerce can easily maintain better relationship with customers. Further, they are also able to get timely feedback for their services and use this information for future reference, after sales services, and customer data maintenance.
5. **Wireless:** In case of mobile commerce, customer need not to be fixed at one place with a fixed infrastructure. But in this form of commerce, he can avail anywhere and anytime business transaction facility. So, it is very convenient for users to operate it.

6. **All time internet:** With new form of mobile technology, user can enjoy all-time internet facilities and can enjoy facilities such as online shopping, online payments, and can also avail several online service facilities.
7. **Mobile app availability:** A number of mobile apps are now available which makes use of mobile commerce easier. These apps are related with online payments, online education, online booking, and online shopping.
8. **Speed facility:** One of the unique features of mobile commerce is that it speeds up transaction process and provides instant satisfaction and contentment to customers. It involves urgency in transaction and speeds up sales through business tactics such as short-term discount, etc.
9. **Possibility of future innovation:** Mobile commerce market is continuously developing over time and thus, will provide opportunity for development in mobile phone market also. Possibility of potential innovation is very high in mobile commerce such as there is need of new mobile apps and innovative payment system.
10. **Affordability feature:** Due to technology development, now smartphones are in reach of lower income group people also. These smartphones can easily be used as internet device for online shopping and payments. (Panneerselvam, 2013).

Further, the main differences between m-commerce and e-commerce are given in Table 3.1.

TABLE 3.1 M-Commerce Versus E-Commerce.

S. no.	Factors	M-commerce	E-commerce
1	Domain	M-commerce is the extended form of e-commerce that involve more innovative technology	Scope of e-commerce is much wider than m-commerce
2	Technology	Smartphones, PDAs	Computers, internet connectivity
3	Convenience	It is easy to handle mobile device. Further, it can be accessed any time and any place	This facility is not possible in e-commerce due to use of fixed infrastructure
4	Security	Due to availability of authentication number it is more secure	It is less secure comparatively

TABLE 3.1 *(Continued)*

S. no.	Factors	M-commerce	E-commerce
5	Internet connectivity	Such boundaries are not there in m-commerce. All time internet facility is now available on mobile phones	It is must in e-commerce to have internet connection and required infrastructure
6	Electricity required	No such compulsion is there in mobile commerce. Only battery backup is required	It is must for smooth functioning of computer and other devices
7	Cost	Comparatively more costly than e-commerce	It is less costly
8	Video conferencing	In this form of commerce video conferencing is possible	In this video conferencing is not possible

PDA, personal digital assistant.
Source: Compiled by author.

3.2 EVOLUTION OF MOBILE COMMERCE

Mobile commerce conducted over mobile or wireless network is very much different from desktop computer-based transactions. It can be differentiated on the basis of presentation, processing, and interactive modalities. The introductory phases in mobile commerce can be explained with the help of mobile revolutions and competitive dynamics of mobile marketplace. Further, this mobile revolution can be divided into three phases and explanation of these phases is given in following sections.

PHASE 1

It was in the year 1997 when WAP forum was formed. Further, in this phase, m-portals were introduced. NTT, DoCoMo's i-mode and Vodafone live (as J-sky) were launched in Japan in the February, 1999 and December 1999, respectively. Another example is Verizon wireless in the United States. Mobile portal was a method for providing mobile apps such as ringtones downloads, weather information, location-based services, and online reservation, etc. In m-portal, walled garden approach was adopted that was closed form of business approach. Moreover, user control and internet openness qualities were not included in m-portal mode. So due to close nature, this method became less useful for users and this led to its withdrawal from mobile business market. Further, more user-friendly technology was introduced over time.

PHASE 2

The second phase started in the year 2000 with introduction of m-internet. Ericsson R380 Smartphone was introduced in the same year. This was the first device with Symbian operating system. In the same year, opera mobile was introduced. This phase is known as open m-commerce market phase. Due to advancement in technology, internet-capable smartphones were introduced with high speed offered by 3G mobile broadband. With introduction of m-commerce business model, now it was compulsory for traditional e-commerce players to modify their websites as per m-internet business model. Now, it was possible to provide location-based services to customers.

PHASE 3

The third phase is known as m-apps phase that started in the year 2007. Due to popularity of m-commerce, new players entered in m-commerce market with advanced technology. In June 2008, Apple's app store was launched. Further, in October 2008, Google launched its Google's Android. In the same line, Apple launched its i-pad and iTunes and Microsoft launched window phones. Again in this phase m-commerce business model shifted from open ended to close ended approach with use of mobile apps. (Kourouthanassis and Giaglis, 2012)

Moreover, it was in the year 1997, in Helsinki area of Finland when first time Coca-Cola Company started mobile commerce activities for payment. In the service sector, Merita bank of Finland was the first bank that adopted mobile banking in the same year. At the same time, smart money concept was introduced in the year 1999, in Philippines. However, in communication sector, Japan's company DoCoMo introduced first mobile internet platform for payment. In India, mobile services started in the year 1998. However, in 1991 with introduction of economic reforms in India, private players also found a way to enter in Indian business. Indian telecom industry got pace in the year 1993–1994 after involvement of private participants. Telecom Regulatory Authority of India (TRAI) was established in the year 1997 and it replaced Department of Telecommunication.

3.2.1 TOOLS FOR MOBILE COMMERCE

By mobile commerce tools, we mean how producers and service providers connect with customers and vice versa. Some of the main tools are as under:

- **Area-based marketing***:* It focused on customers of specific area or locality. Companies locate their customers with the help of Global Positioning System (GPS) and provide goods and services as per requirements of people in that particular area.
- **Short message services:** This is the simplest, cheapest, and widely used method adopted by business houses to sell their products and services. These business houses have data bank of their existing and potential customers and they send bulk messages to their customers with product description and other details such as payment, outlets, and offers.
- **Multi-media message services (MMS):** Color screen mobile phones are basic requirement for the use of multimedia services. Audio, video, and text are used in it. This method of mobile marketing is more convenient than SMS. The reason behind is that product demonstration is possible which proves to be more convenient to customers.
- **Interactive voice response (IVR):** Being a latest one, this method is now widely accepted to provide various services to users. In this system, recorded instructions are there and a customer can benefit by obeying the instructions.
- **Web pages/websites:** With internet connectivity, mobile users can go through various company or product web pages/websites by searching various search engines such as Google and Yahoo, etc. These search engines also provide paid space for advertisement of product and collect handsome amount from these services.
- **Bluetooth:** With proximity of 10 m range, this method is used only by small businesses and shopkeepers such as mall owners and restaurant and café houses (Kaur and Singh, 2016).

3.3 BUSINESS MODEL FOR MOBILE COMMERCE

"M-commerce is an interactive ecology system of people and organizations based on social and technological effects" (Mylonopoulos and Doukidis, 2003).

Like other emerging industries, m-commerce is also a newly developed promising industry with numerous uncertainties related with technology, demand, and strategies. Technical and demand-related uncertainties basically arise at the beginning stage of any industry. Technical uncertainties are mainly related with technological development with acceptable standards. However, demand-related uncertainties are related with development of services as per users' need. Further, there are no established rules and strategies regarding emerging industry like m-commerce. Several opinions are there for development and selection of strategy. But no set criteria are there for the same. It is only repositioning method which can help m-commerce participants to select appropriate strategies for their businesses to achieve competitive position in industry. Every industry has certain characteristics on which any business model depends. Positive externality, mobility, and natural monopolies are the main features of mobile business industry. Cordial relationship with partners or partnership management is the main challenge for m-commerce business. Main features affecting selection of mobile business model are given in Table 3.2.

TABLE 3.2 Mobile Business Market Features Affecting Selection of Mobile Business Model.

Mobility	Connectivity	Monopolies
Ubiquity	Network technology	Rarity of goods
Reach ability	Direct connectivity	Business secrets
Localization	Indirect connectivity	Special privilege
Convenience		Network infrastructures
Quick connection		
Personalization		

Source: Adapted from Camponovo (2002).

From the features given in Table 3.2, it is clear that mobile business market is highly fragmented and requires a large number of business players, mobile phone manufacturers, and network providers. Partnership or acquisitions are certain methods appropriate for mobile business market. Further, the main participants in m-commerce business model are technology, service, network providers, users, regulation, and social context. In technology, we mainly include device manufacturer and equipment vendors. Services include content providers, app providers, and payment agents. Mobile network operators and internet service providers are part of network.

Government, security groups, and regulation authorities are the main regulation authorities to regulate other participants under regulation and social context. Main m-commerce participants along with their subsets are given in Table 3.3.

TABLE 3.3 Mobile Business Model Participants.

Technology	Services	Network	Regulation	Users
Device manufacturer Net equipment vendors Component makers Device retailers Operating system providers Microbrowser producers Development platform providers	Content provider Content owner Content aggregator E-business player Payment agents	Mobile network providers Internet service providers	Government Statutory authority Standardized group	*End users* Corporate Consumers *Vertical players* (*mobility needs*) Travel Logistics Health care Retail

Source: Adapted from Camponovo (2002).

On the basis of above mentioned features and participant, mobile business model is given in Figure 3.1.

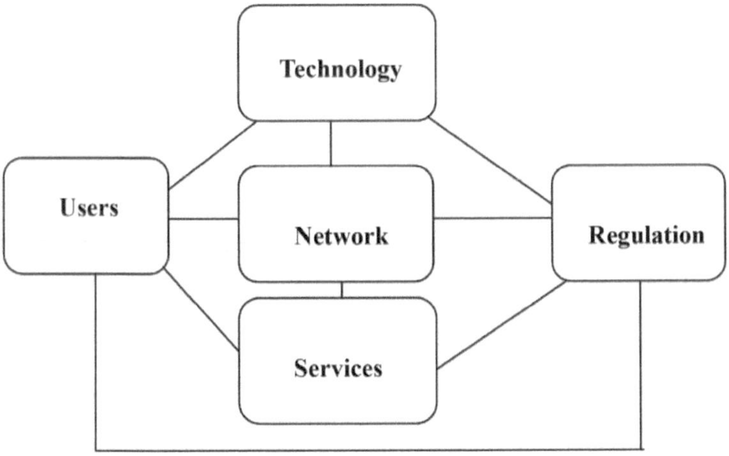

FIGURE 3.1 Mobile commerce model.

Source: Compiled by author.

So, it is clear from above discussion that each player in mobile business market is first of all required to establish their own business model and then club their specific business model with other players in mobile business market. Only after that a complete business model for mobile business can be formulated. But due to newness of mobile business market, it is unpredictable regarding business environment. Further, variety of business models are there and it is very difficult to select appropriate one. Moreover, a business model which provides sustainable and profitable position to mobile business participants is best among all alternatives for mobile business model.

3.3.1 APPLICATIONS OF MOBILE COMMERCE

A number of services can be availed with the help of mobile commerce. These services are featured with any place and any time quality. These services includes money transfer, online booking, information updates, browsing data, mobile banking, brokerage services, content delivery, E-auction, mobile education services and mobile marketing, and advertising. Main services and applications of mobile commerce are as under:

1. **Money transfer:** Money can easily be transferred with the help of mobile commerce. By use of mobile phone both payer and payee can easily transfer money from one place to another.
2. **Online booking services:** It is really a tedious job for most of the people to stand in a queue for air, cinema, bus, railway, and event ticketing. But with the help of mobile ticketing service, one can book, pay, obtain, and validate ticket from any place and any time. The user can get SMS and MMS for confirmation or cancellation of their ticket. Further, all this procedure of mobile ticketing can be adopted by downloading applications for the same.
3. **Information updates:** All important information related with news, stock, traffic, and sports updates which were initially available on Personal Computers only can now be availed on mobile phones.
4. **Browsing data:** Now a large number of upgraded mobile browsers are available, and mobile users can browse World Wide Web at any time and place. They can shop online by operating these browsers. Further, they need not be in present in person to shop goods online.
5. **Mobile vouchers:** Vouchers, coupons, and loyalty cards are represented by virtual tokens. These can be sent to mobile phones and users can avail benefits of these virtual tokens by just showing them

to the concerned person. Moreover, location-based services are also used by big stores to send coupons to their customers when they track them nearby to their stores.

6. **Brokerage services:** Stock market services can easily be provided via mobile device and are very popular these days. Technical name for above service is mobile brokerage and user can take timely and accurate decision related with the stock by using this mobile brokerage service.
7. **Mobile banking:** Banks provide mobile banking facilities to their customers for transfer and remitting money. Further, they can purchase stock with use of this service. Notifications, mini statements of bank accounts, balance checking, Personal Identification Number provision, and cash in and out transactions on an Automated Teller Machine are some other facilities provided by banks and financial institutions to its customers.
8. **Content purchase and delivery:** Due to innovative technology, now smartphones are in much use these days. These mobiles have quality of portable audio and video convergence. Further, high-speed 4G networks are there. Due to high speed and uninterrupted connectivity of internet, mobile users can purchase full audio/video contents online. Ring tones, wall papers, and games are mainly sold and delivered through mobile commerce.
9. **Mobile marketing and advertising**: Marketing and advertising of product and services with the help of mobile commerce is in trend these days. This is very convenient and less expensive method for business houses to expand their brand. Further, due to recent evolution of this concept, few rules and regulations have been formulated till date. However, security of customer data is another problem faced by mobile marketer and advertisers.
10. **E-auction:** Now, electronic auction can be possible with the help of mobile commerce. In it, bidding for goods and services can be performed with the help of electronic implementation. Here, first of all deals are finalized between buyer and supplier and only after that physical transaction takes place. This method of auction is less time consuming and convenient for both suppliers and buyers.
11. **Education services:** Now, it is possible to access various information related with all subjects by students, academicians, researchers, NGO's, and funding agencies with the help of browsing of related data on their mobile devices. Online journals and books are also

available easily. Further, online access to libraries is also possible by subscribing to these.

12. **Location information:** Knowledge of location of the mobile phone user is one of the important information for vendors during business transactions. This facility is only possible in case of mobile commerce. Some of the location-based services include local offers, local map, people tracking and monitoring, and weather information.

3.4 TRENDS IN MOBILE COMMERCE

Mobile commerce is very popular these days due to easy availability and reasonable rate of mobile phones. Further, in the year 2015, the number of mobile phone users worldwide was 7 billion. In India, it was above 1.15 billion as compared to 1.06 billion in China. Moreover, India scored first rank in mobile phone users, followed by China, the United States, Brazil, and Russia. Globally, up to the year 2015, only 40% people have internet connection. However, in India 26% people have internet connection as compare to China (50.30%), the United States (65%), Brazil (59.08%), and Russia (73.41%). Also, in India 17% people were smartphone users in 2015, and India ranks 127th in internet users (Global Attitudes Survey by Pew Research Centre, 2015). As per data provided by Statista in 2014, the number of smartphone users worldwide was 1.57 billion that will increase up to 2.32 billion in 2017 and 2.87 billion in the year 2020. Further details are given in Table 3.4

TABLE 3.4 Mobile Phones and Internet Users Ranking.

Country	No. of mobile phones (in billions)	Ranking	% of internet users	Ranking
World	7		40	
India	1.15	1	26	127
China	1.06	2	50.30	90
United States	0.33	3	65	40
Brazil	0.28	4	59.08	71
Russia	0.25	5	73.41	43

Source: Adapted from http://en.m.wikipedia.org/wiki/list-of-countries-by-number-of-internet-users (accessed Feb 9, 2017.)

Globally, in the year 2015, around 35% of e-commerce transactions took place via mobile Phones. As per report by HubSpot in year 2015, smartphones surpass computer and tablets regarding number of Google search queries. HubSpot predicts that mobile commerce will grow by 31% in 2017 as compare to e-commerce which is predicted to grow by 15%. So, m-commerce grows 200% faster than e-commerce. However, China is number one mobile shopper with 56% share and 562 million dollar revenue (Kolowich, 2015). Details are given in Table 3.5.

TABLE 3.5 Percentage Share of Mobile Shoppers.

Country name	Rank	% share of mobile shoppers
China	1	56
USA	2	77
UK	3	66
Japan	4	42
Germany	5	27

Source: Adapted from http;//www.criteo.com>media>criteo-state-of-mobile-commerce-q1-2015-ppt (accessed Feb 9, 2017).

There are approximately 1 million e-commerce sites and keep on increasing day by day. Out of these, 65,000 sites use mobile commerce app for their websites due to favorable cost benefit analysis. Some of the best mobile commerce apps have been given in Table 3.6.

TABLE 3.6 Best Mobile Apps Worldwide.

S. no.	Apps	S. no.	Apps
1	Amazon	6	Best buy
2	Walmart	7	Flipkart
3	Ebay	8	Shop.com
4	Alibaba	9	Groupon.com
5	Target	10	Ikea.com

Source: Adapted from https://www.trendingtopmost.com/worldpopular-list-top10/2017-2018-2019-2 (accessed July 3, 2017).

3.4.1 M-COMMERCE GROWTH IN INDIA

India is the fastest growing m-commerce market, and it is evident from the fact that in the year 2012, in India, m-commerce growth rate was 30.3% which increases to 62.9% in the year 2014. It was followed by Taiwan with 28.2% in 2012 and 62.6% in 2014 and Malaysia with 25.4% in 2012 and 45.6% in 2014. India is the second largest internet user base with more than 300 million internet users. These days "Mobile only internet" is in trend in India. Due to slow connectivity of 3G in India, Wi-Fi is mainly used by the mobile users. As smartphone users are increasing in India, the demand for m-commerce apps also increased and it registered a growth rate of 131% in the year 2015. In the year 2012, around 1.56 billion apps were downloaded which increased up to 9 billion in the year 2015 with 75% cumulative annual growth rate. Top 10 mobile apps used in India in the year 2015 have been given in Table 3.7.

TABLE 3.7 Top 10 Mobile Apps in India.

Ranking	Mobile apps	Ranking	Mobile apps
1	Flipkart	6	Ola
2	Myntra	7	Uber
3	Amazon	8	Jabong
4	Snapdeal	9	ShopClues
5	Paytm	10	Meru

Source: Adapted from Dalal, M. https://www.livemint.com/consumer.NdIEK83hrwa640qc-Qoes5k/India-top-10e-commerce-apps (accessed July 3, 2017).

As per Forbes, due to popularity of mobile commerce, one of the top fashion e-commerce Myntra decided in the year 2015 to shut down its website and become mobile app only. Flipkart is also thinking to experiment in the same way.

3.4.2 SWOT ANALYSIS OF M-COMMERCE IN INDIA

High use of technology is an important indicator of economic growth in developing country like India. In developed countries, most of the activity is seen in B2B segment but for developing countries, B2C segment is more feasible. The reason behind this is lack of basic infrastructure facilities such as power, road, transport, and bandwidth, etc. Improper connectivity of internet is the main reason behind less use of mobile commerce facility in rural area in India. Privacy, protection of consumer interest, and data protection are some other concerned areas in m-commerce. To develop a strong m-commerce base, it is essential to know the strengths, weaknesses, opportunities, and threats of m-commerce in India. Strength, Weakness, Opportunity, and Threat (SWOT) analysis matrix of m-commerce in India is given in Table 3.8.

TABLE 3.8 SWOT Analysis of Indian M-Commerce Market.

Strength	**Weakness**
Convenient	Lack of high speed
Wireless	Interrupted
Personalized	Internet connectivity
No geographical limitation	Complicated
Real-time availability	Hacking
All time internet facilities	Lack of awareness
Mobile app availability	Not suitable for small business
Opportunities	**Threat**
Increase in mobile users	Lack of confidence
Availability of apps in local language	Security issues
Improvement in technology	Lack of proper internet laws
Cashless India initiatives	Lack of standards
Availability of internet in rural areas	Product specific
	Difficulty in reverse transactions

Source: Compiled by author.

On the basis of SWOT analysis of Indian mobile commerce market, it is clear that to increase the growth rate of Indian mobile commerce market, it is essential for Indian government to build required infrastructure for proper internet connectivity. Some of the main areas where immediate attention is required are as under:

- educate people about use of mobile commerce,
- provide proper security for mobile payment,
- formulate clear rules and regulations for mobile commerce business,
- to promote mobile business, make mobile devices more affordable,
- try to improve mobile internet connectivity in rural areas, and

- to increase the popularity of mobile commerce, try to reduce internet tariffs.

3.4.3 DEMONETIZATION AND MOBILE PAYMENT SYSTEM IN INDIA

Mobile payment is a hybrid form of payment. It includes element from other forms of payments such as credit card, prepaid card, and telephone bills. Mobile payment system works as an intermediary between buyer and seller and payment has to be made with the help of mobile device. Some of the main methods used for mobile payment include:

- Mobile Web Payments
- Mobile Phone-based Payments
- Direct Mobile Billing
- Card-based Mobile Payments
- Mobile Wallets

The main highlights of demonetization decision and its impact on Indian Payment System is as follows:

1. The main reason behind demonetization decision taken by Indian government in November 2016 were:
   a) Demolish counterfeit currency
   b) To unearth black money
   c) To put check on corruption activity
2. As per figures provided by RBI in March 2016, Rs. 500 and Rs. 1000 notes constituted a gigantic share in Indian currency. The currency in circulation at that time was 16,415 billion. Out of this, participation of Rs. 500 notes was 47.8% and Rs. 1000 notes was 38.6%, which constitute approximately 86% of total currency.
3. To strengthen Indian economy against other developing and developed countries, Indian government took this decision of demonetization. The main emphasis was to convert Indian economy in cashless economy where more and more use of digital mode and less use of cash in transactions is possible. Moreover, digital technology will boost growth, expand opportunities, and improve service delivery in any economy.
4. In India, paper currency is the main reason behind evils such as black money, tax evasion, terrorist activities, and existence of informal

economies. Further, all these evils lead to corruption. Increased use of paper money leads to more corruption in economy. It is clear from the Table 3.9.

TABLE 3.9 Status of Cashless Payments and Corruption Perception in Selected Countries.

Countries	% noncash payment[a]	Corruption perception index[b]	Rank in corruption perception index (2015)[b]
Belgium	93	77	15th
France	92	70	23rd
Canada	90	83	9th
UK	89	81	10th
Sweden	89	89	3rd
Australia	86	79	13th
The Netherlands	85	87	5th
USA	80	76	16th
Germany	76	81	10th
South Korea	70	56	37th
India	22[c]	38	76th

[a]http://www.worldatlas.com/article/which-are-the-world-s-most-cashless-countries.html (accessed Dec 15, 2016).

[b]https:://www.iaca.int/images/news/2016/corruption_Perception_Index_2015_report.pdf (accessed Dec 20, 2016).

[c]https://www.equitymaster.com/5minWrapUp/charts/index.asp?date=11/14/2016&story=1&little=Just-22-of-Indias-transactions-Are-Non-Cash (accessed Dec 20, 2016).

Source: Patnaik (2017).

5. In India, at present (up to December 2016), 15 lakh point of sales terminals provide card-based payments. Up to March 31, 2017, an additional 1 million new point of sales terminal would be installed. Further, to promote digital payment system, government has exempted these devices and also raw material used for these devices from excise duty up to March 31, 2017.
6. In Indian economy, cash circulation is very high and it constitutes 13% of India's GDP. More than 95% of transactions take place in cash form only, and nearly 92% of ATM cards are used for cash withdrawal only. To encourage digital transaction culture Jan Dhan Aadhar Mobile (JAM) has been introduced by Indian government. With this scheme, every remote area in India can be covered to

promote digital payment system. Further, JAM is used by Indian government departments to transfer government funds in rural area.

7. Indian economy is rural economy and around 68.84% of people live in villages. As per data provided by Reserve Bank of India in July, 2016, around 697.2 million debit card and 25.9 million credit cards had been issued by banks which does not equal number of individuals holding those cards because of multiple card holding by single person. Further, it also reveals from report that regions with higher female population and more remote area show lower level of financial inclusion. To promote e-payments, plastic transactions, and cashless payments, Indian government has reduced service charges on transaction through debit/credit cards.
8. To promote cashless economy and encourage digital payment system, a number of schemes were introduced by Indian Government after demonetization:

a) **Aadhar Payment App:** The main feature of this app is that it can be used with and without phones. In it, Aadhar card number of holder is linked with his/her bank account number and with the help of biometric reader, holder enters his Aadhar card number and selects a bank for the transaction.

b) **RuPay***:* Like Visa or Mastercard, this is a form of debit or credit card. This scheme is basically for poor people and lower income people to attract to cashless economy. Under Jan Dhan Scheme, RuPay was launched by National Payment Corporation of India. The person who adopted this scheme and opened an account with bank, got rupee 1 lakh accidental insurance.

c) **Bharat Interface for money (BHIM):** BHIM is basically an e-wallet used for online transactions. In it, mobile number is attached with the bank account where holders link their Aadhar number. Further, BHIM is Aadhar-based mobile payment application and only works with UPI-enabled bank accounts.

d) **Digi Dhan Vyapar Yojana** and **Lucky Grahk Yojana** were also launched by Indian government after demonetization to promote digital payment. Digi Dhan Vyapar Yojana is for merchants and Lucky Grahak Yojana for consumers. These are basically digital payment schemes for consumers and merchants.

e) **Vittiya Saksharta Abhiyan Campaign***:* To encourage people to adopt digital payment methods and cashless economy. The ministry of human resource development appealed private and public institutions to develop a cashless campus by use of digitally enabled cashless economic system for transactions.

3.4.4 ADVANTAGES OF CASHLESS ECONOMY TO M-COMMERCE

Being a fastest growing economy, Indian economy suffers from different socio-political and economical issues such as corruption, terrorism, black money, etc. Further, in India only 10–15% people are those who ever used noncash method of payment. So, number of users is very few. Moreover, a number of advantages are attached with cashless payment system that directly helps mobile business market. These advantages are as under:

- Reduction in transaction cost
- Cover more rural customers due to electronic payment compulsion.
- Assurance of sound security system for cashless payments by the government.
- Formulation of concrete rules and regulations for the safety of both customers and vendors.
- Accelerate the growth of development of basic infrastructure facilities required for mobile commerce.
- Awareness campaigns about cashless transaction by the government help to build confidence in users. It directly helps to promote mobile commerce.

3.5 CONCLUSION

Worldwide, mobile commerce is growing rapidly. Increased competition, customer awareness, and different market strategies are certain factors those directly contribute in mobile commerce growth. Further, mobile commerce as a game changer in online market is very much customer friendly and give opportunities to customers to shop anytime at any place. Moreover, with introduction of 3G and 4G technologies, now future of mobile commerce is very bright. It provides value for money to the users by providing convenience, timely, and personalized services. It is evident from marketing experiences that ever-growing technology will definitely boost mobile commerce market if proper marketing techniques are clubbed with strategic initiatives. So, the future of mobile commerce is bright with tremendous opportunities.

KEYWORDS

- **m-commerce**
- **evolution**
- **trends**
- **SWOT analysis**
- **demonetization**
- **cashless payment system**
- **business model**

REFERENCES

Camponovo, G. Mobile Commerce Business Models. Work Supported by Swiss National Science Foundation. *Version 0.5* **2002**.

Kwon, O. B.; Sadeh, N. Applying Case-based Reasoning and Multi-agent Intelligent System to Context-aware Comparative Shopping. *Decis. Support Syst.* **2004,** *37*, 199–213.

Kourouthanassis, P. E.; Giaglis, G. M. Introduction to the Special Issue on Mobile Commerce: The Past, Present and Future of Mobile Commerce Research. *Int. J. Electron. Commer.* **2012,** *16* (4), 5–17.

Kaur, R.; Singh, S. Mobile Commerce: Indian Perspective. *Int. J. Innov. Res. Comput. Commun. Eng.* **2016,** *4* (3), 4320–4326.

Myonopoulos, N. A.; Doukidis, G. I. Introduction to the Special Issue: Mobile Business: Technological Pluralism, Social Assimilation and Growth. *Int. J. Electron. Commer.* **2003,** *8* (1), 5–22.

Panneerselvam, K. M. Mobile Commerce—A Mode of Modern Business. *Asia Pacific J. Market. Manag. Rev.* **2013,** *2* (7), 141–149.

Sharma, D.; Sundar, K. D.; Murthy, V. R. Evaluating the Impact of Government Policies and Regulations on M-commerce in India: A System Dynamic Modeling Approach. *Int. Bus. Manag.* **2012,** *7* (23), 54–80.

Satinder; Niharika. The Impact of Mobile Commerce In India: A SWOT Analysis. 2nd International Conference on Science, Technology and Management, University of Delhi, New Delhi, India, 2015. http://data.conferenceworld.in/ICSTM2/P2503-2513.pdf (accessed Feb 9, 2017).

Yan, L. The Analysis on How China Turn to be the Global Power in M-commerce with the Advantage of Backwardness. *Manag. World* **2005,** *7*, 162–165.

List of Countries by Number of Internet Users. http://en.m.wikipedia.org/wiki/list-of-countries-by-number-of-internet-users (accessed Feb 9, 2017).

State of Mobile Commerce Growing Like a Weed, Q1-2015. http;//www.criteo.com>media>criteo-state-of-mobile-commerce-q1-2015-ppt (accessed Feb 9, 2017).

Top Ten Best Online Shopping Websites in World. https: //www.trendingtopmost.com/world-popular-list-top10/2017-2018-2019-2 (accessed Sept 3, 2017).

Dalal, M. India's Top Ten Mobile E-commerce App. http://www.livemint.com/consumer/NdIEK83hrwa64OcQoeS5K/India-top-10e-commerce-apps (accessed Sept 3, 2017).

Kolowich, L. How is Mobile Commerce Growing Around the World? (Infographic). http://blog.hubspot.com/marketing/mobile-commerce-growth-global, 2015 (accessed Feb 9, 2017).

Profile of Indian Internet Users. Global Attitudes Survey. Pew Research Centre, 2015. http://www.pewresearch.org/fact-tank/2016/04/06/global-tech-companies-see (assessed Feb 9, 2017).

SUGGESTED READINGS

Agarwal, A.; Bhatawal P. M-commerce in India: Promise and Problems. *Int. J. Res. Comput. Commun. Technolo.* **2015,** *4* (4), 273–276.

Deshmukh, P. S.; Deshmukh, P.; Thampi, G. T. Transformation from E-commerce to M-commerce in Indian Context. *Int. J. Comput. Sci. Issues* **2013,** *10* (4), 55–60.

Dhingra, K.; Bhardwaj, A.; Aggarwal, A. M-commerce. *Int. J. Eng. Res. Gen. Sci.* **2015,** *3* (2/2), 704–706.

Golden, R. A.; Regi, S. B. Mobile Communication in Modern Business Era. *Int. J. Curr. Res. Acad. Rev.* **2013,** *1* (4), 96–102.

Gupta, S.; Vyas, A. Benefits and Drawbacks of Mobile Commerce in India—A Review. *Int. J. Adv. Res. Comput. Commun. Engender.* **2014,** *3* (4), 6327–6329.

Jahanshahi, A. A.; Mirzaie, A.; Asadollahi, A. Mobile Commerce Beyond Electronic Commerce: Issues and Challenges. *Asian J. Bus. Manag. Sci.* **2011,** *1* (2), 119–129.

Mishra, A.; Medhavi, S.; Mohd, K. S.; Mishra, P. C. Scope and Adoption of M-commerce in India. *Int. J. Adv. Res. Comput. Commun. Engender.* **2016,** *5* (8), 231–238.

Ngai, E. W. T.; Gunasekaran, A. A Review for Mobile Commerce Research and Application. *Decis. Support Syst.* **2005,** *43*, 3–15.

Nawara, M. A. M-commerce in India. *Int. J. Innov. Res. Comput. Commun. Eng.* **2016,** *5* (4), 231–238.

Pattanaik, B. K. Demonetization, Cashless Economy and Development. *Yojan* **2017,** 52–57.

Singh, D. Electronic Commerce: Issues of Policy and Strategy for India. *Indian Council Res. Int. Econ. Relat.* **2002,** 1–66.

Shafer, S. M.; Smith, H. J.; Linder, J. C. The Power of Business Models. *Bus. Horiz.* **2005,** *48* (3), 199–207.

Shettar, R. M. Services and Applications of Mobile Commerce in India: An Empirical Study. *J. Res. Human. Soc. Sci.* **2016,** *4* (11), 94–100.

Tsalgatidou, A.; Pitoura, E. Business Model and transaction in Mobile Electronic Commerce: Requirements and Properties. *Comput. Netw.* **2001,** *37* (2), 221–236.

Shettar, R. M. Services and Applications of Mobile Commerce in India: An Empirical Study. *J. Res. Human. Soc. Sci.* **2016,** *4* (11), 94–100.

Tsalgatidou, A.; Pitoura, E. Business Model and Transaction in Mobile Electronic Commerce: Requirements and Properties. *Comput. Netw.* **2001,** *37* (2), 221–236.

CHAPTER 4

SUCCESS FACTORS OF M-COMMERCE: A CUSTOMER PERSPECTIVE

ANU G. AGGARWAL[1*], GUNMALA SURI[2], and HIMANSHU SHARMA[1]

[1]*Department of Operational Research, University of Delhi, Delhi 110007, India*

[2]*University Business School, Punjab University, Chandigarh, India*

**Corresponding author. E-mail: anuagg17@gmail.com*

ABSTRACT

With the advancement in technology and the introduction of internet in 1940s, the traditional electronic data interchange (EDI) has transformed to internet-dependent commerce, electronic commerce (e-commerce). E-commerce has enabled users to access services and make transactions from anywhere and anytime. Nowadays e-commerce is moving toward mobile commerce (m-commerce). M-commerce was inculcated as a part of e-commerce in the 1990s. This allowed the users to do commercial activities while they were moving. Although many studies related to m-commerce have existed in literature, they were focused on specific regions or areas or from specific viewpoints. Researchers have distinguished the success factors into critical success factors (firm's perspective) and customer-oriented success factors (customer's perspective). It has been proved previously that m-commerce when compared with e-commerce share some similar points but they are not analogous. The major point of difference is the mobility criterion possessed by m-commerce. With over 7.7 billion mobile connections and 4.7 billion unique mobile subscribers in the world, which comes out to be more than the world population, there exists a need to study m-commerce especially from the customers' perspective. The aim of this chapter is to define the factors that affect success of m-commerce from customers' point of view, evaluate these factors, and rank some predetermined mobile applications on the basis of these factors. Previous studies were centered on finding the system success factors and less work has concentrated toward customer satisfaction. The success of e-commerce firms depends on the perception of online customers toward their website. But since most of the e-commerce firms are extending to m-commerce, the success of these firms depends on the satisfaction that the customers obtain while using the mobile application. This chapter discusses in detail the factors that impact the success of m-commerce. Here, we extend the factors defined by Kabir and Hasin (2011), by adding a criterion, namely, "online customer feedback." The factor is incorporated so as to take care of the impact of feedbacks on the potential purchasers of that product/service. Also the model proposed in this chapter combines the Multi Criteria Decision-Making (MCDM) techniques of Analytical Hierarchy Process (AHP) and Preference Ranking Organization METHod for Enrichments Evaluation (PROMETHEE). The AHP is used to obtain the criteria weights through a structured survey and PROMETHEE is used to obtain the final ranking of the m-commerce applications concerned. A numerical illustration is given to show the efficiency of the proposed model. The illustration validated the significance of the two criteria: online customer feedback and personalization. This showed the importance of feedbacks/reviews and personalized products/services among the new-age tech-savvy customers.

4.1 INTRODUCTION

Until the early 1990s, the commercial activities were conducted using electronic data interchange (EDI) which was assumed to be a closed and standardized form of communication between computers. But with the advent of internet technology, they evolved into electronic commerce (e-commerce). Thus, with the development of internet in 1948, many "brick-and-mortar" firms transformed to purely click or "brick-and-click" firms. E-commerce can be classified on the basis of the nature of their transaction into business-to-business (B2B), business-to-consumer (B2C), peer-to-peer (P2P), consumer-to-business (C2B), etc. In the 1990s, mobile commerce (m-commerce) was accepted to be a part of e-commerce. This led to more advancement of firms as they could now move from "website firms" to "application firms." The credit of the popularity of m-commerce goes to an increasing trend in the number of smartphone users in the world, especially in a developing country such as India, which has shown largest increments in the number of mobile phone users over the years. In India, the retail m-commerce revenue increased from US$ 7.78 billion in 2015 to US$ 15.27 billion in 2016, is expected to reach US$ 63.53 billion by 2020 (Statista, 2017), and is represented by Figure 4.1.

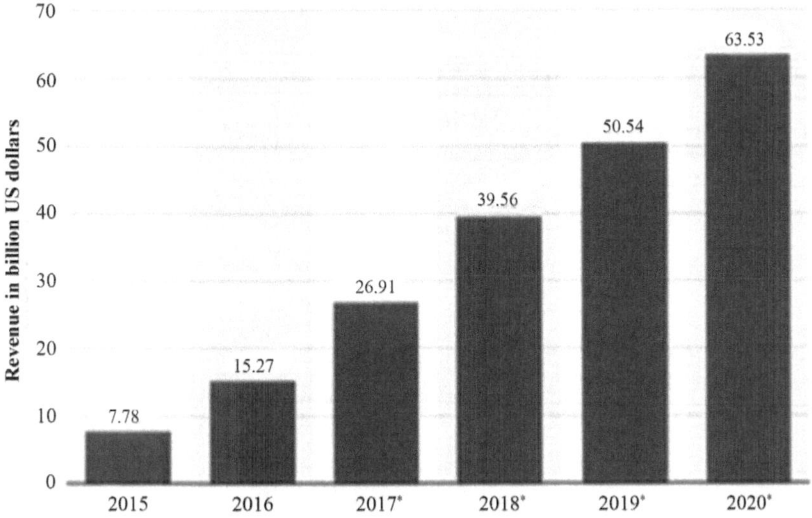

FIGURE 4.1 Mobile commerce revenue generated in India over the years.
Source: Adapted from Statista (2017).

Almost two-third of the world population possesses a mobile phone out of which more than half of the world population is smartphone users (Statista, 2017). The combination of faster mobile connections and improved access to smartphones has made it possible that more than half of all the web pages are now served by mobile phones. Mobile's share of web traffic is up 30% year-on-year, with the majority of this increase coming from the world's developing economies. There are four major forces fuelling m-commerce's growth. (1) Low transaction cost: m-commerce has enabled the users to pay online for their purchased products/services anytime and anywhere and also perform online banking. (2) Low communication cost: this gives the opportunity to m-commerce users to communicate with their peers both locally and globally, especially the use of social applications. (3) Increase in smartphone users: the increase in the number of smartphone users shows that now more people have access to advanced technologies and resources to use mobile applications. (4) Increase in mobile internet users: it has been observed that there has been a dramatic increase in the number of people accessing web via mobile phone other than tablets and laptops/PCs (Hootsuite, 2017). These forces are shown diagrammatically in Figure 4.2.

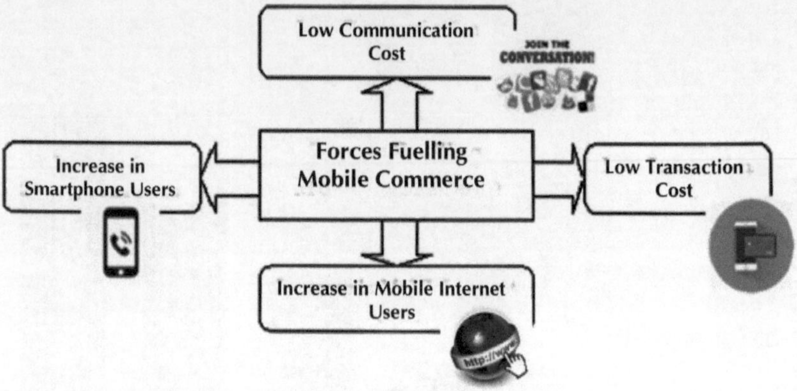

FIGURE 4.2 **(See color insert.)** Forces fuelling mobile commerce.
Source: Compiled by authors.

Earlier, the companies that wished to engage in m-commerce focused on the processes that customers preferred such as purchase and delivery, alternative evaluation, evaluation after purchase, information search, and identification (Turban et al., 2002). Despite the fact that the m-commerce is gaining abundant market, previous researchers have focused on its limited aspects. This necessitates a need for defining a success factor. Many researchers have divided the success factors into two categories: success factors from the firm's perspective (called the critical success factors) and success factors from the view point of customers (called the customer-oriented success factors). Many studies related to success factors with regard to e-commerce exist in literature. While some studies focused on Information System (IS) success (DeLone and McLean, 1992), whereas other studies concentrated toward the customer satisfaction side (Molla and Licker, 2001). These studies showed that despite the similarity between e-commerce and m-commerce in some aspects, they are different in many other features. Thus, it is considered impractical to evaluate m-commerce success on the basis of success factors related to e-commerce. It was observed that the customer-oriented success criteria defined (Molla and Licker, 2001) for e-commerce can be extended to define the customer-oriented success factors for m-commerce by adding the two criteria: mobility and personalization (Kabir and Hasin,

2011). The mobility takes care of the difference in the computer website and mobile applications, whereas the personalization criterion covers the advancements in technology made over the past 10 years. In this study, we add another factor, the online customer feedback. The online customer feedback represents the effect of social communication between the online users. This factor explains the impact of online buzz on the buying behavior of customers. Thus, this chapter proposes a model that incorporates the factors such as system quality, content quality, use, trust, support, mobility, personalization, and online customer feedback.

The model developed in this chapter is consistent with the latest advancements in the technology and also gives importance to the aware customers. Since customer satisfaction is the prime objective for an m-commerce application to become successful, there exists a need for a Multicriteria Decision-Making (MCDM) model to assess the ambiguous and uncertain nature of customers. Here, we use the MCDM techniques such as Analytical Hierarchy Process (AHP) and Preference Ranking Organization METHod for Enrichment of Evaluations (PROMETHEE). AHP is a powerful tool for handling both qualitative and quantitative multicriteria factors in decision-making problems, developed by Saaty (1980). Through this method, a complex problem can be converted into a hierarchical process which can be solved easily. It allows evaluating, numerically, the tangible and intangible elements of the problem. PROMETHEE was introduced by Brans and Vincke (1990). The method enables the decision-maker to obtain both partial and final ranking. The method is useful when it is applied to problems which involve complex computations incorporating multicriteria decisions related to human perception and judgments which have a long impact.

Thus, the main objective of this chapter is to find the best m-commerce application evaluated on the basis of customer-oriented success factors using AHP and PROMETHEE, where the AHP technique is used to provide criteria weights which are a prerequisite for computation of PROMETHEE, which is used for the final ranking of the mobile applications considered in the study. Also, to validate our model, a numerical illustration is performed based on a survey taken up in the Delhi NCR region in India. The remainder of this chapter is structured as follows: Section 4.2 covers the theoretical background relevant to our study. The customer-oriented m-commerce success criteria are discussed in Section 4.3. In Section 4.4, the AHP and PROMETHEE techniques are presented. Section 4.5 describes the proposed model. Section 4.6 covers a numerical illustration based on the proposed model. The discussion and conclusion is covered in Section 4.7. Section 4.8 discusses the limitations and future scope of the study.

4.2 THEORETICAL BACKGROUND

The advent of internet has enabled the firms to market goods and services to their purchasers and provides the facility to process business transactions such as online ordering and online payments. The early 1990s witnessed the introduction of e-commerce with the deployment of internet-based technologies and standardized form of computer-to-computer communication. The e-commerce systems proved to be a better substitute for the traditional EDI. Though earlier the EDI and the e-commerce were treated to be synonymous, it was later possible to differentiate between the two. With this digital revolution, firms attempted to place a wide range of products on the web, through their databases for the convenience of their online customers. Thus, e-commerce has become a necessity for better performance of an enterprise in long term, rather than being just an option.

The era of 1990s witnessed the inclusion of m-commerce as a type of e-commerce. The major reason for the popularity of m-commerce was its distinguishing feature of using and accessing the desired information while moving (anywhere, anytime). Many studies have contradicted the assumption that the factors for evaluating m-commerce and e-commerce might be same (Molla and Licker, 2001). This transformation of wire-connected internet to mobile internet access brought about by the technological advancements has affected the customers' needs and their purchasing pattern. With the technologies changing each day, the study about e-commerce and its subsidiary m-commerce are gaining momentum. The studies related tom-commerce applications are still new and have a wider research perspective for researchers in this domain. Although, m-commerce is now gaining a large-scale market, earlier researches have focused on its limited application. If the m-commerce tackle these limitations in advance and overcome them successfully, it may result in the firm being able to cope with the changes in market more easily. This demonstrates the importance for the study of success factors to the companies/users and what other alternatives they have.

Success is a factor consisting of various dimensions which can be measured at various levels such as individual, group, technical, and organizational, making use of numerous economic, financial, or perceptual criteria (Molla and Licker, 2001). Numerous studies have been done in the field of IS success measure, but still these researches prove to be a speculative one. The combined studies of Mason (1978) and Shannon and Weaver (1949) yielded the six main determinants of information success: System Quality, Information Quality, Use, User Satisfaction, Individual Impact, and Organizational Impact. The System Quality concentrated on the characteristics of the information system itself. The Information Quality focused on the characteristics of information product such as accuracy, timeliness, and meaningfulness. The Use or the User Satisfaction analyzed the interactions of the information system with its recipients/users. The Individual Impact studied the influence that information product has on the management decisions. DeLone and McLean (1992) proposed an IS success model which assessed many individual dimensions for information system success. According to DeLone and McLean (D&M) Model of IS success, the major success determinants were System Quality, Information Quality, Use, and User Satisfaction. The D&M IS Success Model stated that there exist three different levels to assess the success of e-commerce systems such as individual, system, and organizational. Kim focused on the effects of system quality toward e-commerce systems, which help in establishment of web presence by companies (Kim, 1999). Moreover, the content quality, which is different from technical quality, affects the e-commerce systems (Tarasewich, 2003). E-commerce researchers also noticed the usability criterion by assessing the use level and satisfaction of users with e-commerce systems. Though the topic has been researched and significant contributions have been made in this field, less emphasis is put on integrating the dependent variable (customer satisfaction) and the independent variables. According to D&M model, use and user satisfaction were considered to be the dependent variables in the e-commerce system.

Later, an extended and comprehensive e-commerce success model was proposed by Molla and Licker (2001). The model incorporated the importance and online transaction phases of e-commerce systems. The criteria considered by Molla and Licker for e-commerce system success is: e-commerce system quality, use, content quality, support, and trust. The changes in criteria made by this model with comparison to D&M model were: system quality replaced by e-commerce system quality, information quality replaced by content quality, and user satisfaction replaced by customer e-commerce satisfaction. These changes were made to bring into the customer satisfaction agenda, which is centric to the modern marketing management policy. According to them, on the basis of nature of e-commerce model, the customer satisfaction can be differentiated into two types: (1)

core product and service satisfaction and (2) satisfaction with the system and process used to deliver these products and services.

Customer feedback mechanism is a social communication process of providing opinion about a product or service by the users. Social communication is a powerful tool that influences the purchasing decision of the potential customers. This is important because the people have trust in others rather than the communications with the marketers (Goldsmith, 2008). The concept of word-of-mouth (WOM) has been well described in literature by many researchers. It is considered to be the oldest medium of exchanging opinions on various goods and services offered by the markets (Goyette et al., 2010). It is believed to be more effective than other promotion mix factors (Katz and Lazarsfeld, 1955). Many researchers have shown the positive correlation between the WOM concept and clients' trust (Bergeron et al., 2003), service quality (Parasuraman et al., 1988), satisfaction (Anderson, 1998), and clients' intention to purchase (Crocker, 1986). The recent up gradation in technology shows the relevance of WOM in online context and thus represents a modern version of WOM (Godin, 2001). Here, we define the factor "online customer feedback" to incorporate internet based WOM. This factor consists of online reviews and star ratings. It is believed that good reviews and higher star ratings do have a positive influence over potential customers' buying decision (Chavalier and Mayzlin, 2006).

Customer satisfaction is considered to be a major element for long-term success of any business. The basic method of assessing customer satisfaction is SERVQUAL (Zeithamal et al., 1990). According to them, "SERVQUAL measures the gap between the perception of customers, the potential of improvement, and the level of service provided." Another model to measure customer satisfaction was proposed by (Naumann, 1995) that incorporated attributes such as product quality, product design, and value. Despite the extensive researches found in academic literatures of management and marketing on customer satisfaction, the main focus of the studies were on the importance of products and services based on the interpersonal relationship between the service providers and customers, while ignoring the investigation of customer interactions with the m-commerce applications. Also, existing works related to User Information Satisfaction, Customer Satisfaction, and SERVQUAL, have a limited scope and provide a gap between the existing studies and the comprehensive success model.

In this chapter, we will discuss in detail the customer-oriented success factors for m-commerce. Here, we will take the factor: online customer feedback, along with the ones defined by Kabir and Hasin (2011). Their study was motivated by the Molla and Licker Model of E-commerce Success based on customer-oriented factors, with adding two factors: mobility and personalization, to give a taste of m-commerce and also to incorporate technological advancements. Personalization is a factor specific to the customer preferences and the alternatives, benefits, and other incentives that a potential customer expects from the online retailer. Mobility of devices and applications is their ability of being used anywhere while moving. These variables prove to be a major driving force for satisfying customer needs and wants, which in turn increase the market share of m-commerce firms. Also, the proposed model will evaluate the best m-commerce success factors on the basis of customer satisfaction and also rank some predefined applications with respect to these factors. For the empirical results, we have used the multicriteria techniques of AHP and PROMETHEE. The AHP technique has previously been applied to numerous applications such as drugs selection (Vidal et al., 2010), weapon selection (Deng and Shen, 2006), etc. Also, PROMETHEE technique has previously been used for various applications such as water resource planning (Hajkowicz and Collins, 2007), selection of lean manufacturing systems (Anand and Kodali, 2008), etc. The next section will discuss in detail the criteria taken up for the study.

4.3 CUSTOMER-ORIENTED SUCCESS FACTORS FOR M-COMMERCE

Here, we consider the following criteria: System Quality, Content Quality, Use, Trust, Support, Mobility, Personalization, and online customer feedback as represented by Figure 4.3.

4.3.1 SYSTEM QUALITY

This criterion helps in identifying whether the online customer is satisfied with the application's performance or not, taking into account the hardware and software factors. The potential competitors in m-commerce are just a click away and any dissatisfaction caused due to mobile application site's failure, impacts the m-commerce firm largely. Thus, it can be inferred that an important factor for a successful m-commerce firm is a seamless and smooth performance of the application's website. Varshney and Vetter (2002) stated the attributes such as reliability of the system and online response time, under the system quality. However, additional variables were discussed by Guo and Shao (2005) under this criterion such as online response time, 24-h availability, page loading speed, and visual appearance. Finally, the items which were found to be suitable under system quality from m-commerce perspective were: online response time, 24-h availability, page-loading speed, and visual appearance. "Online response time" represents the efficiency of the interface system of the mobile application website. By "24-hour availability" we mean that whether the m-commerce application is operable the whole day or there are some off hours. The time the application's site takes to reload or refresh and to migrate from various search results represents "page-loading speed." By "visual appearance" we mean how attractive the mobile application website is and also the readability of its content.

4.3.2 CONTENT QUALITY

Content quality is a prerequisite for firms to attract the online customers toward their website. In e-commerce, the products and the services without content are purely valueless and thus a major source of value is content (DeLone and McLean, 1992). The way of presenting the information on the website is termed as content quality. Despite information being considered as an asset to businesses, much of the emphasis of e-commerce has been made on elevated content such as the information, data, experience, and knowledge to higher significant levels (Molla and Licker, 2001). According to him, numerous attributes such as accuracy, up-to-datedness, comprehensiveness, understandability, completeness, timeliness, reliability, relevancy, and preciseness have been identified as determinants of user satisfaction in respect to content. It has been observed that the m-commerce can be judged on exactly the same attributes. The quality of content and its capability to meet the needs and expectations of the customers affect the success of the organizations and determine customers' conversion and retention. The content available on the firm's website influences the customer buying decision. The most appropriate attributes under this criterion are found to be the one defined by D&M model. According to this model, the attributes of content quality, which are taken up in this study are: up-to-datedness, understandability, timeliness, and preciseness. By "up-to-datedness" we mean the efficiency of the mobile application to inform the customer about the product inventory and latest arrivals/departures of goods. By "timeliness" we mean the visitor obtains the information about the goods/services instantly. "Understandability" means how better the data on the application is displayed and is also recognizable. The correct and exact information without any discrepancies is termed as "preciseness."

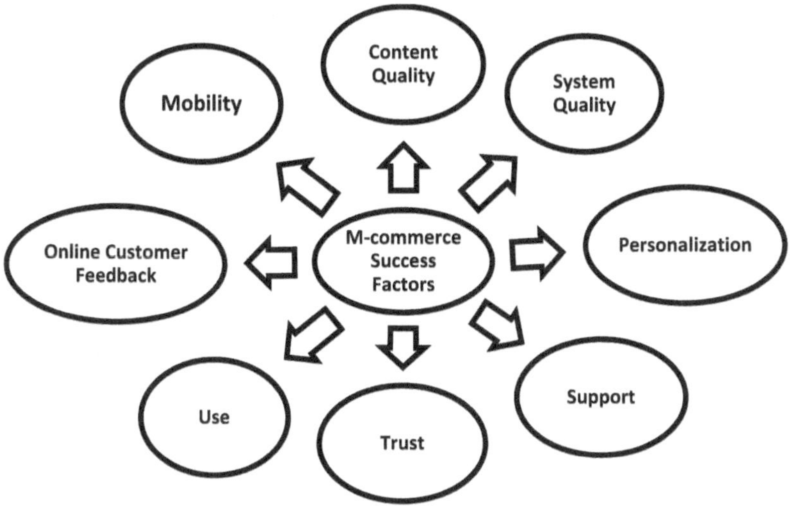

FIGURE 4.3 Mobile commerce success factors.
Source: Compiled by authors.

The success of an e-commerce website depends on the extent to which the website is useful to the customer. The market share and reach of m-commerce firms are indicated using use levels as captures through "hits" and "visits." Based on the purposes of e-commerce systems, they can be divide into informational, transactional, or customer service. According to Tarasewich (2003), two components make up an e-commerce system: information and transaction. Here again, "information" represents the significance of the content available on the application's website, and the "transaction" represents the how smoothly and effectively the payment process is carried out.

4.3.3 TRUST

Trust is an important criterion for m-commerce success since customer's insecure feelings may affect their disposition toward privacy issues and security. Trust helps in customer conversion and retention toward am-commerce site. There are mainly two attributes: security and privacy, which affect the future of e-commerce systems (Warrington et al., 2000). He stated that when providing online information, the level of security presents the major concern for customers. "Security" represents the capabilities of am-commerce system to ensure that there is no breach of customers' personal information while making a transaction. The ability of an m-commerce system to keep the personal information of the customers confidential and to protect the customers' data from being included in various databases is termed as "privacy."

4.3.4 SUPPORT

Another criterion which is of interest to online customers is support. This is also a customer-oriented criterion that helps in customer retention. The support and services provided by the operators are highly valued during all phases of transactions. According to Tarasewich et al. (2000), the criterion includes the following attributes: tracking order status, payment alternatives, and Frequently Asked Question (FAQ). Another variable included is the account maintenance on the m-commerce application's website. "Tracking order status" is related to the service provided by the retailer to track the purchased product. Also, with improvements in the technology, various online payment options have been introduced and can be provided by the m-commerce firms termed as "payment alternatives." "Frequently asked questions" represents the query handling potential of the firm's customer care. These days m-commerce provides the customer an opportunity to download their applications at minimal costs and create an account on their website to avail various incentives and exclusive offers. This is the advantage of "account maintenance" in m-commerce.

4.3.5 MOBILITY

Mobility is the property by which the services can be availed from anywhere and also at any time at the discretion of the user. This is the major distinguishing feature between the e-commerce system and the m-commerce system. This is due to the fact that the e-commerce needs an infrastructure which is not movable, whereas the m-commerce depends on mobile phones which are portable and can be accessed from anywhere. Thus, m-commerce enables the customers to employ their services and make transactions anywhere. The two major attributes under this criterion are: device and application (Tarasewich, 2003). According to him, "mobility of device and application raises the issue of their suitability for the user under some circumstances." By "device" it is meant by the mobile phones used. This is a very critical attribute since some of the users may not have the latest technology smartphones to support the applications in their system. By "application" it means the interface on the mobile phones through which we can have access to the product/services we look forward to have.

4.3.6 PERSONALIZATION

The term personalization means to provide the products or services as per customer taste. From the m-commerce perspective, this is the most critical success factor (Tarasewich et al., 2000). Since mobile devices have major issues related to the battery capacity and its size and configuration, to increase their usability there is a need for personalization. The attributes under this criterion are: location, time, and individual preferences. The ability to have access to the m-commerce application websites from any part of the world and at the customers' suitable time gives the importance of the attribute "location" and "time," respectively. "Individual preferences" means that the customer can build their own shopping kart and select the products/service by considering various filter searches based on their constraints.

4.3.7 ONLINE CUSTOMER FEEDBACK

The offline WOM has been considered to be a major determinant of consumers' buying decision. WOM is considered to be an exchange of idea or information or communication, which can be both informal and formal in nature and has a greater influence over the customers (Goyette et al., 2010). However, the introduction of internet has enabled the consumers to offer their consumption-related experience (Hennig et al., 2004), which in this study is termed as "online customer feedback." This is most prominent in m-commerce where the viewers can have access to the feedback provided by the previous purchasers. Two major attributes are considered under this criterion such as: online reviews and star ratings. By "online reviews" we mean the linguistic comments written by the past purchasers on the mobile applications for the product/service taken by them. Some of the applications provide the consumers the option to rate by selecting stars, called the "star ratings," which is consistent with the Likert scale used by researchers.

4.4 PRELIMINARIES

4.4.1 ANALYTICAL HIERARCHY PROCESS

The AHP is based on the following steps:

Step 1: Constructing a Hierarchical Structure

MCDM models are initially structured as a hierarchy of interrelated decision elements to simplify the problem. All the criteria, subcriteria, alternatives, and the overall objective of the problem are transformed into a hierarchical structure. The hierarchy consists of the goal at the topmost level of hierarchy, the criteria and the subcriteria at the middle level, and the alternatives at the bottom level of hierarchy (Wang and Yang, 2007), as represented by Figure 4.4.

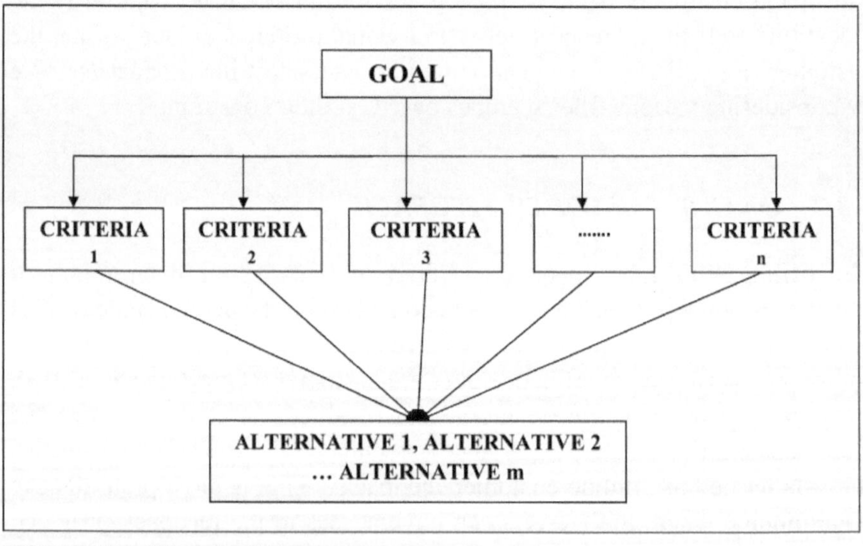

FIGURE 4.4 Hierarchical structure.
Source: Compiled by authors.

Step 2: Construct Pair-wise Comparison Matrix

The next step is to construct a pair-wise comparison decision matrix to obtain relative weight of the criteria. The criteria are compared pair-wise according to their influence and based on the specified criteria in the higher level. The pair-wise comparisons are done based on a comparison scale of five levels as shown in Table 4.1. Suppose $D = \{D_j \mid j = 1, 2 \dots n\}$ be the set of criteria. The matrix can be obtained, consisting of elements $p_{ij} \mid i, j = 1, \dots, n\}$ representing the relative weights of the comparison, as illustrated:

$$P = \begin{pmatrix} p_{11} & \cdots & p_{1n} \\ \vdots & \ddots & \vdots \\ p_{n1} & \cdots & p_{nn} \end{pmatrix} \tag{4.1}$$

where p_{ij} satisfy the following condition:

$$p_{ij} = \frac{1}{p_{ji}} \; p_{ii} = 1, p_{ij} > 0 \tag{4.2}$$

Step 3: Calculate the Eigenvector of the Matrix

Using the formula

$$PH = \beta_{\max} H \tag{4.3}$$

The $\beta_{\max}$ can be obtained from above equation. If the $\beta_{\max}$ is equal to n and the rank of matrix P is n, then P is consistent. The weight of each criterion will be calculated by normalizing any of the rows or columns of the matrix P.

Step 4: Check Consistency

The consistency of the matrix P depends on two parameters: the consistency index (CI) and the consistency ratio (CR). They are defined as:

$$CI = \frac{\beta_{\max} - n}{n - 1} \tag{4.4}$$

$$CR = \frac{CI}{RI} \tag{4.5}$$

where RI is the random index, given by Table 4.2.

TABLE 4.1 Comparison Scale of Five Levels.

Definition	Values
Equal importance	1
Importance	3
Extreme importance	5
Intermediate values	2, 4

Source: Compiled by authors.

TABLE 4.2 RI Value.

N	1	2	3	4	5	6	7	8	9
RI	0	0	0.58	0.90	1.12	1.24	1.32	1.41	1.45

Source: Compiled by authors.

If CR is less than 0.10, the result is acceptable and matrix P is consistent. Otherwise, we return to Step 1 and repeat the procedure again.

4.4.2 THE PROMETHEE METHOD

PROMETHEE is an outranking approach used for MCDM problems and was introduced by Brans and Vincke (1990). Compared to other outranking approaches, this method is simple and easily computable. The major assumption for applying this method is that the number of alternatives (sometimes conflicting), that are to be ranked, must be fixed (Anand and Kodali, 2008). The method starts with an evaluation matrix that evaluates the alternatives on the criteria. Other than this, two more inputs are required for this method: (1) the criteria weights and (2) the type of function of each criterion considered in the problem. The criteria weights can be obtained through various methods (Mergias et al., 2007). In this chapter, we will be using the AHP method to obtain criteria weights.

The PROMETHEE converts the decision-making problem to following:

$$\max\{f_1(a), f_2(a), \dots, f_n(a) \mid a \in A\} \tag{4.6}$$

where A is the finite set of all alternatives, f_j denotes the "n" criteria to be maximized, and $f_j(a)$ is an evaluation of the alternative a in terms of criterion j. Since, obtaining the preferences of the alternatives is the sole objective of this method; therefore, there exists a need to formulate a preference function.

When two alternatives (say "a" and "b") are being compared, the preference function represents the difference of "a" and "b" in the range of 0–1 in terms of particular criterion. Let

$$Q_{j(a,b)} = G_j\left[f_j(a) - f_j(b)\right] \tag{4.7}$$

$$0 \leq Q_{j(a,b)} \leq 1 \tag{4.8}$$

be the preference function associated to the criterion, $f_j(i)$ where G_j is a nondecreasing function of the observed deviation (d) between $f_j(a)$ and $f_j(b)$. According to (Brans et al., 1986), there are six types of preference functions, namely, usual function, U-shape function, V-shape function, level function, linear function, and Gaussian function. Each function does not require more than two thresholds of $p,q,$ or s to be determined (Wang and Yang, 2007). Indifference threshold q is the largest deviation to consider negligible on that criterion and is a small value with respect to the scale of measurement. Preference threshold p is the smallest deviation to consider decisive in the preference of one alternative over another and is thus a large value with respect to scale of measurement. Gaussian threshold s is used only with the Gaussian preference function and is usually an intermediate value between indifference and a preference threshold.

Using PROMETHEE, we can compute following quantities for each alternative a and b:

$$\pi(a,b) = \frac{\sum_{j=1}^{n} w_j Q_j(a,b)}{\sum_{j=1}^{n} w_j} \tag{4.9}$$

$$\varphi^+(a) = \sum_{x \in A} \pi(x,a) \tag{4.10}$$

$$\varphi^-(a) = \sum_{x \in A} \pi(a,x) \tag{4.11}$$

$$\varphi(a) = \varphi^+(a) - \varphi^-(a) \tag{4.12}$$

For each alternative a, which belongs to the set of all alternatives A, $\pi(a,b)$ is an index which represents the overall preference of a over b. The value $\varphi^+(a)$ is a leaving flow and measures how a dominates all the other alternatives. Similarly, the value $\varphi^-(a)$ is the entering flow and measures how a

is dominated by all other alternatives. The final value $\varphi(a)$ represents the net flow, where a higher value represents a greater dominance of alternative a. Now, for the purpose of solving the problem, the emphasis is laid on two major PROMETHEE methods which are: (1) the partial ranking given by PROMETHEE I and (2) the complete ranking given by PROMETHEE II.

PROMETHEE I gives the preference (*aPb*), or indifference (*aIb*), or the incomparability (*aRb*) situations, which are formulated under the following criteria,

$$\begin{aligned} &aPb \text{ if}: \varphi^+(a) > \varphi^+(b) \text{ and } \varphi^-(a) < \varphi^-(b); \text{or} \\ &\varphi^+(a) > \varphi^+(b) \text{ and } \varphi^-(a) = \varphi^-(b); \text{or} \\ &\varphi^+(a) = \varphi^+(b) \text{ and } \varphi^-(a) < \varphi^-(b); \end{aligned} \tag{4.13}$$

$$aIb \text{ if}: \varphi^+(a) = \varphi^+(b)^+ \text{ and} \varphi^-(a) = \varphi^-(b) \tag{4.14}$$

$$\begin{aligned} &aRb \textit{ if}: \varphi^+(a) > \varphi^+(b) \text{and } \varphi^-(a) > \varphi^-(b); \text{or} \\ &\varphi^+(a) < \varphi^+(b) \text{ and } \varphi^-(a) < \varphi^-(b) \end{aligned} \tag{4.15}$$

PROMETHEE II gives the complete ranking of the alternatives from the best to the worst. For this purpose, the value of net flow is used. Since PROMETHEE I provides the decision-maker with partial ranking, it cannot equal the solution of complete ranking provided by PROMETHEE II. Still, there have been some cases in literature where both the rankings are equal.

4.5 METHODOLOGY

The proposed model for ranking m-commerce applications on the basis of customer-oriented success factors is composed of two approaches, namely, AHP and PROMETHEE. The phases of the model are discussed below and represented by Figure 4.5.

Phase 1: Criteria Identification

In this phase, the m-commerce applications and the criteria which will be used in ranking are determined, and the decision hierarchy is formed. AHP model is formulated with the objective at the top level, the success attributes at the middle level, and the m-commerce applications at the bottom level.

Phase 2: Criteria Weight Calculation
In this phase, pair-wise comparison matrices are constructed to obtain criteria weights. The elements in the matrices are filled using the ratings provided in Table 4.1. Then the arithmetic means of the values in the pair-wise matrices are computed, and we obtain the normalized matrix, which is the final matrix. The weights of the criteria are calculated based on this final matrix.

Phase 3: Rank Applications Using PROMETHEE
Ranking of the m-commerce applications is done using the PROMETHEE technique. First, the partial rankings are obtained using PROMETHEE I and then the final rankings are obtained using PROMETHEE II.

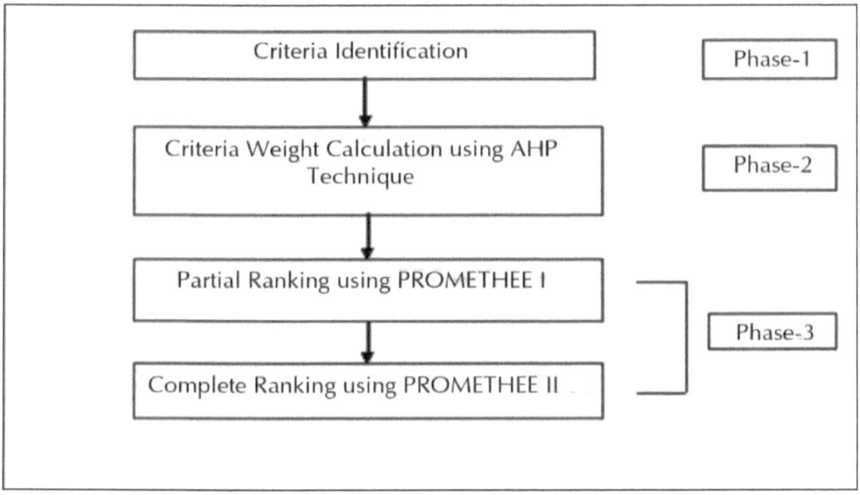

FIGURE 4.5 Phases of the proposed model.
Source: Compiled by authors.

4.6 NUMERICAL ILLUSTRATION

In this section, we validate the proposed model with the help of a case analysis. The objective of the problem is to rank the m-commerce applications on the basis of customer-oriented success factors. The proposed model combined the MCDM techniques of AHP and PROMETHEE to obtain the required results. First, we required the pair-wise comparison matrix for the set of criteria to evaluate the AHP process. This requirement was fulfilled with the help of a survey consisting of respondents who frequently used mobile applications for shopping and searching purpose. The survey was conducted in Delhi NCR region and through it, we successfully obtained the pair-wise comparison matrix and also the top five most preferred applications by the respondents, which will be considered as the alternatives. Also, the experts associated with the m-commerce businesses were involved to construct the PROMETHEE evaluation matrix. The whole process is composed into three phases as discussed in previous section.

Phase 1: Criteria Identification

In this phase, a hierarchical structure of the problem is constructed consisting of three levels, as shown in Figure 4.6. The topmost level comprises the overall goal of the problem, that is, Best m-commerce Application. The next level comprises the criteria such as: System Quality (C1), Content Quality (C2), Use (C3), Trust (C4), Support (C5), Mobility (C6), Personalization (C7), and Online Customer Feedback (C8). Alternatives consisting of the m-commerce applications such as: A1, A2, A3, A4, and A5, form the lowest level of the hierarchy.

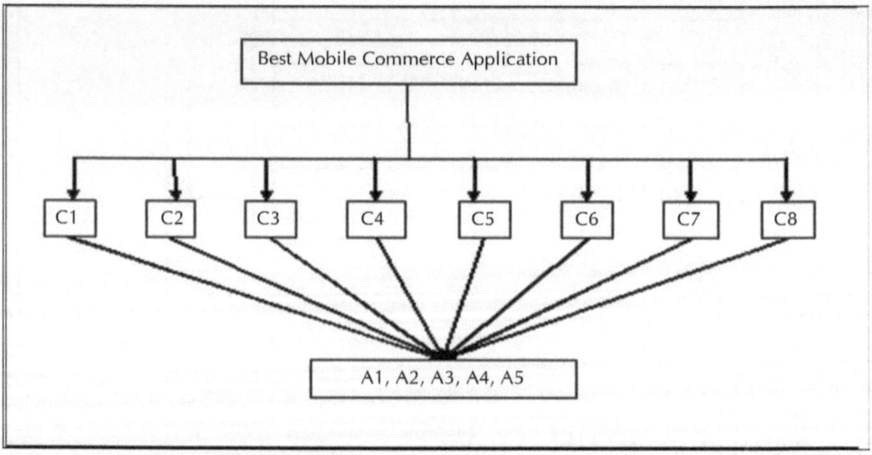

FIGURE 4.6 Hierarchy of the numerical illustration.
Source: Compiled by authors.

Phase 2: Criteria Weight Calculation

In this phase, the feedback from the respondents of the survey was used to obtain pair-wise comparison matrix, as represented by Table 4.3.

TABLE 4.3 Pair-wise Comparison Matrix.

	C1	C2	C3	C4	C5	C6	C7	C8
C1	1	2	0.33	0.25	0.33	0.25	0.2	0.25
C2	0.5	1	0.33	0.33	3	0.33	0.33	0.2
C3	3	3	1	0.33	3	0.33	0.25	0.33
C4	4	3	3	1	3	1	1	1
C5	3	0.33	0.33	0.33	1	0.33	0.25	0.25
C6	4	3	3	1	3	1	1	0.33
C7	5	3	4	1	4	1	1	1
C8	4	5	3	1	4	3	1	1

Source: Compiled by authors.

Next, the element of each column was divided by the sum of the respective column to obtain a normalized matrix. Taking arithmetic means row-wise, we get the criteria weights and also check for the consistency of the matrix. The values for β_{max}, CI, and CR are obtained using eqs 4.3–4.5. The results are represented in Table 4.4. Also, Figure 4.7 shows the AHP results graphically.

TABLE 4.4 AHP Results.

Criteria	**C1**	**C2**	**C3**	**C4**	**C5**	**C6**	**C7**	**C8**
Weight	0.045	0.057	0.089	0.176	0.053	0.157	0.195	0.229
β_{max}	–	–	–	8.764	–	–	–	–
CI	–	–	–	0.109	–	–	–	–
RI	–	–	–	1.41	–	–	–	–
CR	–	–	–	0.077	–	–	–	–

APH, analytical hierarchy process; CI, consistency index; CR, consistency ratio, RI, random index.

Source: Compiled by authors.

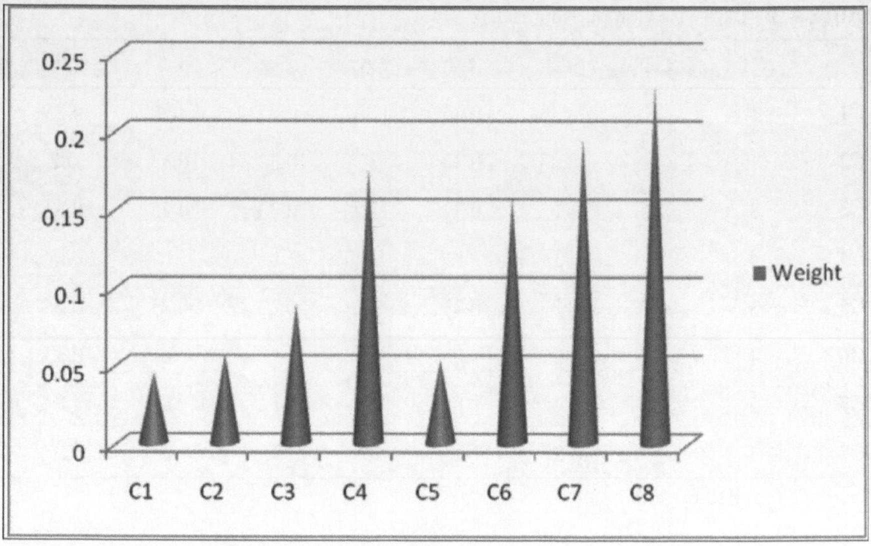

FIGURE 4.7 Criteria and their AHP weights.
Source: Compiled by authors.

Since the CR value is less than 0.1, we imply that the comparison matrix is consistent and the weights obtained can be used for ranking purpose in the next phase.

Phase 3: Rank Applications Using PROMETHEE

To get the evaluation matrix, some of the m-commerce experts were requested to rate the criteria on the basis of each alternative. To keep the study consistent, the rating scale was taken to be the 5-point Likert scale (1= not satisfied; 5= extremely satisfied). Also to get the solution, we have made use of the preference function of type: Level Function, which is defined as:

$$q(x) = \begin{cases} 0x \leq q \\ \frac{1}{2} q < x \leq p + q \\ 1x > P + q \end{cases} \tag{4.16}$$

The level type of preference function is taken here, as it is considered to be best to tackle qualitative data, especially the data is the type yes/no or Likert scale (Turcksin et al., 2011). The evaluation matrix is given by Table 4.5.

TABLE 4.5 Evaluation Matrix.

DM	C1	C2	C3	C4	C5	C6	C7	C8
Weights	0.045	0.057	0.089	0.176	0.053	0.157	0.195	0.229
A1	3	3	3	1	2	3	3	2
A2	4	4	2	3	3	2	4	3
A3	3	5	4	4	4	3	5	4
A4	2	2	3	3	4	4	2	3
A5	3	3	2	5	4	3	5	5

Source: Compiled by authors.

Next we consider the criterion C1:

Let T1 be the distance matrix for S1 obtained using the formula for Euclidean distance (*A–B*) and is shown in Table 4.6. The negatives values are included because of their significance in further solving the problem.

TABLE 4.6 Distance Matrix for Criterion C1.

T1	A1	A2	A3	A4	A5
A1	0	−1	0	1	0
A2	1	0	1	2	1
A3	0	−1	0	1	0
A4	−1	−2	−1	0	−1
A5	0	−1	0	1	0

Source: Compiled by authors.

Then, the corresponding Preference matrix is obtained using eq 4.16 and using the value of $p = 2$ and $q = 1$, and is represented by Table 4.7.

TABLE 4.7 Preference Matrix for Criterion C1.

Q1	A1	A2	A3	A4	A5
A1	0	0	0	0	0
A2	0	0	0	0.5	0
A3	0	0	0	0	0
A4	0	0	0	0	0
A5	0	0	0	0	0

Source: Compiled by authors.

Similarly, we can obtain the Preference matrix for each criterion. Next using formula given in eq 4.9, we get the values of $\pi(a,b)$, as represented by Table 4.8.

TABLE 4.8 $\pi(a,b)$ Values for Alternatives.

$\pi(a,b)$	A1	A2	A3	A4	A5
A1	0	0	0	0	0
A2	0.088	0	0	0.149	0
A3	0.355	0.045	0	0.098	0.073
A4	0.115	0.079	0	0	0
A5	0.415	0.203	0	0.3	0

Source: Compiled by authors.

Next by using eqs 4.10–4.12, we formulate the leaving flows, entering flows, and the net flow for each alternative. This gives us the final ranking (PROMETHEE II) as represented by Table 4.9.

TABLE 4.9 PROMETHEE Results.

	A1	A2	A3	A4	A5
φ^+	0	0.059	0.143	0.049	0.23
φ^-	0.243	0.082	0	0.137	0.018
φ	−0.243	−0.023	0.143	−0.088	0.212

Source: Compiled by authors.

From Table 4.9, we obtain the final ranking of the m-commerce applications in descending order as: A5, A3, A2, A4, and A1, and is represented by the flow value using Figure 4.8.

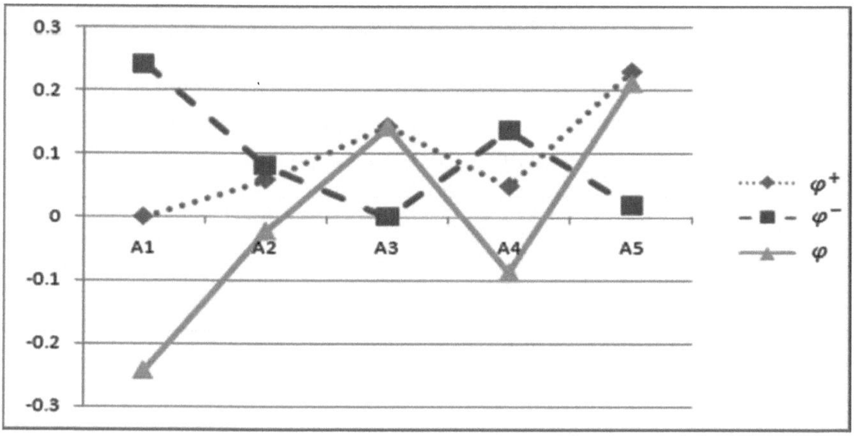

FIGURE 4.8 PROMETHEE results with flow values.
Source: Compiled by authors.

The AHP results give higher weights to Online Customer Feedback (C8), Personalization (C7), Trust (C4), and Mobility (C6). The high weight of criterion C8 represents the importance of the online social communication among the potential customers. Their purchasing decision is highly affected by the online reviews and star ratings provided by the previous consumers on the mobile application. Since, this is the criterion added in the proposed model; its higher value validates our assumptions of the model. Next is the criterion C7, and this tells the necessity for the m-commerce retailers to offer individual customized products/services, according to their taste. Also, customers prefer to purchase their desired goods/services at their own location and at any time which is convenient to them. Another criterion of importance is the trust. For am-commerce application to be successful, it should provide an interface to the users which is secure and encrypts theirs privacy. Customers prefer a secure platform when they pay for their goods/services online, and that their details will be kept hidden from other users. Mobility is another main feature which distinguishes m-commerce from e-commerce. It is impractical to carry laptops/PCs everywhere, and this is where the need arises for movable devices, and this is shown by high weight of the criterion mobility. Thus, this validates the proposed model which is consistent under a customer-centric approach.

From the PROMETHEE results, we can infer that out of the five alternatives under study, A5 is the best alternative and A1 is the worst alternative. The m-commerce application A5 has been rated above average for the abovementioned higher weighted criteria. This shows that the online reviews and the star ratings provided by the past purchasers have a positive impact on the potential customers and this displays their satisfaction level with the application. Also, the application provides various features related to the products/services demanded and helps the purchasers to get products according to their own preference, anytime, and anywhere. Moreover, the application provides an excellent platform for the transaction and payment processes, and people are comfortable with the arrangements of security and privacy provided by the m-commerce application system. Obviously, mobility is an influential criterion in m-commerce and the ones that provide this feature the best are the one ranked higher, which is the case with alternative A5. The alternative A1 is ranked the lowest because it is rated below average for the major criteria: online customer feedback and trust, along with being averagely rated for personalization and mobility. Thus, it portrays a behavior opposite to A5, and is inconsistent with the above reasoning.

4.7 DISCUSSION AND CONCLUSION

Nowadays, e-commerce is moving toward m-commerce, and is an area of research interest to many researchers. This chapter proposes a model which incorporates "online customer feedback" along with the ones defined by Kabir and Hasin (2011). Thus, the overall criteria under study are: content quality, system quality, use, trust, support, mobility, personalization, and online customer feedback. The model is constructed by keeping in view the importance of customer online behavior and latest advances in the technology. The objective of the chapter was to rank the m-commerce applications based on the customer-oriented success factors. Therefore, a model was proposed which inculcated the MCDM models such as AHP and PROMETHEE. The AHP technique was used to obtain the criteria weights and PROMETHEE was used to obtain the final ranking of the m-commerce applications. Here, AHP is preferred over Fuzzy AHP (FAHP). Though FAHP is used while dealing with uncertainty and vagueness of linguistic values, but this approach has some major limitations arising due to large number of pairwise comparisons to be made and it may also result in unfitting scores to some criteria (Kabir and Sumi, 2014; Wang and Yang, 2007). PROMETHEE is preferred over other outranking approaches, such as ELECTRE, as it is considered a flexible algorithm, which is suitable for an integrated assessment analysis (Kabir et al., 2013) and to be more stable than ELECTRE (Brans et al., 1986). To validate our model, an illustration was provided using the eight criteria and five alternatives. A customer survey was conducted to get the inputs for

AHP process and feedback from the experts was used to obtain inputs for the PROMETHEE process. Overall analysis showed that the customers are mostly affected by the online customer feedback, personalization, and trust. The proposed model is a flexible model and can be further used for ranking and preference of attributes within a set of variables.

4.8 LIMITATIONS AND FUTURE SCOPE

In this chapter, we have proposed a model combining the techniques of AHP and PROMETHEE. However, AHP has some limitations. Since the method is based on accurate measurements and crisp evaluation, most of the selection parameters cannot be calculated precisely (Kabir and Hasin, 2011). PROMETHEE also has some limitations as it does not lay any proper rules on how to measure weights (Gervasio and da Silva, 2012). It assumes that the decision-maker is able to weigh the criteria appropriately, no specific guidance is provided for its determination (Kabir and Sumi, 2014). The numerical illustration is based on the criteria instead of subcriteria to reduce the computational complexity and also that the objective of the paper is to rank m-commerce applications, which is contrary to the other studies which focused just on the evaluation of criteria and subcriteria (Kabir and Hasin, 2011). Many future studies can be conducted using various MCDM techniques such as Shannon's entropy, ELECTRE, VIKOR, etc. Also, since this was a study of success factors from customers' perspective, future studies can focus on the success factors of the service providers.

REFERENCES

Anand, G.; Kodali, R. Selection of Lean Manufacturing Systems Using the PROMETHEE. *J. Modell. Manag.* **2008,** *3* (1), 40–70.

Anderson, E. W. Customer Satisfaction and Word-of-Mouth. *J. Serv. Res.* **1998,** *1* (1), 5–17.

Bergeron, J.; Ricard, L.; Perrien, J. Les Determinants de la fi Délitédes Clients Commerciaux dans L'industrie Bancaire Canadienne. *Can. J. Adm. Sci.* **2003,** *20* (2), 107–120.

Brans, J. P.; Vincke, P. H. A Preference Ranking Organization Method. *Manag. Sci.* **1990,** *26* (4), 8–14.

Brans, J. P.; Vincke, P. H.; Mareschall, B. How to Select and How to Rank Projects: The PROMETHEE Method. *Eur. J. Oper. Res.* **1986,** *14,* 228–238.

Chavalier, J. A.; Mayzlin, D. The Effect of Word of Mouth on Sales: Online Book Reviews. *J. Mark. Res.* **2006,** *43* (3), 345–354.

Crocker, K. E. The Influence of the Amount and Type of Information on Individuals' Perception of Legal Services. *J. Acad. Mark. Sci.* **1986,** *14* (4), 18–27.

DeLone, W. H.; McLean, E. R. Information Systems Success: The Quest for Dependent Variable. *Inf. Syst. Res.* **1992,** *3* (1), 60–95.

Deng, Y.; Shen, C. Evaluating the Main Battle Tank Using Fuzzy Number Arithmetic Operations. *Def. Sci. J.* **2006,** *56* (2), 51–257.

Gervasio, H.; da Silva, L. S. A Probabilistic Decision-making Approach for the Sustainable Assessment of Infrastructures. *Expert Syst. Appl.* **2012,** *39*, 7121–7131.

Godin, S. *Les Secrets du Marketing Viral: le Boucheà-oreille à la Puissance 10!* Maxima Laurent DuMesnil Éditeur: Paris, 2001; p 197.

Goldsmith, R. E. Electronic Word-of-mouth. In *Electronic Commerce: Concepts, Methodologies, Tools, and Applications*; IGI Global: USA, 2008; pp 2143–2149.

Goyette, I.; Ricard, L.; Bergeron, J.; Marticotte, F. E-WOM Scale: Word-of-mouth Measurement Scale for e-Services Context. *Can. J. Adm. Sci.* **2010,** *27* (1), 5–23.

Guo, S.; Shao, B. *Quantitative Evaluation of e-Commercial Websites of Foreign Trade Enterprises in Chongquing*, Proceedings of International Conference on Services Systems and Services Management, Chongquing, China, 2005; IEEE; pp 780–785.

Hajkowicz, S.; Collins, K. A Review of Multiple Criteria Analysis for Water Resource Planning and Management. *Water Resour. Manag.* **2007,** *21* (9), 1553–1566.

Hennig, T. T.; Gwinner, K. P.; Walsh, G.; Gremler, D. D. Electronic Word-of-mouth via Consumer-opinion Platforms: What Motivates Consumers to Articulate Themselves on the Internet? *J. Interact. Mark.* **2004,** *18* (1), 38–52.

Hootsuite. 2017. https://www.wearesocial.com (accessed Feb 16, 2017).

Kabir, G.; Hasin, M. A. A. Evaluation of Customer Oriented Success Factors in Mobile Commerce Using Fuzzy AHP. *J. Ind. Eng. Manag.* **2011,** *4* (2), 361–386.

Kabir, G.; Sumi, R. S. Integrating Fuzzy Analytic Hierarchy Process with PROMETHEE Method for Total Quality Management Consultant Selection. *Prod. Manuf. Res.* **2014,** *2* (1), 380–399.

Kabir, G.; Sadiq, R.; Tesfamariam, S. A Review of Multi-criteria Decision Making Methods for Infrastructure Management. *Struct. Infrastruct. Eng.* **2013**, *10* (9), 1176–1210.

Katz, E.; Lazarsfeld, P. F. *Personal Influence*; Free Press: Glencoe, IL, 1955.

Kim, E. In *A Model of an Effective Web*. Proceedings of the Fifth Americas Conference on Information Systems; Haseman, W. D., Nazareth, D. L., Eds.; Association for Information Systems, 1999; pp 523–526.

Mason, R. O. Measuring Information Output: A Communication Systems Approach. *Inf. Manag.* **1978,** *1* (5), 219–234.

Mergias, I.; Moustakas, K.; Papadopoulos, A.; Loizidou, M. Multi-criteria Decision Aid Approach for the Selection of the Best Compromise Management Scheme for ELVs: The Case of Cyprus. *J. Hazard. Mater.* **2007,** *147* (3), 706–717.

Molla, A.; Licker, P. S. Ecommerce System Success: An Attempt to Extend and Respecify the DeLone and McLean Model of IS Success. *J. Electron. Commer. Res.* **2001,** *2* (4), 131–141.

Naumann, E. *Creating Customer Value: The Path to Sustainable Competitive Advantage*; Thomson Executive Press: Cincinnati, 1995.

Parasuraman, A.; Zeithaml, V. A.; Berry, L. L. SERVQUAL: A Multiple-item Scale for Measuring Customer Perceptions of Service Quality. *J. Retailing* **1988,** *64* (1), 12–40.

Saaty, T. L. *The Analytic Hierarchy Process*; McGraw-Hill: New York, 1980.

Shannon, C. E.; Weaver, W. *The Mathematical Theory of Communication.* University of Illinois Press: Urbana, 1949.

Statista. 2017. https://www.statista.com (accessed Feb 16, 2017).

Tarasewich, P. Designing Mobile Commerce Applications. *Commun. ACM* **2003,** *46* (12), 57–60.

Tarasewich, P.; Nickerson, R. C.; Warkentin, M. Issues in Mobile e-Commerce. *CAIS* **2000,** *8*, 41–64.

Turban, E.; King, D.; Lee, J. K.; Warkentin, M.; Chung, H. M. *Electronic Commerce 2002: A Managerial Perspective,* 2nd ed.; Prentice Hall: New Jersey, 2002.

Turcksin, L.; Bernardinia, A.; Macharisa, C. A Combined AHP-PROMETHEE Approach for Selecting the Most Appropriate Policy Scenario to Stimulate a Clean Vehicle Fleet. *Procedia Soc. Behav. Sci.* **2011,** *20*, 954–965.

Varshney, U.; Vetter, R. Mobile Commerce: Framework, Applications and Networking Support. *Mobile Netw. Appl.* **2002,** *7* (3), 185–198.

Vidal, L. A.; Sahin, E.; Martelli, N.; Berhoune, M.; Bonan, B. Applying AHP to Select Drugs to be Produced by Anticipation in a Chemotherapy Compounding Unit. *Expert Syst. Appl.* **2010,** *37* (2), 1528–1534.

Wang, J. J.; Yang, D. L. Using a Hybrid Multi-criteria Decision Aid Method for Information Systems Outsourcing. *Comput. Oper. Res.* **2007,** *34* (12), 3691–3700.

Warrington, T. B.; Abgrab, N. J.; Caldwell, H. M. Building Trust to Develop Competitive Advantage in E-Business Relationships. *Competitiveness Rev.* **2000,** *10* (2), 160–168.

Zeithamal, V. A.; Parasuraman, A.; Berry, L. L. *Delivering Quality Service: Balancing Customer Perceptions and Expectations;* The Free Press: New York; 1990.

CHAPTER 5

DRIVING FORCES OF M-COMMERCE

KULVINDER KAUR BATTH*

Department of Commerce, K. C. College, University of Mumbai, Mumbai, India

**E-mail: kulprofessional@gmail.com*

ABSTRACT

M-commerce is not an extension of e-commerce but it is a new dimension in the era of new-age marketing. M-commerce has created a new definition in modern-day marketing. It has set new parameters and designed a path for marketing revolution in the future. The ease, comfort, accessibility, and convenience have broadened the outlook of the participants in the industry and widened the scope of business. This new dimension is alarming and rising at a pace much faster than the speed of the jet planes.

Driving forces for m-commerce are the factors which will lead to expansion of the technology to a wider market segment. It will help the companies to walk toward a new technological revolution. It will lead to a better connectivity between buyers and sellers, connectivity which will also encompass Business to Business, (B2B), Business to Consumer (B2C), Consumer to Business (C2B), and Consumer to Consumer (C2C).

M-commerce will provide customers' comfort, convenience, and cost-effective ways to fulfill their requirements. It will lead to better linkage between the buyers and sellers. The sellers will be able to keep in better touch with the buyers of their goods. Customer satisfaction will be high due to fulfillment of wants and a 24/7 approach with 360 degree reach.

M-commerce on one hand is a separate segment from e-commerce and on the other hand is a technology which has overcome its challenges and weaknesses. It is an extension of e-commerce while developing features which provide much better advantages. M-commerce aims to provide ease, convenience, and comfort to its users in every possible manner. The benefit of buying and selling goods and enjoying services anywhere and anytime is the key to the success of m-commerce.

5.1 INTRODUCTION

Communication is the key to generation of ideas. All living beings communicate through different modes of communication. Communication leads to expression, sharing of ideas, and fostering new developments. With the advent of technology, new forms of communication also developed. The lines of communication have gone through a drastic change. The relationships have transformed into more viral mode. Celebrations are in real time in the world with friends and dear ones participating from all the nooks and corners of the world (Cotlier, 2000; Darling, 2001).

M-commerce can be defined as the wireless mode by which buying and selling can be undertaken. The transactions are held through wireless medium without any need for plugging in. the technology provides the comfort and convenience to the users (Fitchard, 2001). Mobile communication is device which stands apart from all the historical communication methods. It is a new chapter and has set multidimensional parameters. Mobile is considered to be most powerful medium of communication. Technology has reached at a very fast rate to every strata and segment of the world. It has superseded all the demographical, geographical, or sociological boundaries in the world (Feldman, 2000).

Mobile has in true sense led to technological inclusion. It has incorporated each and every individual under its ambit. This technology has created platforms for sharing of knowledge, expression, ideas, and innovations and is building awareness among the masses.

There is no rocket science to get adapted to this great technology. It is user friendly, accommodating, economical, and has bypassed all the barriers of communication. A technology which provides solution to all communication deterrents, language barriers, and above all, it wins over all the special needs. The latest software's have incorporated the demands of the people born with special needs. The technology is not only supporting them but also complementing their requirements. People with different learning disabilities can make use of the technology and a large number of apps, which instill confidence in their abilities.

5.2 COMPONENTS OF M-COMMERCE

The financial components of m-commerce include mobile banking, mobile financial services, and mobile payment. M-commerce also includes transportation services, gaming services, household services, house-help services, online groceries, and so on. Mobile technology has been so far successful with the convenience of carrying on all buying and selling activities comprising anything from anywhere, without the need of the usage of desktop. The extension of m-commerce should encompass all the commerce and business transactions to be carried on the internet. M-commerce applications should be user friendly providing the ease and convenience to its users. Mobile financial services are an inherent contributory factor for the success of m-commerce (Gordon and Gutiorrez, 2007). Financial services include the money transfer, banking transactions, payment of insurance premium and mutual fund payments, and so on. Mobile payments comprise payment for the usage of services online through application-based services by transferring the funds from the online bank accounts to the mobile payment wallet. Such as if a person has used the online cab service, the payment for the same can be done from the bank account directly to the service providers account through the mobile without the need for carrying cash. Growing demand for mobile banking has created large number of opportunities for the banks to target the account holders for providing mobile banking services (Clarke, 2008). It provides ease and comfort of getting the information related to the banking transactions with respect to your account. Mobile financial services need a large amount of security or an enhanced finger recognition system so as to provide authenticity and thus, increasing the number of users (Vanathi et al., 2016).

When all the financial transactions can be handled through the mobile, it provides convenience as well as the comfort of handling the financial transactions. The 24/7 access to the account is considered to be a safe option. In case of any emergency, the customer will not have to wait for the next day for branch to open to know the robbery or a fraud from his account. The earlier information will lead the account holder to keep himself abreast of transaction on his account. M-commerce provides an opportunity to banks to target their tech-savvy customers and be in constant touch with them by providing variety of schemes and offers. Banks can also reach to new customers by offering technology as a strong point and the efforts taken at their end to ensure safety and security as front-end parameters.

Mobile banking could have a brighter future as compared to online banking, if the banks and their technology departments can upgrade and implement the high-end skills to reach out to customers in every possible way. M-commerce requires safety and security as strong phenomena. It also needs the innovative strategies to beat the tough competitive wars and create retention among the existing users. Cost-effective plans can be a possible tool to attract new customers as well.

5.3 EXTENSION OF E-COMMERCE

The rise of e-commerce has led to the progress toward m-commerce. Technological revolutions have created a push toward ease and convenience in the most cost-effective manner. Donegan (2000), Schwartz (2000), and Liebmann (2000) discuss that people consider e-commerce and m-commerce as two synonymous terms.

Nohria and Leestma (2001) state that the advantage of mobility which comes along with m-commerce is unbeatable. Skourletopoulos et al. (2016) discuss the transition toward fifth-generation (5G) mobile networks. The research on fifth-generation mobile-enabled services will conceptualize by 2020 with the one thousand times more network traffic than it is today. Internet access and web browsing on mobiles have become an inseparable part of mobile. Mobiles come with an added advantage of touch screen technology, which makes it easy to handle anywhere and anytime. It also incorporates all the comforts and conveniences satisfying the needs of the consumers. Changes in the technology have given the m-commerce world a better push and a great support to carry on transactions more smoothly (Turban et al., 1999).

Harter (2000) analyzes that one of the important driving force for the m-commerce technology is the easy access of internet on mobiles. Consumers are happy to enjoy the merging of two technologies at the top of communication world—mobile and internet. When internet is offered free on mobile, they are willing to make use of it. But a great resistance is shown toward the payments involved in the usage of internet services still by a large number of users (Jones, 2000).

M-commerce is an extension of e-commerce with the provision of using of a large number of apps without the need for a desktop (Cotlier, 2000). Customers should be presented with the ideology of a brand new medium in the m-commerce technology (Ramakrishnan, 2001). Nohria and Leestma (2001) argue that the participants of m-commerce technology have to develop a change in the thinking abilities and a shift in their approach while using this technology. Ramakrishnan (2001) endures that the m-commerce technology has to imbibe the needs and desires of the consumers, marketers, and all the other authorities participating. M-commerce and e-commerce should be considered as two different tools in the technological inventions. Their roles have to be categorized and understood in isolation. They both vary with their own advantages and disadvantages (Cotlier, 2000).

Zhang et al. (2002) emphasize on the three main drivers of m-commerce. The three main drivers mentioned are value chain, technology, and consumer demand, which are the main reasons contributing for the mass revolution of m-commerce.

5.4 COST

The cost is higher due to the network and other associated systems. The investment rises as it has to be made available while the customer is on move. Cost also depends on the data usage, downloads, files accessed, and the services provided. Mobile communication companies have tried to lure consumers by providing free internet usage on mobiles. It has proven a good amount of success for them at the initial stage since the number of people opting for the communication services has considerably risen up. But the big question would be that something which was freely available all this while, will it still be popular when the cost factor is added to it (Carrigan, 2000).

Marketers have understood the fact that the two binaries, that is, mobile and internet are now inseparable from the consumers. In fact, they are more of an addiction for the consumers. The service which was coming free to them for a period of 6 months, customers might will pay a small fee for it rather than forgo the usage of internet addiction which they cannot imagine living without (Dezoysa, 2001, 2002).

Pricing strategies should be carefully drafted and efficiently delivered, without the consumer realizing the burden of it. The consumers should feel excited about the brand which they have accessed for this while without paying any charges and a nominal fee might be just a token of appreciation for them to repay.

5.5 SOCIAL MEDIA MARKETING

Entertainment industry is a big industry and growing leaps and bounds. Mobile communication systems have changed the entertainment industry as well (Kuoppamaki et al., 2017). Movies can be downloaded, TV serials could be uploaded, movie tickets could be booked online, and utility payments could be made. Mobiles have provided convenience, comfort, and above all, it has caused enormous changes in the business dynamics. Mobile marketing and mobile ticketing are the two terms which have got added in the mobile revolution.

Music, movies, TV serials, games, and so on, mobile technology provides a wider range of entertainment services to its users. The technology aims to provide 24/7 entertainment channel to the consumers in the form of mobile, the one of major drivers for the China's mobile economic boom (Winston, 2016). It is an entertainment outlet which is not dependent on indoor connectivity but also extends to outdoors through a wireless medium. It is available everywhere, with or without internet connectivity (Kuoppamaki et al., 2017).

Mobile entertainment also covers videos and audio messages which are travelling through the reins of social media. They are reaching out to a larger number of audiences. Audio and videos once downloaded on the mobiles can be listened and watched any number of times without the requirement of any kind of connectivity or services. Mobile entertainment services provide mobility, which helps during travelling, vacations, or out of station. These services are very helpful for the users to fulfill their leisure requirements.

Social media has created impetus in the spread of mobile entertainment. Real or unreal videos keep on getting uploaded on the mobiles and the users ensure the video gets viral faster than any mode of communication.

5.6 DRIVING FORCES FOR M-COMMERCE

Mobile applications have been a boom for all the businesses. Every business organization establishes a mobile application to provide services in the viral world (Skourletopoulos et al., 2016). Technology aspires to be accessible all the time. The generations of today, no more waits to log on to the desktop or a laptop. The handiest option is a mobile. Whether it is making payments, transferring of funds, shopping, ordering goods, or any other online transaction, the fastest mode is mobile (Vanathi et al., 2016).

Mobile technology, due to its ease, accessibility, and convenience has taken over all the other electronic and other gazettes from us. Whether it is a diary, calendar, clock, alarm, calculator, camera, walkman, radio, albums, newspapers, or contacts collection, it has taken over almost every other thing. The mode of communication now remains mobile, whether through chat, messenger, mails, social media, video calling, or anything else (Gordon and Gutiorrez, 2007; Clarke, 2008).

Medium of communication for today's generation is largely mobile; therefore, a vast amount of time is spent on it. Marketers realized that to survive the tough marketing warfare, they have to have mobile advertising as an important element of IMC, that is, Integrated Marketing Communication. It has changed our lives, the way we communicate, travel, chat, explore, shop, order, express, celebrate, wish, and handle our relationships.

The new social quotient formulas are derived from the number of Facebook friends, Twitter followers, Whatsapp groups, and so on. Social circles are getting viral with the clubs evolving in it, in the form of Whatsapp groups. Psychological satisfaction is now defined by the number of likes, comments on Facebook, comments on Instagram, and comments posted on walls of social media.

The changing times have changed the era of communication, the way we express, share, or post anything. The words are replaced now by emoticons available aptly for each and every possible situation (Keen, 2001; Kelly, 2001).

Recent changes in the business environment have brought immense changes in the marketing strategies and have streamlined broader business definitions.

Some of the driving forces for m-commerce are discussed in following sections.

5.6.1 COMPETITION

Three main organic objectives of business are survival, growth, and image. These organic threefold objectives have articulated the businesses to strategize their marketing policies.

Marketing warfare has set new boundaries in the industry. Competition among the brands is omnipresent. The cutthroat competition has crossed all the boundaries and parameters of doing businesses. Every other brand strives for retention and creating reminders in the minds of the customers. Therefore, there is a dire need to be where the customers are, that is, the world of mobiles. Resultantly, most of the brands have created and placed their applications online. In fact, they have gone a step further to motivate the consumers to download the app by luring them to attractive offers, ranging from initial free orders, discounts, or fee waivers. Consumers have to download the app on their mobiles and the services begin along with the monetary incentives (Herman, 2000; Hooper, 2001).

Competition between OLA and Uber has been through a tough journey. The two brands had to first compete with the existing market of taxis. And then the brands had the competition among themselves. As per Darwin's theory—Survival of the fittest, they knew that the competition war had to be won at all cost. Therefore, when both the brands introduced their apps, the challenge which lied ahead was to make these apps popular. Marketing teams of both the companies got the idea of popularizing the apps by adding a monetary incentive to it. The consumers got the monetary benefit of either the first ride free or Rs. 250 off on their first ride.

5.6.2 TECHNOLOGY

Advancements in technology have created a new era in the world of marketing and advertising. Brands are vying for consumer's attention by all means. On one hand, high-end technology has created innumerable opportunities for the brands to make their products and services accessible and believable for the consumers but on the other hand, technology also brings resistance from the users (Heinz et al., 2017).

Technology works all the ways whether it is to involve the customers in the production process or at any other stage of delivery of product. It leads to providing products to customers, especially customized ones by fitting in their desires and expectations (Heinz et al., 2017).

Technology has made possible a delivery system which has never been imagined before, while ensuring that the products and services are being rendered in most shortened time frame. It is an era where the physical delivery of goods could be virtually tracked and exact delivery time can be estimated.

The app world makes it easier for the marketers to be available to the customers 24/7. It makes the concept of accessibility and convenience true in the real-time frame. It wins over the barriers of geography as well as demography. The markets are widened when the expansion takes place and the consumer base increases. Companies like Reliance has introduced plans such as "Jio," which aims to cater, cover, and reach the internet technology revolution to the last person of the country. Technology aims to provide network at zero cost at the fastest possible speed. The company's initially package which came up with free internet services has been extended to a longer period, looking at the popularity of the plan. The free internet offer has been so attractive that millions of users have joined the network and are part of the "Jio" family.

E-dark is the term used to identify with a period when a mobile internet user has no data package left to access internet. A large number of popular social media networks have extended their services and applications which can be accessed even without the availability of internet.

5.6.3 GLOBAL VILLAGE

In the current scenario, the world is one big stage with all kinds of products in a big market inclusive of buyers and sellers. The challenge lies for the developing nations, as they have to meet the standards laid by the developed nations. The world is walking on the lanes of cutthroat technology. Every nation and every other company has to imbibe and inculcate the technological advancements in its functioning behavior. Global village has on one hand provided immense opportunities to marketers by creating wider markets. On the other hand, it has laid m-commerce as a challenge (Molina and Merono, 2017).

Emerging markets are opportunities in front of the marketers all over the world to be grabbed. They have the advantage of selling the goods to the customers all over the world without any kind of geographical barriers. The wider markets have enhanced the scope of their businesses.

M-commerce is an opportunity which stratifies and simplifies the concept of global village more comfortably. Touch screen technology provides the marketers ease as well as opportunity to carry on the business all over the world while sitting at one corner of their place of business. This is a technology which knows no boundaries and makes the world one small village by integration of economies. Marketers can benefit from the differentiated prices and trade on equity while sailing and reaping the profits of the trade (Young, 2001).

In this big village, there are no entry or exit barriers. There is also no cost being imposed on entry or exit of goods. The goods do not have to be confined to any selected markets anymore. They are free to be sold in any part of the world to any customer (Molina and Merono, 2017).

5.6.4 SIMPLICITY

M-commerce has been successful, magnificent, and powerful tool due to the presence of inherent simplicity in it, a technology which does not require training or a particular skill (Papadopoulou, 2017). It is also not limited to the concept of literacy; therefore, the number of users is inclusive or exhaustive. It cuts across all the boundaries of age, gender, language, education, income, occupation, and so on. It is customer friendly as well as standardized.

The aim of any revolution should be inclusion to succeed and to excel. Inclusion should be true and real, encompassing each and everyone under its ambit. M-commerce has been truly inclusive with the technology reaching to each and every section of society (Papadopoulou, 2017).

Whether it is the domestic helpers in our houses, vegetable vendors, street sellers, hawkers, daily wage earners, there is a choice for each one due to the ease of price ranges offered in the mobile segment as well as telecommunication technology (Pelet and Taieb, 2017).

Idea mobile subscribers have definitely loved all the advertisements of the company with the single theme of inclusion. The marketing team really is worth mentioning with all the Idea advertisements diligently crafted with social messages and effectively delivered (Papadopoulou, 2017).

The dream of becoming self-sufficient by equipping the required entrepreneurial skills among the simple villagers with the help of technology ensuring that there goods reach foreign markets is what Idea advertisements have been showcasing. These concepts are instilling the confidence among the population to excel in their respective skills (Pelet and Taieb, 2017).

5.6.5 VALUE CREATION

Value creation entitles the consumer's ideology about the brand to be positive and supportive. Google maps can be used on the mobiles without any internet connection or any data plan. Facebook has extended a range of service for its users in India, which extends its services without the need for any internet access. Rebtel.com, Freekall.com, and FireChat apps offer free calling services to its users without any internet connection. DD (Doordarshan) channels can be watched on the mobile phones without the usage of any internet connection. Games can be played on the mobiles without any internet access (Whinston et al., 2001).

There is a need for partnering between the sellers and the telecommunication industry. The strong association will provide a large variety of services to its customers without charging them for each and every second of internet usage. This will lead to creation of lot of value, which will create brand equity. Brand equity will create a positive brand image developing brand loyalty among the consumers for ever, toward the products of the company (Brewin, 2001).

The websites such as Wikipedia and Twitter can be accessed without internet connectivity. Latest news can be read, FM channels can be listened, bus and train schedules can be planned, spiritual apps can be read, and existing mails can be opened without any connection to internet.

Reliance initially offered Jio services for free, but the popularity, number of users, and the usage have driven the company to introduce a plan which will be at a reasonable cost. Large customer base which has now become used to the services of the company and fast internet speed will not like shutting down of the services. They rather might go ahead and pay the reasonable charge to benefit from the continuous internet connectivity.

YouTube offers an offline mode, wherein the users can watch the videos offline. The users get the advantage of watching the videos, which might have been watched earlier, or the videos which have been saved on mobile in offline mode.

5.6.6 SAFETY AND SECURITY

Advent in technology has introduced safety systems so as to detect, capture, and prevent the mobiles from any kind of virus attacks. Increase in the security and safety develops confidence among the users and therefore, the sense of trust builds (Rockhold, 2001).

Innovative software have been introduced which prevent any kind of malpractices, frauds, or any cheating cases in the m-commerce transactions. The purpose is to prevent the users from any kind of unexpected frauds.

Companies have introduced safety devices in their systems and consumers are being warned from time to time about the possible frauds. They are also educated about the cybercrimes and the methodology to prevent such crimes themselves (Swartz, 2001a).

Government should take necessary steps to ensure safety and security in the mobile communication network. Stringent laws with strict punishments and severe penalties will create a sense of fear and prohibit the culprits to be averse to such frauds and malpractices (Swartz, 2001b).

5.6.7 INNOVATION

Technological innovations have been high end, with the slimmer handsets, sleek designs, elegant looks, competitive camera clarity, expandable text, unimaginable memory sizes, and so on. There has been no boundary for the revolution taking place in the new era. The current scenario is highly competitive enforcing the companies to change before the change by grabbing the cutting-edge technology in their processing (Jainschigg and Grigonis, 2001).

The latest software downloaded on the smartphones allows the users to click pictures with clarity as close to DSL camera. It also provides the option to the user to photoshop the pictures and the required editing. Software also provides the advantage of transferring the color movies and sharing it with the friends in the real time.

Location-based customized services are provided to the customers wherever they are. The consumers get assistance of any kind if they are at a new place and wish to get accustomed. Innovations have reached to an extent where even if the mobiles are lost, it can be easily tracked by the authorities through a unique number, which provides exact location, even if the robber switches it off (Schwartz, 2000; Secker, 2001; Shaffer, 2000; Shim, 2001).

5.6.8 CONSUMERS

The success of m-commerce lies in its ability to satisfy the demands, changing revolutionary desires and expectations of the consumers. Adequate fulfillment of consumer's value-added services will lead to value creation. Consumers of today have choices, due to which they continuously compare among the brands and look for profitable opportunities (Cohn, 2001).

India already has been rated as a country with highest amount of minutes of usage (MOU) among the mobile users. The success of smartphones in the country will provide a boost to the m-commerce industry to sail in this winning segment and reap the fruits (Chanay, 2001).

The days have gone when the consumers use phones only for making or receiving calls. Customers demand has touched undemanding heights with the inherent changes in the current marketing environment.

Consumers are keen on doing m-commerce transactions. The interest and eagerness ratio is very high among the existing mobile users and the presence of rising interest in the new users. Consumers are adaptive and receptive to the technology in the smartphones. The ease and convenience which are inseparable part of the revolution have touched the right chords across all the users (Schooler, 2001).

The approach of m-commerce technology is user friendly and customized. Older generation is also not behind and has readily accepted the new addition. They believe that it can make life simpler and easier.

5.7 CONCLUSION

A systematic effort needs to be in place to ensure the success of m-commerce technology. This will entail incorporating all the stakeholders in the technological innovation process (Feldman, 2000). The internet service providers on the one end, the application developers on the other end, and most importantly, the users of the m-commerce technology should be encouraged to play a participative role (Schneiderman, 2001). M-commerce is a technology which is ready to use, explore, and succeed in the present era. It provides all the advantages to the generations whether "X," "Y," or "Z." It is a technology which has adequate business security solutions. It is ready to revive and learn from its previous experiences and is willing to change every second. It is quickly grabbing all the technological innovations happening all over the world (Carrara, 2000).

Latest smartphones have finger print recognition system software. Therefore, unless the user itself tries to access the phone, no one else can do so. Digital biometrics will not only increase the security of the phone but such software will also prevent any kind of data theft or frauds.

The success of m-commerce will depend on the effective merging of the stakeholders, incorporating marketers, as well as telecommunication companies (Schwartz, 2000). Value-added services will be an added advantage to effectively satisfy the stakeholders and create high-end brand loyalty among the users.

REFERENCES

Brewin, B. M-commerce Hits Snag as Cell Earners. *Computerworld,* 2001, Vol. 35 (30), p 18.

Carrara, J. Who Holds the Keys? *Wireless Review,* 2000, Vol. 17 (23), pp 68–70.

Carrigan, T. Licensing Places Big Cost Burdens on M-commerce. *Marketing,* 2001, Vol. 17 (16).

Chanay, X. Boom or Bust for SIM. *Telecommunications,* 2001, Vol. 35 (6), pp 109–112.

Clarke, I.III. Emerging Value Propositions for M-commerce. *J. Bus. Strat.* **2008,** 41–57.

Cohn, M. Full Beam. 2001. http://www.mcommerceworld.com/articles/article (accessed April 23, 2016).

Cotlier, M. Wide Wireless World. *Catalog Age,* 2000, pp 16–17.

Darling, A. Waiting for the M-commerce Explosion. *Telecommunications,* 2001, Vol. 35 (2), pp 34–38.

Dezoysa, S. The Cost of M-commerce. *Telecommunications,* 2001, Vol. 35 (7), p 10.

Dezoysa, S. M-commerce Payment Architecture. *Telecommunications,* 2002, Vol. 35 (8), p 10.

Donegan, M. The M-commerce Challenge. *Telecommunications,* 2000, Vol. 34 (1), p 58.

Feldman, S. Mobile Commerce for the Masses. *IEEE Internet Comput.* **2000**. http://compui:er.org/internet/ (accessed May 2, 2016).

Fitchard, K. Mapping Next Generation Wireless. *Telephony,* 2001, Vol. 241 (18), p 26.

Gordon, X.; Gutierrez, J. A. An Exploratory Study of "Killer Applications" and Critical Success Factors in M-commerce. In *Web Technologies for Commerce and Service*; 2007; pp 232–235.

Harter, B. Motient's M-commerce Movements. *Wireless Review,* 2000, Vol. 17 (19), pp 10–12.

Heinz. J.; Thomann, M.; Fischer, P. Ladders to M-commerce Resistance: A Qualitative Means-end Approach. *Comput. Hum. Behav.* **2017,** *73* (2), 362–374.

Herman, J. The Coming Revolution in M-commerce. *Bus. Commun. Rev.* **2000,** *30* (10), 24–25.

Hooper, M. Billing the 3G Extravaganza. *Telecommunications,* 2001, Vol. 35 (9), pp 77–80.

Jainschigg, J.; Grigonis, R. M-Commerce Alternatives. *Computer Telephony,* 2001. http://www.cconvergence.com/ (accessed May 7, 2016).

Jones, J. Vendors Walk Thin Line. *InfoWorld,* 2000, Vol. 22 (50), pp 1–27.

Keen, P. G. W. Go Mobile-now. *Computerworld,* 2001, Vol. 35 (24), p 36.

Kelly, S. M-commerce Slower than Expected. *Communications News,* 2001, Vol. 38 (7), p 10.

Kuoppamaki, S. M.; Taipale, S.; Wilska, T. A. The Use of Mobile Technology for Online Shopping and Entertainment Among Older Adults in Finland. *Telemat. Inform.* **2017,** *41* (5), 110–117.

Liebmann, L. Help for Building Sticky Web Sites. *Information Week*, 2000, Vol. 815, pp 158–164.

Molina, E. J., Merono, A. L. Drivers of Mobile Application Acceptance by Consumers. In *Mobile Commerce: Concepts, Methodologies, Tools and Applications*; 2017; Vol. 1 (1), pp 42–58.

Nohria, N.; Leestma, M. A Moving Target: The Mobile-commerce Customer. *MIT Sloan Manag. Rev.* **2001,** *42* (3), 104.

Papadopoulou, P. Exploring M-commerce and Social Media: A Comparative Analysis of Mobile Phones and Tablets. In *Research Paradigms and Contemporary Perspectives on Human Technology Interaction; IGI Global*, 2017; pp 1–21.

Pelet, J. E.; Taieb, B. From Skeuomorphism to Flat Design: When Font and Layout of M-commerce Websites Affect Behavioral Intentions. In *Advances in National Brand and Private Label Marketing*; Springer Proceedings in Business and Economics, Springer: Cham, Switzerland, 2017; pp 95–103.

Ramakrishnan, J. WWWireless Wonder. *Corporate Location*, 2001, pp 60–61.

Rockhold, J. Find Your Location Reality. *Wireless Review,* 2001, pp 6–8.

Schneiderman, C. Are You Billing in Real-time. *Telecommunications,* 2001, Vol. 35 (7), pp 71–72.

Schooler, J. Give Them What They Want. *Credit Union Management,* 2001, Vol. 24 (8), pp 42–45.

Schultz, B. The M-commerce Fallacy. *Network World,* 2001, Vol. 18 (9), pp 77–82.

Schwartz, E. Mobile Commerce Takes Off. *InfoWorld,* 2000, Vol. 22 (25), pp 1–32.

Secker, M. Does M-commerce Know Where It's Going? *Telecommunications,* 2001, Vol. 35 (4), pp 85–88.

Shaffer, R. A. M-commerce: Online Selling's Wireless Future. *Fortune,* 2000, Vol. 142 (2), pp 262.

Shim, R.; Rice. V. How to Unwire Your Business. *Technology Review*, 2001, pp 46–54.

Skourletopoulos, G.; Mavromoustakis, C. X.; Mastorakis, G.; Batalla, J. M.; Sahalos, J. N. An Evaluation of Cloud Based Mobile Services with Limited Capacity: A Linear Approach. *Soft Comput.* **2016,** *21* (16), 4523–4530.

Swartz, N. Taking It to the Street. *Wireless Review,* 2001a, Vol. 18 (5), pp 54–58.

Swartz, N. Hot & Cold: M-commerce Opportunities. *Wireless Review,* 2001b, Vol. 18 (6), pp 32–38.

Turban, E.; Lee, J.; King, D.; Chung, H. M. *Electronic Commerce: A Managerial Perspective;* Prentice Hall: Englewood, 1999; pp 436–439.

Vanathi, B.; Shanmugam, K.; Uthairaj, V. R. A Secure M-commerce Architecture for Service Provider to Improvise Quantity and Quality of the Product Using Fingerprint Authentication and Gender Classification. *Asian J. Inf. Technol.* **2016,** *15* (2), 232–242.

Whinston, A.; Barna, A.; Shutter, J.; Wilson, B., Pinell, J. Measuring the Internet Economy. University of Texas, 2001. www.intemetindicators.com (accessed June 4, 2016).

Winston, M. Mobile Entertainment. In *China's Mobile Economy*; Wiley and Sons, 2016. http://do;org/10.1002/9781119321392.ch6 (accessed June 6, 2016).

Young, D. Will Security Power M-commerce? *Wireless Review*, 2001, Vol. 18 (8), p 12.

Zhang. J. J.; Yuan. Y.; Archer. N. Driving Forces for M-commerce Success. *J. Intern. Commer.* **2002,** (454), 81–104.

ADDITIONAL READINGS

Kalakota, R.; Whinston, A. B. *Frontier of Electronic Commerce*; Addison-Wesley: Reading, MA, 1996.

Parsons, R. How Should We Market WAP Services? *Telecommunications,* December 2000, Vol. 34 (1), pp 95–96.

Prior, M. M-commerce Infancy no Day in the Park. *Retailing Today,* 2001, Vol. 40 (16), pp 1–35.

Sliwa, C. M-commerce Awaits Killer Apps. *Computerworld,* **2001,** *35* (22), 16.

Sweeney, T. Tech Obstacles Keep Wireless E-retailers Grounded. *Information Week*, 2001, pp 72–76.

Taaffe, J. Organization for Economic Cooperation and Development. *The OECD Observer,* 2001, pp 29–33.

Tsalgatidou, A.; Pitoura, E. Business Models and Transactions in Mobile Electronic Commerce: Requirements and Properties. *Comput. Netw.* **2001,** *37,* 221–236.

CHAPTER 6

ENCUMBRANCES IN M-COMMERCE: A DEMOGRAPHIC CONTEXT

NIDHI SINGH* and SHUBHI AGARWAL

Department of Business Administration, Faculty of Commerce, University of Lucknow, Lucknow, Uttar Pradesh, India

**Corresponding author. E-mail: nidhiks7@gmail.com*

ABSTRACT

Mobile commerce or m-commerce, which depends on technology, is predicted to be the next big phase in the society after E-commerce era. As we are moving in the present age of globalization and trying to touch the height of development, there are root factors which have been overlooked; thus, everyone is not benefitted by the advancement. The main question is whether the development is for all or only for few. However, it could be seen that m-commerce adoption and level of use is low as compared to the smartphone usage. The objective of this chapter is to compare the usage of m-commerce with respect to smartphones in India, and will look out what are the various demographic factors which are obstructing m-commerce usage. M-commerce is complex as it has emerged recently. There are challenges related to it. Its adoption is far from its full potential. It has not grown according to the predictions. Internet usage is increasing as well as usage of wireless devices is increasing but m-commerce is not being used at the same rate. Therefore, there is a need to explore various factors hindering its growth. This chapter is concerned with internet usage, smartphone usage, and m-commerce usage. Usage of smartphone is increasing day by day yet comparatively, m-commerce is not increasing with the same pace; therefore, a comparison showing its use in detail is discussed. The factors which are hindering m-commerce usage are identified and main emphasis has been given on demographic factors. This chapter will contribute to the existing literature related to m-commerce usage in India. It shows the current status of m-commerce usage, and compares it with the usage of smartphone. In-depth analysis of the hindrances will help in overcoming these and therefore will increase the rate of usage of m-commerce, and it will provide an environment where people will be more oriented toward its use.

6.1 INTRODUCTION

As a rapidly maturing global technology, internet is having an increasing impact on the world's economic, political, and social affairs. While electronic commerce (e-commerce) impacts the global business environment overpoweringly, the focus of technology and wireless has shifted more on mobile computing and the wireless web. With this trend comes a new set of issues specifically related to mobile e-commerce.

Mobiles help tremendously in dealing with customers thereby increasing number of transactions. The rise of mobile has been rapid in past years and it has transformed the trends in marketing, customer experience, and all other segments of business.

But this rapid advancement has benefitted many but not all. Every technology faces heterogeneous environment. Introduction of m-commerce have made many promises regarding free flow of information but developing nations like India face many hurdles in deriving the whole benefit. Security, connectivity, performance, etc., has been issues which have been widely explored in the past. There are various other reasons in the demographic context which can be explored. The chapter discusses about the history of internet as well as m-commerce. The usage statistics of internet has been shown with respect to urban and rural areas. The uses statistics has been shown with respect to the devices which have been used more for the internet uses, purpose for which internet is used and the overall statistics of smartphone usage. The perspective of noninternet users has been discussed and it could be clearly understood that if internet uses do not exist then how we could think of m-commerce usage.

6.2 HISTORY OF INTERNET IN INDIA

For the present generation, and even the septuagenarian, it is almost impossible to think back of a time when the world had no internet. August 1995, marked the introduction of internet in India and it was launched by Videsh Sanchar Nigam Limited (VSNL), which was then a public sector enterprise and now part of Tata Communications, and it primarily focused on providing oversees communication. People in six Indian cities could access the internet through dial-up services, which promised a speed range of 9.6–128 Kbps on premium leased lines at costs ranging from INR 5000 to INR 2,000,000 per annum. In 6 months, VSNL was able to acquire more than 10,000 customers, many of whom were prominent business leaders (IAMAI and KANTAR, 2016).

After a year, in 1996, India's first cyber café opened in Mumbai. Over the next few years, cyber cafes became a common phenomenon. People had to wait for their turn before being assigned an available computer. That time, internet was slow in speed and computers had less random access memory but that was not any obstacle because use of internet was mostly confined to content browsing, sending an e-card, or enjoying a chat session. Having internet connection at home was considered a luxury, and therefore uncommon. Dial-up connections were the only option available at an average speed of 10 Kbps (NASSCOM, 2016).

The late 1990s saw internet expanding its reach and computing becoming a part of people's lives. Internet adoption spread across business segments and heralded the dotcom boom. Just as people and businesses in India and the world were getting to accommodate with the internet better and leveraging its potential, an Indian entrepreneur started his webmail company called Hotmail. Within 2 years of its launch, Sabeer Bhatia's Hotmail had millions of registered users. In 1997, Microsoft acquired Hotmail.com for $400 million, its largest acquisition at that time.

A key breakthrough in the growth of internet in India is the opening up of the sector by the Government to allow private internet service providers (ISPs) to set up internet infrastructure. This was in November 1998. A decade ago, the industry had already witnessed the launch of National Association of Software and Service Companies (NASSCOM) to promote the Information Technology (IT) industry in the country. NASSCOM led to tremendous growth in India's software and services exports, and enabled the IT sector to make a significant impact to the country's growth domestic product.

From Yahoo (recently acquired by Verizon) and MSN launching their Indian sites in 2000, to Wikipedia adding Indian regional languages and first broadband option introduction by BSNL in 2004—there are many key milestones in the internet's journey in India in its early years. Figure 6.1 clearly shows the history of internet in India.

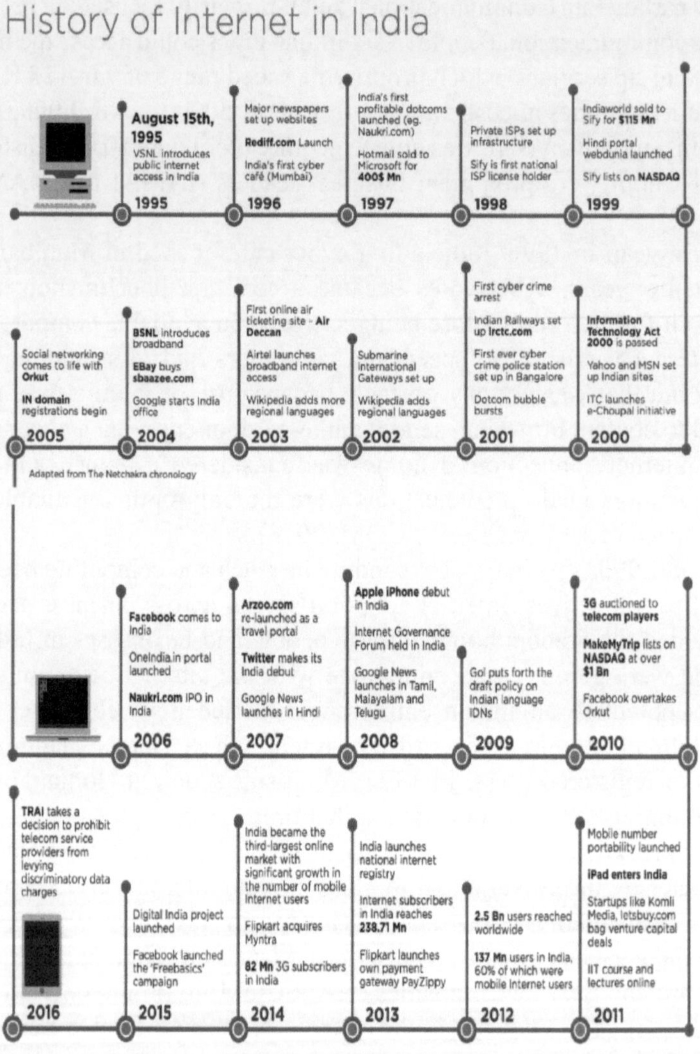

FIGURE 6.1 (See color insert.) Internet history in India.
Source: Adapted from NASSCOM. The Future of Internet in India; 2016. http://www.nasscom.in/sites/default/files/Article_News/PR_NASSCOM_AKAMAI_Rise%20of%20Internet_17082016.pdf (accessed Mar 10, 2017).

6.2.1 CHANGING INDIA: ONE CONNECTION AT A TIME

Internet has transformed our lives way we learn, shop, work, and even the way we connect. Gone are the days when people had to wait in long queues at telephone booths, more popularly known as Public Call Offices, to talk to family and friends stationed overseas for a steep amount once a week. But, that was the reality two decades ago. With the introduction of mobile phones and Reliance overhauling the playing field, mobile telephony became affordable to the common man. Even today, India has some of the lowest calling rates in the world. With the proliferation of Wi-Fi and low-cost data plans, Skype, Whatsapp, and other VOIP options are the preferred modes of calling, especially for long-distance calls. Today, internet video calling is not only virtually free, but it also ensures that people are connected to family and friends all the time. Distances of hundreds and thousands of kilometers seem to have phenomenally reduced. Everything, from e-commerce services and products, advertising, online content, devices, connectivity, as well as private infrastructure and the government, as a whole, make up India's internet economy (NASSCOM, 2016).

Today, we can buy anything online, from clothing to groceries, even motorcycles and cars. Online businesses are generating a large amount of revenues annually. Consumers are rapidly and intuitively adapting to new modes of online sales, for instance, in 2014, the sale of Chinese smartphone manufacturer's Xiaomi phones through exclusive flash sales on India's home grown e-commerce platform—Flipkart—did not last more than 5 s. That is a powerful example of how the internet has changed consumer behavior in the country—it is also testament to how speeds have accelerated (averaging 3.5 Mbps in Q1 2016). India is also home to the world's second-largest user base for social media giants such as Facebook and LinkedIn. And, it took less than a decade for companies like Facebook to tap the large user base, since its debut in 2006, followed by Twitter in 2007 (Muthukumar and Muthu, 2015)

In addition, the internet has changed the way we bank today. A visit to the bank to deposit the monthly salary, or waiting for at least 4 days for an outstation check to be cleared, or even having to pay to get outstation check cleared, are all a thing of the past.

6.3 M-COMMERCE HISTORY

M-commerce was first used in Helsinki where a Coca-Cola vending machine operated through mobile phone was set up. First phone-based banking service was introduced in Marita Bank in 1997 which enabled payment through text messages and thus, m-commerce found its way (Sharma, 2016). In 1998, downloadable ringtones were started in Finland through which, the first sales of digital content were done. In 2000, increase of services related to m-commerce could be witnessed. Mobile parking payment was started in Norway. Austria offered train ticketing through mobile. Japan offered mobile purchase of airline tickets. Cellular phones have become so popular that mobiles have become means to interact with customers for many businesses. In order to exploit the potential of m-commerce market, mobile phone manufactures such as Sony, Blackberry, and Apple are working with carriers such as AT wireless, Telenor, Vodafone, NTT, DoCoMo, Bharti Airtel, Sprint Nextel Corporation, and Reliance Communications and are developing Wireless Application Protocol-enabled smartphone and are offering complete range of telecom service. Smartphones offer services for email and fax. There are new trends transforming the world m-commerce is sometimes described as a wireless extension of wired electronic commerce which is easily accessible anytime from anywhere. Products and services such as m-shopping, m-money transfer, mobile banking, mobile ATM, location-based services, etc., are the factors making m-commerce so popular. M-commerce is innovative way of doing business as it allows making any transaction anytime from anywhere. M-commerce is dependent on the availability of mobile connectivity. As we have become accustomed of making mobile phone calls from anywhere at any time, in the same way, consumers have become used to shop using a handheld device, personal digital assistant (PDA), mobile and tablet, and smart wireless devices. M-commerce offers multiple advantages which include flexibility and distribution, instant connectivity, ubiquity, personalization, and immediacy. There are many ways in which consumers of India can be benefited from advancement of m-commerce. M-commerce connects mobile users irrespective of their varied geographical locations (Mishra et al., 2016).

Mobility has the benefit of portability where users can carry mobile everywhere with them and do anything, anywhere in real time using mobile devices because it has no geographical constraint. They can order any product online from any corner of the country and get it delivered to their place safely and conveniently. M-commerce is based on wireless mobile communication system, which utilizes cellular technology. M-commerce enables the user to use services such as mobile entertainment, travel ticketing, booking, games, product locating, searching, wireless reengineering, mobile financial applications, healthcare services, and shopping almost everything online using apps easily.

6.3.1 CHARACTERISTICS OF M-COMMERCE

The reason behind adoption of m-commerce is its unique characteristics. The two fundamental characteristics are mobility and reachability (BenMoussa, 2003; Camponovo and Pigneur, 2003; Ng-Kruelle et al., 2002; Turban et al., 2002).

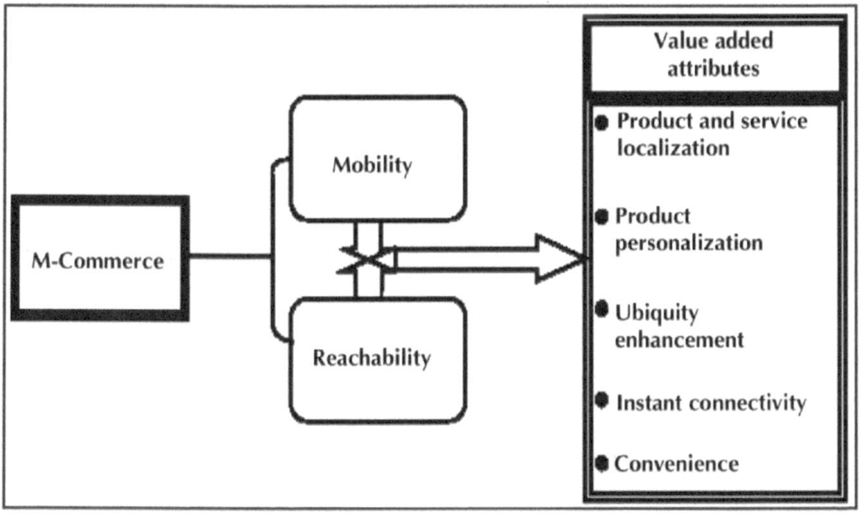

FIGURE 6.2 M-commerce characteristics.
Source: Reprinted from Sang Hyun Kim. Impact of Mobile-commerce: Benefits, Technological and Strategic Issues and Implementation. *J. Appl. Sci.* **2006,** *6* (12), 2523–2531. Published under the Creative commons attribution License.

Mobility characteristics (as shown in Fig. 6.2) of m-commerce relate to uses of mobile devices at any time and any location. Mobile devices users can be reached instantly. Aforesaid two characteristics remove the geographical and time obstacles for a particular situation. Other characteristics are ubiquity, user identity, location awareness and interactivity, security, and instant connectivity (Muller-Veerse, 1999). Studies (Kenny and Marshall, 2000; Kannan et al., 2001; Tarasewich and Warketentin, 2002; Vetter, 2001) had concluded that ubiquity is the most obvious advantage of m-commerce because it can satisfy the need for real-time information and communication anywhere, regardless of a user's location.

6.3.2 APPLICATIONS OF M-COMMERCE (FIG. 6.3)

M-shopping: Mobile shopping is the process of purchasing goods and services through mobile devices. Almost all big retailers offer online shopping options to their customers. Presently, the highest boom is in this sector. People spend more time on mobile phones than desktop computers. Main reason is its mobility and portability feature. Shopping using mobile phone has become easy and smooth. Best quality, trendy, and reasonable products are easily available through m-shopping on mobile with secured mobile wallet payment system (Sharma, 2016; Sandhu, 2012).

Entertainment: Entertainment is also one of the important applications of m-commerce. Playing online games, watching online videos, and listening songs on smartphone devices played an important role in development not only for entertainment industry but also for m-commerce industry. Entertainment applications are a segment with maximum share in m-commerce market, and it is predicted to become the leading part of m-commerce. India being the country of youth, features of m-commerce and its applications are popular among young population.

Education: Education is also among the main applications of m-commerce. One can access lot of contents without any time and location constraint using handheld devices. Many of the online journals, books, research papers, and articles have their dedicated applications and help the students in their education and learning (Sharma, 2016; Sandhu, 2012).

Travel and ticketing: With the introduction of internet-enabled fast-speed smartphones, mobile users have started booking online tickets easily. Booking tickets has become easy these days. Be it for railways, road, or air travelling, through mobile ticketing apps, tickets for any medium of transport can easily be booked. Indian Railways launched official mobile application which helps consumers to check train schedule, availability, booking, cancellation, and other related functions. There are many apps for road transport booking such as Ola cab, Uber, Jugnoo, etc. Almost all airlines have their mobile applications for various mobile platforms to provide facility to their customer (Sharma, 2016; Sandhu, 2012).

Banking apps: Mobile banking is an integral part of m-commerce. Many banks provide online banking facility. Mobile banking services can be classified in two categories: transaction based and nontransaction based. Online facilities such as Fund transfer, Micro payment, Bill payment, Tax payment, Request for check, Request for mini statement, various statements, etc., can be availed using mobiles. Mobile banking services are provided by all leading banks such as Axis Bank, Bank of Baroda, and SBI, etc. (Sharma, 2016; Sandhu, 2012).

6.3.3 INDIAN M-COMMERCE: AT A GLANCE

Internet + wireless + e-commerce = m-commerce

- 1505 CAGR till 2016
- 200% jump in flipkart cell visit
- 30% B2B, trade India through cell
- 35% eBay traffic through cell
- 60% revenue through cell for Snapdeal
- 800% growth through mobile
- Indian eCommerce market $4b

(Accel partners study)

FIGURE 6.3 Indian m-commerce at a glance.
Source: Adapted from http://www.digit.in/telecom/mobile-commerce-in-india-ready-to-hit-mainstream (accessed Jan 1, 2017).

Online stores: Presently, the lifestyle of the people is extremely busy especially in cities as male and female both are working. As both are busy with their work and commitments, they are left with only a little spare time. With wide availability and usage of smart handheld devices, people find internet as the easiest medium to fulfill their needs and requirements. M-commerce makes the shopping procedure much easy as all the sites are available there on mobile, just one touch away, which gives users' facility to shop online anything, anytime, and anywhere. All these features have made online shopping so immensely popular. Few best online shopping apps are mentioned below:

1. Amazon.in
2. Flipkart.com
3. Snapdeal.com
4. Koovs.com
5. Yepme.com
6. Jabong.com
7. Myntra.com

8. Askmebazaar.com
9. ebayinc.com
10. Voonik.com

Mobile banking: Mobile banking service is provided by a bank or other financial institution on a 24-h basis. Transactions of financial and nonfinancial nature are performed in mobile banking. The types of financial transactions which a customer may transact through mobile banking include obtaining account balances and list of latest transactions, electric and other bill payments, fund transfer, and many other functions. Some mobile banking service providers in India are:

1. hdfcbank.com
2. icicibank.com
3. onlinesbi.com
4. axisbank.com

Mobile wallets: Mobile wallet is mobile payments mechanism under which mobile user opens a mobile wallet account in a partner bank and deposits some money in the account which can be used for making online payments. Issuing bank registers the mobile number and gives a mobile wallet account number to the user. While making payment, the user enters the account number in the mobile phone and sends it to the bank via SMS, bank checks available balance in mobile wallet account. If balance is sufficient account is debited for the requested amount. Mobile wallets are gaining popularity day by day in India. One important feature of mobile wallets is that user need not to have a bank account, especially in developing countries like India where still many citizens do not have a bank account. They can open their account in mobile wallet partner banks and can make easy online and offline payments. Following are a few mobile wallet companies in India.

1. Paytm
2. Momoe
3. PayUmoney
4. MobiKwik
5. Citrus
6. State Bank Buddy
7. Citi MasterPass
8. ICICI Pocket
9. HDFC Chillr
10. LIME

6.4 INTERNET USAGE STATISTICS IN INDIA

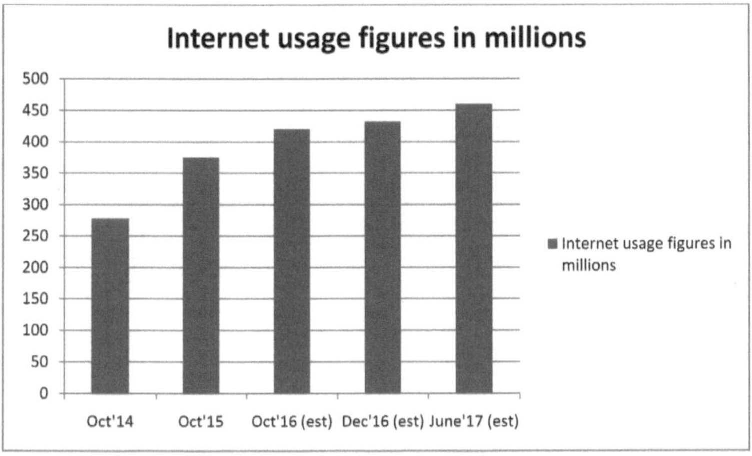

FIGURE 6.4 Internet usage in India.
Source: Adapted from An IAMAI & KANTAR IMRB Report. *Internet in India 2016*; 2016. http://static.bestmediainfo.com/wp-content/uploads/2017/03/Internet-in-India-2016.pdf (accessed Apr 4, 2017).

Figure 6.4 clearly shows as on December 2016, India had estimated 432 million internet users. This however, does not take into account the impact of demonetization. It is estimated that by 2017, internet users in India are most likely to be in a range of 450–465 million. The report finds that the overall internet penetration in India is around 31% presently.

6.5 INTERNET PENETRATION IN INDIA

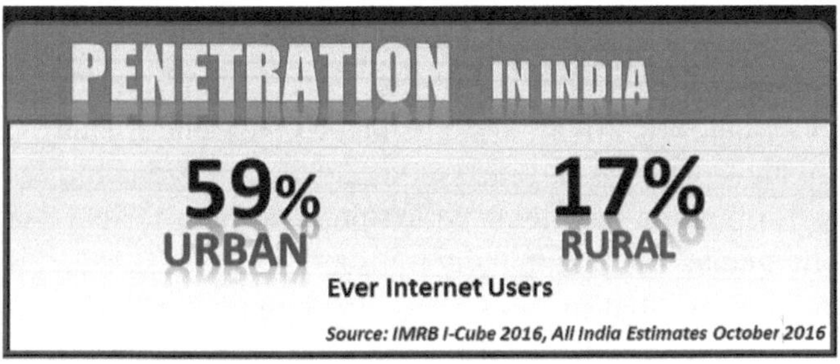

FIGURE 6.5 Penetration in India of internet users.
Source: Adapted from An IAMAI & KANTAR IMRB Report. *Internet in India 2016*; 2016. http://static.bestmediainfo.com/wp-content/uploads/2017/03/Internet-in-India-2016.pdf (accessed Apr 4, 2017).

Figure 6.5 clearly defines the internet penetration in India which is 59% in Urban India and 17% in rural India, so it could be seen that the difference is too much.

6.6 PURPOSE OF INTERNET ACCESS

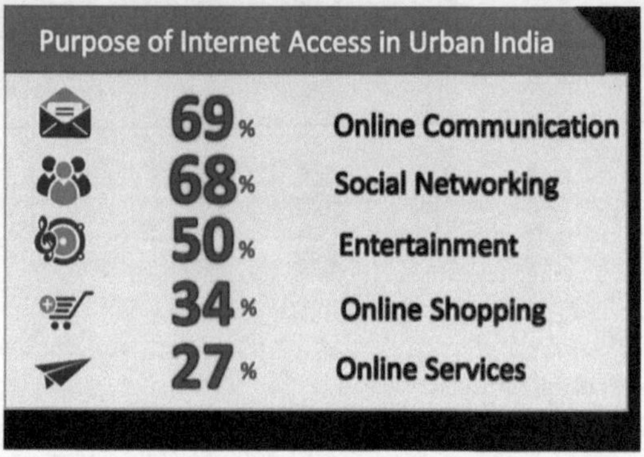

FIGURE 6.6 Purpose of internet usage in urban India.
Source: Adapted from An IAMAI & KANTAR IMRB Report. *Internet in India 2016*; 2016. http://static.bestmediainfo.com/wp-content/uploads/2017/03/Internet-in-India-2016.pdf (accessed Apr 4, 2017).

All the fields of daily human life have been touched by internet (Fig. 6.7 clearly explains it).

Figure 6.6 clearly defines that in urban areas, it is used widely for online communication. People can communicate to a person overseas thus removing the distance barrier. Another common use is social networking which is 68% in which means for users to interact over internet are such as e-mail, instant messaging, and online forums. In total, 50% users use it as source of entertainment by playing games, watching movies, free music, and chatting. Because of unavailability of time, people have started using mode of online shopping; thus, there is no distance constraint for any product. It is growing widely with a fast rate. About 27% users are using it for online services say for making payments which prevent them for carrying cash or avoiding long queues for making any transaction.

Figure 6.7 defines that in rural areas, 39% people are using internet for entertainment, whereas 34% use it for social networking and 31% use it for communication. Online ticketing is another use which is 12%. In rural areas, awareness, knowledge, and sense of security is a big concern so uses and users are limited in number; therefore, only 4% people are using it for online shopping.

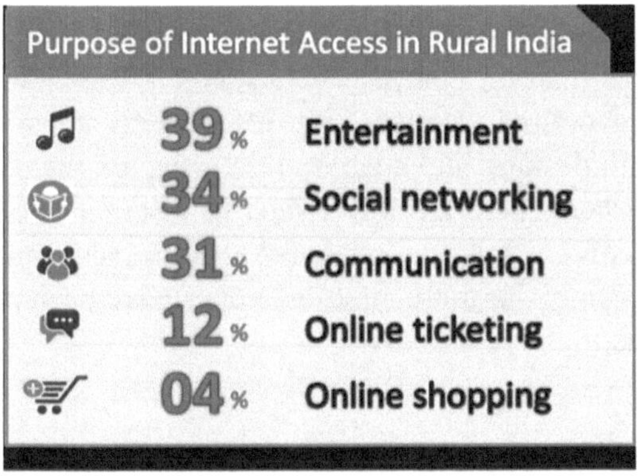

FIGURE 6.7 Purpose of internet usage in rural India.
Source: Adapted from An IAMAI & KANTAR IMRB Report. *Internet in India 2016*; 2016. http://static.bestmediainfo.com/wp-content/uploads/2017/03/Internet-in-India-2016.pdf (accessed Apr 4, 2017).

So when it is analyzed, it could be clearly figured out that the difference between urban and rural is too high and the percentage of online shopping and other online service is lowest among all.

6.7 DEVICE USED FOR INTERNET ACCESS

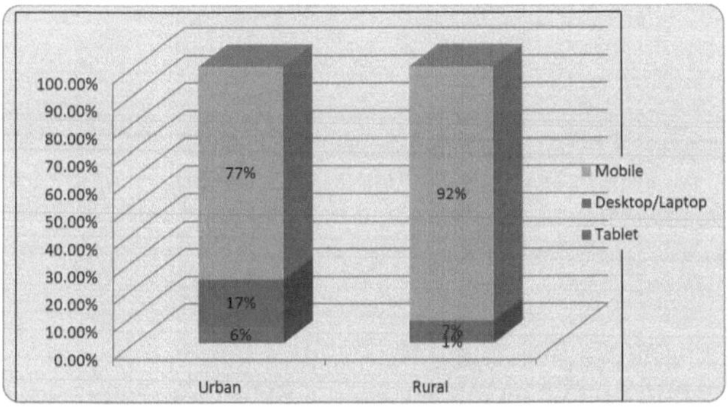

FIGURE 6.8 **(See color insert.)** Device used for internet usage.
Source: Adapted from An IAMAI & KANTAR IMRB Report. *Internet in India 2016*; 2016. http://static.bestmediainfo.com/wp-content/uploads/2017/03/Internet-in-India-2016.pdf (accessed Apr 4, 2017).

Figure 6.8 clearly defines, be it rural area or urban area, the internet use on mobile has surpassed the use of desktop and tablet because 77% of urban users and 92% of rural users consider mobile as the primary device for accessing the internet due to ease, comfort, and portability factor in mobile. In 2015, Google officially confirmed that more Google searches take place on mobile devices than on computers. On-screen keyboards make tablets as a less preferred option, which is only 1% in rural areas. Mobile has the advantages of both desktop and a tablet. It is expected the usage of desktop will go down further. There are low-priced handsets in the markets. More and more companies are entering into handset manufacturing. Internet usage rates have been falling; therefore, mobile phones have been preferred choice for internet access.

6.8 POINT OF INTERNET ACCESS IN URBAN INDIA

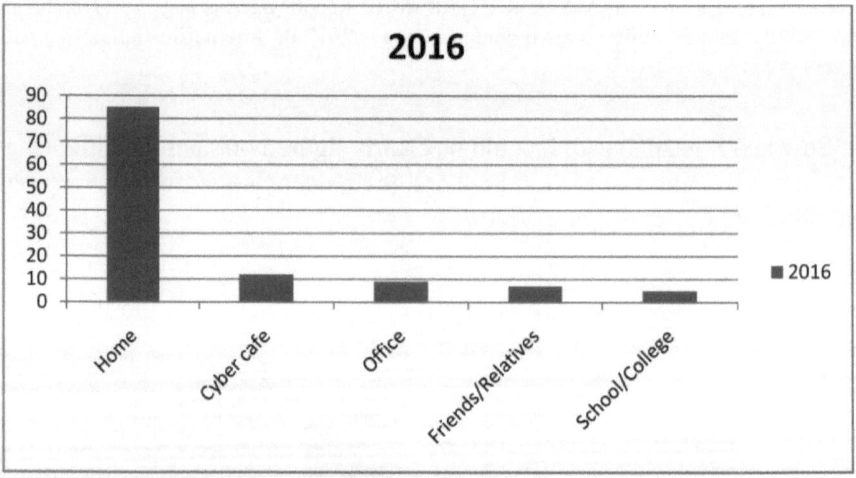

FIGURE 6.9 Point of internet access in urban India.
Source: Adapted from An IAMAI & KANTAR IMRB Report. *Internet in India 2016*; 2016. http://static.bestmediainfo.com/wp-content/uploads/2017/03/Internet-in-India-2016.pdf (accessed Apr 4, 2017).

Figure 6.9 explains that with the declining rates of internet connection, it is affordable and people have easy access to it. Its usage is not confined to cyber café or any fixed location. People are free to access it from anywhere they feel comforable;, only for that reason, all internet-provided services are used so easily by all in urban India.

6.9 POINT OF INTERNET ACCESS IN RURAL INDIA

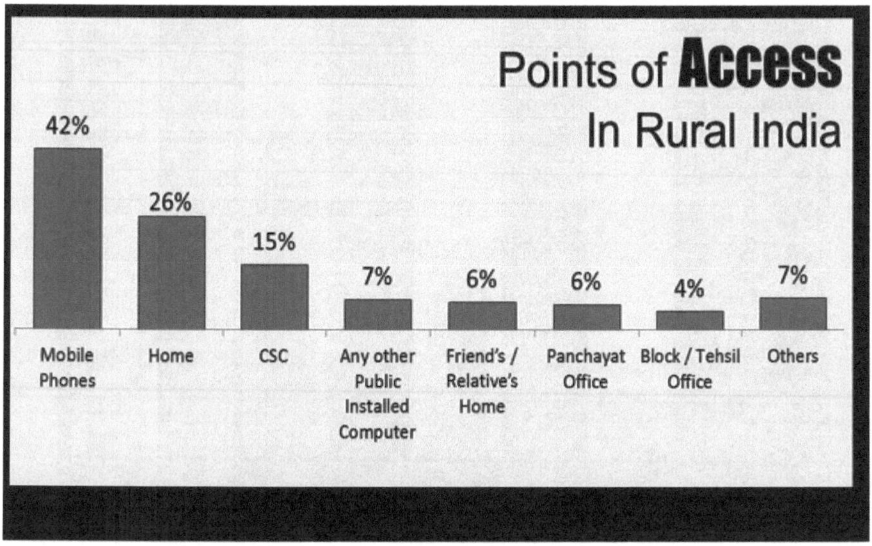

FIGURE 6.10 Point of internet access in rural India.
Source: Adapted from An IAMAI & KANTAR IMRB Report. *Internet in India 2016*; 2016. http://static.bestmediainfo.com/wp-content/uploads/2017/03/Internet-in-India-2016.pdf (accessed Apr 4, 2017).

Looking at Figure 6.10 it could be seen that as in rural areas, main reasons for internet access are entertainment and social networking; therefore, it is more viable to use internet on mobile phones at home as it provides more privacy to the users. Users are not using it for any other reasons therefore usage for other options are very low in percentage. As mobile serves as a cheaper and easily available to all option, they prefer it and the affordable internet connection makes it easier to access internet from their location.

6.9.1 SMARTPHONE MARKET IN INDIA

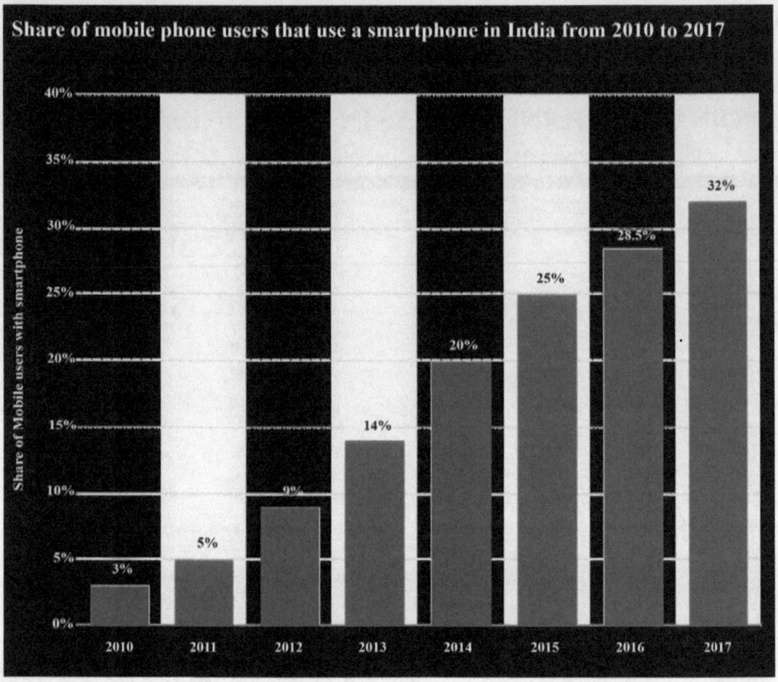

FIGURE 6.11 Smartphone penetration.
Source: Adapted from Statista (2017). https://www.statista.com/statistics/257048/smartphone-user-penetration-in-india/ (accessed Apr 10, 2017).

Looking at Figure 6.11, it could be clearly identified that the usage of smartphone is increasing day by day and so the use of m-commerce should be increased but the scenario is not so. In comparison of smartphone, the usage of m-commerce is not increasing with the same pace as people are still carrying the old mindset.

6.10 ENCUMBRANCES IN THE USAGE OF M-COMMERCE IN INDIA

Anything which is in existence has positive and negative aspects and its acceptance do have to face many challenges. Noninternet usage by the people of urban and rural India has been seen and there are different opinions. If internet usage is not there among people in India then how can we speak about m-commerce usage because everything is dependent on internet usage only. There are many hurdles in the nonusage of m-commerce.

(Gupta et al., 2016) The barriers that are discussed by different authors are as follows:

- lack of awareness and personal touch,
- technical limitations and low internet connectivity,
- usability problem and language barrier,
- security issues, and
- internet infrastructure.

6.10.1 BRIEF DESCRIPTION OF ENCUMBRANCES DISCUSSED BY DIFFERENT AUTHORS

Security: The very main issue related to m-commerce is security. Users have doubt in mind while making online payment as they need to share the information related to debit/credit card over the internet. Another threat is of hacking and virus. Hackers may hack the device and steal all personal and important information from mobile. Viruses also damage the handset if the security measures are not proper.

Language barrier: In India, most of people are not well versed with English language, so for them it is not possible to do transactions through mobile due to this language barrier. This is also one of the major issues.

Lack of awareness: M-commerce is popular in cities only not in rural areas. Even today, many people in India are not aware of mobile phones and internet. Those who are aware are still hesitant regarding buying online, making payment online, and how to return if not satisfied with delivered product. Moreover, people still have suspicions and reservations regarding services offered by m-commerce.

Data transmission rate: Speed is one of the biggest issues of m-commerce. Common users still get the speed of 2 mbps for 3G and 4G network mobile phones, although data transfer at 14.5 mbps can be attained but the charges for such service are very high.

Lack of network coverage: Network coverage is also a big issue. Many time users do not get connectivity at many geographical locations. Without coverage, using m-commerce is just not possible.

Low graphical resolution: Products and offers are not very clearly visible on mobile in comparison to desktop or laptop. Youngsters are fine with it but middle- and old-aged people do not find the graphical resolution very convenient to look upon and work.

But as per the internet usage, it could be clearly identified that there are demographic barriers which are hindering m-commerce usage. These barriers are as follows:

- by income or lack of means,
- by education or lack of knowledge,
- by gender, and
- by age.

The data show no internet usage due to lack of income and education, so we could conclude that if there is no internet usage, how can we think of m-commerce usage. As m-commerce is still new to people, data is unavailable on m-commerce showing the exact statistics, so internet has been used as a means to understand the m-commerce hurdles. The statistics shown clearly give a brief description about the mindset of people who do not use internet in India.

Among the nonusers of internet, the reasons behind not accessing internet services can be grouped into following categories:

- lack of knowledge of internet,
- lack of means (or infrastructure), and
- beliefs (the opinions held by the nonusers).

6.10.2 NONINTERNET USERS IN URBAN INDIA

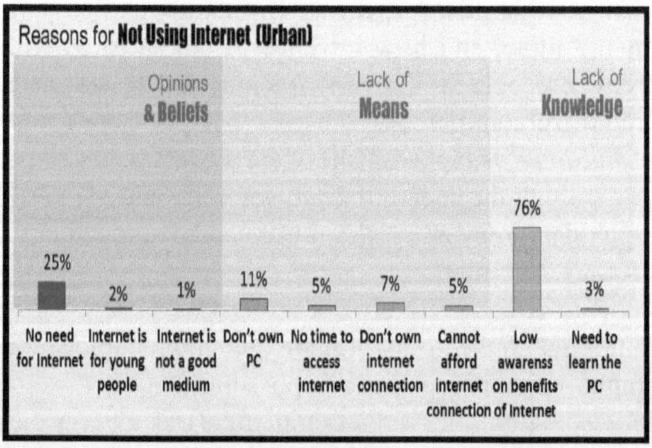

FIGURE 6.12 (See color insert.) Noninternet users in urban India.
Source: Adapted from An IAMAI & KANTAR IMRB Report. *Internet in India 2016*; 2016. http://static.bestmediainfo.com/wp-content/uploads/2017/03/Internet-in-India-2016.pdf (accessed Apr 4, 2017).

In urban areas, the most dominating reasons amongst urban nonusers are as follows:

- *Lack of knowledge*
  - *Low awareness on benefits of internet* as they have their own feeling regarding the use of internet which is 76%. It clearly denotes that people fail to understand the benefit of using the internet.
  - *Need to learn the personal computer (PC)*: It is unbelievable that still in this techno-savvy age, there are people who need to learn PC usage and that percentage is 3%.
- *Lack of means*
  - *Do not own PC*: There are 11% of people who do not own PC that is why they do not use internet.
  - *No time to access internet*: 5% of people have reason that they do not have time to use internet.
  - *Do not own internet connection*: 7% of people living in urban areas do not own internet connection which is very surprising in this digital age.
  - *Cannot afford internet connection*: 5% of people who reside in urban areas cannot afford the internet connection which is an unexpected reality.
- *Opinion and beliefs*
  - *No need of internet*: 25% of people living in urban areas do feel that there is no need of internet in their life.
  - *Internet is for young people*: 2% of people carry a thought that internet is for young people.
  - *Internet is not a good medium*: 1% of people feel that internet is not a good medium for them to use.

6.10.3 NONINTERNET USERS IN RURAL INDIA

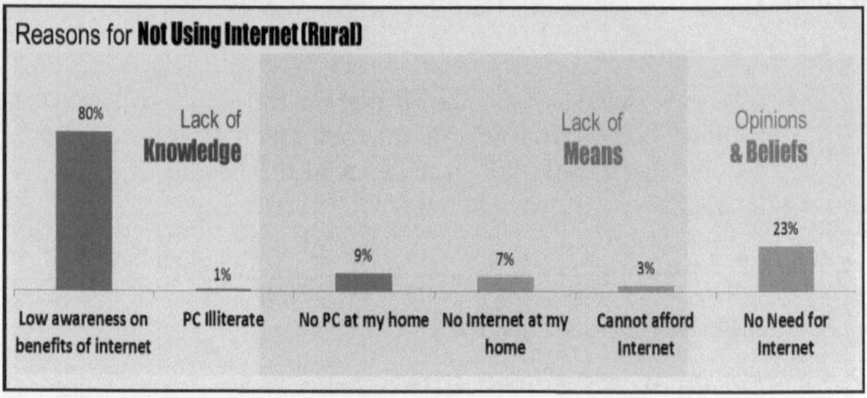

FIGURE 6.13 (See color insert.) Noninternet users in rural India.
Source: Adapted from An IAMAI & KANTAR IMRB Report. *Internet in India 2016*; 2016. http://static.bestmediainfo.com/wp-content/uploads/2017/03/Internet-in-India-2016.pdf (accessed Apr 4, 2017).

In rural areas, the most dominating reasons amongst rural nonusers are as follows:

- *Lack of knowledge*
  - *Low awareness on benefits of internet* as they have their own feeling regarding the use of internet which is 80%. It clearly denotes people fail to understand the benefit of using the internet.
  - *PC illiterate*: it is unbelievable that still in this techno-savvy age there are people who need to learn PC usage and that percentage is 1%.
- *Lack of means*
  - *No PC at my home*: There are 9% of people who do not own PC that is why they do not use internet.
  - *No Internet at my home*: 7% of people living in rural areas do not own internet connection of their own which is very surprising in this digital age.
  - *Cannot afford internet connection*: 3% of people who reside in urban areas cannot afford the internet connection which is an unknown reality.
- *Opinion and beliefs*
  - *No need of internet*: 23% of people living in rural areas do feel that there is no need of internet in their life.

6.10.4 INTERNET USAGE BY GENDER

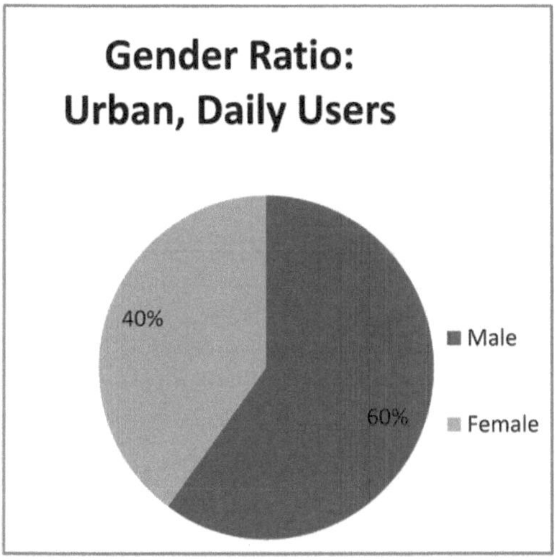

FIGURE 6.14 Gender-wise usage in urban India.
Source: Adapted from An IAMAI & KANTAR IMRB Report. *Internet in India 2016*; 2016. http://static.bestmediainfo.com/wp-content/uploads/2017/03/Internet-in-India-2016.pdf (accessed Apr 4, 2017).

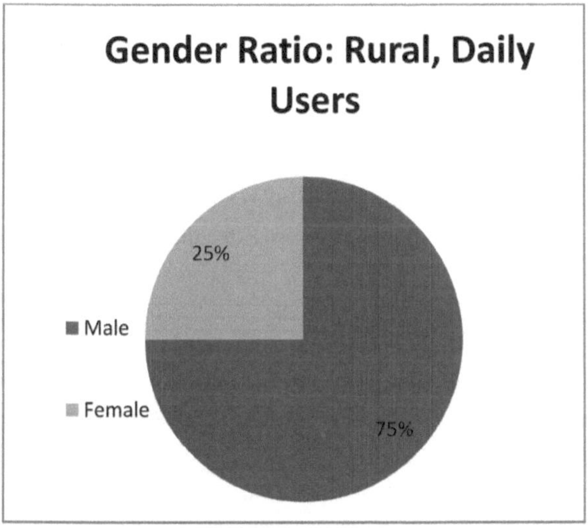

FIGURE 6.15 Gender-wise usage in urban India.
Source: Adapted from An IAMAI & KANTAR IMRB Report. *Internet in India 2016*; 2016. http://static.bestmediainfo.com/wp-content/uploads/2017/03/Internet-in-India-2016.pdf (accessed Apr 4, 2017).

From Figures 6.14 and 6.15, we can say that gender difference exist in urban areas; gender ratio is slightly better in urban areas as compared to rural areas due to various reasons such as education levels are higher in females in urban areas. In urban areas, more females work as compared to rural areas, so we could trace it out that if gender is a parameter in the use of internet, it will be the hurdle for m-commerce usage too.

6.10.5 *INTERNET USAGE BY AGE*

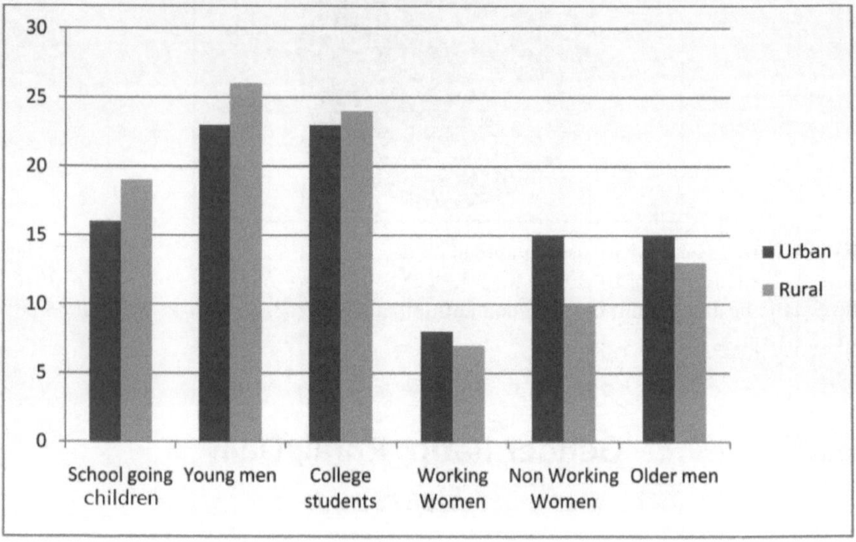

FIGURE 6.16 Age group wise internet usage.
Source: Adapted from An IAMAI & KANTAR IMRB Report. *Internet in India 2016*; 2016. http://static.bestmediainfo.com/wp-content/uploads/2017/03/Internet-in-India-2016.pdf on (accessed Apr 4, 2017).

When we go through the statistics of Figure 6.16, it could be clearly seen that the school-going kids, young men, and college students are using more internet as compared to other age groups and the difference is there in the urban and rural population. When we look at Figures 6.12 and 6.13, we could clearly see that the belief is also there that young people use internet. The internet usage defines the m-commerce statistics and it is also the thought that everything is limited to a specific age group.

A report from Regalix Research (2016) says Indian retail customers are taking strongly to m-commerce, with nearly 83% of people owning a smartphone shopping online on their mobile phones. In particular, customers in the age group of 25–34 are taking to their mobile phones to shop online, with 90% of customers in this group doing so. These are the findings from the "State of M-Commerce 2016" survey which was conducted by Regalix Inc. Mobiles are used heavily by customers to shop online but still online shopping has a potential to grow. Only 53% of those surveyed said that they had shopped online within the last month. Further, only 25% of respondents said that they shopped on their phones at least once a week.

There seems to be a gender divide in the frequency of online shopping, with more men (63%) doing so at least once a month than women (40%).

The lion's share of online retail is divided between three platforms—Flipkart (44%), Amazon (32%), and Snapdeal (19%). Interestingly, preference between these platforms is segmented by age, with Flipkart the preferred retailer for 49% of respondents in the 18–24 age group, but only 35% in the 25–34 age group. Snapdeal received more support in the 25–34 age group (25%) than in the 18–24 group (only 13%) (RapidValue Solutions, 2014).

Each product can be clearly divided into categories of male and female. Overall, gadgets and electronics was the most popular product category at 60%, followed distantly by clothing and accessories at 20%. While 66% of men shopped for gadgets online, only 31% of women did so. On the other hand, more women (65%) shopped online for clothing, accessories, and home products as compared to men (27%).

Although the number of retail customers for shopping online is increasing at a very fast pace, many still prefer to explore and search online and purchase the product offline. In total, 42% of respondents said that they preferred to do so while purchasing gadgets and electronics, while 33% said that they bought clothing and accessories this way. Interestingly, clothing and accessories was the one category in which offline purchasers outnumbered online shoppers across all age and gender groups.

The two biggest factors influencing a customer's decision to shop online are cash-on-delivery (CoD) (34%) and free delivery (34%). CoD was the preferred payment mode for the majority of customers, with 62% preferring this over net banking, credit/debit cards, or mobile wallets.

Mobile apps are the preferred way for customers to shop on their phones, with an overwhelming 94% declaring they preferred apps to mobile websites.

The study also found that while 81% of respondents said they were unaffected by mobile advertisements, a higher percentage of women (25%) said that advertisements influenced their shopping behavior than men (17%).

Online shopping is only set to grow with 31% of those who have not shopped online so far saying they planned to do so in the future. What has kept 75% of such shoppers away is concern over product quality.

6.11 CONCLUSION

M-commerce is dependent on internet. Although m-commerce has penetrated the Indian market, still it is at the nascent stage and evolving with each passing day. But some barriers can hinder the growth of m-commerce in India. Thus, overcoming these barriers will help m-commerce to expand its growth and help the people to connect to global business. It is required to overcome these barriers for its universal acceptance. We need to focus on the demographic aspects which have been neglected so far. Thus, to provide equality and its usability to all, it is an imperative to explore these demographic barriers. Then only it will lead to development of all. By studying heterogeneous population and its characteristics, most of the issues to overcome their problems about internet usage could be sorted out. This chapter provides the insight into the various barriers which hinders its growth and acceptance. This chapter clearly gives insights into how the lack of knowledge, lack of income, age, and gender are creating hurdles in the use of internet. When we see the statistics provided by the IAMAI (2016), it could be clearly figured out that demographic factor are the biggest hurdle in development and acceptance of any new thing and if internet is not being used, how could we think of m-commerce usage. The main contribution of this chapter is the development of the country-level adoption barriers.

KEYWORDS

- **internet**
- **m-commerce**
- **demographic factors**
- **smartphone**
- **knowledge**
- **income**
- **gender**

REFERENCES

An IAMAI & KANTAR IMRB Report. *Internet in India 2016*; 2016. http://static.bestmediainfo.com/wp-content/uploads/2017/03/Internet-in-India-2016.pdf on (accessed Apr 4, 2017).

BenMoussa, C. In *Workers on the Move: New Opportunities Through Mobile Commerce*, UKAIS Conference, University of Warwick, Apr 9–11, 2003.

Camponovo, G.; Pigneur, Y. Business Model Analysis Applied to Mobile Business. 5th International Conference on Enterprise Information System (ICEIS), Angers, Apr 23–26, 2003.

Gupta, C.; Chandhok, A.; Gupta, M. Hardship of M-commerce in India: Problems, Issues & Challenges. *IOSR J. Bus. Manag. (IOSR-JBM)* **2016,** *18* (1), 22–26 (e-ISSN: 2278-487X; p-ISSN: 2319-7668).

http://www.digit.in/telecom/mobile-commerce-in-india-ready-to-hit-mainstream (accessed Jan 1, 2017).

Statista. 2017. https://www.statista.com/statistics/257048/smartphone-user-penetration-in-india/ (accessed Apr 10, 2017).

Kannan, P. K.; Chang, A.; Whinston, A. B. In *Wireless Commerce: Marketing Issues and Possibilities*, Proceedings of the 34th Hawaii International Conference on System Sciences Piscataway; Sprague, R. H. Jr., Ed.; IEEE: New Jersey, 2001.

Kenny, D.; Marshall, J. F. Contextual Marketing: The Real Business of the. *Harvard Business Review*, 2000.

Kim, S. H. Impact of Mobile-commerce: Benefits, Technological and Strategic Issues and Implementation. *J. Appl. Sci.* **2006,** *6* (12), 2523–2531.

Mishra, A.; Medhavi, S.; Mohd, S. K.; Mishra, P. C. Scope and Adoption of M-commerce in India. *Int. J. Adv. Res. Comput. Commun. Eng.* **2016,** *5* (8), 231–238 (ISO 3297:2007 Certified).

Muller-Veerse, F. *Mobile Commerce Report*; Durlacher Research Ltd: London; 1999.

Muthukumar, S.; Muthu, N. The Indian Kaleidoscope: Emerging Trends in M-commerce. *Int. J. Adv. Res. Comput. Commun. Eng.* **2015,** *4* (1), 7.

NASSCOM. *The Future of Internet in India*; 2016. http://www.nasscom.in/sites/default/files/Article_News/PR_NASSCOM_AKAMAI_Rise%20of%20Internet_17082016.pdf (accessed March 10, 2017).

Ng-kruelle, G.; Swatman, P. A.; Rebne, D. S.; Hampe, F. The Price of Convenience: Privacy and Mobile Commerce. *Q. J. Electro. Comm.* **2002,** *3*, 273–285.

RapidValue Solutions. *Internet, Smartphone & Social Media Usage Statistics*; 2014; p 13. https://www.google.co.in/social-media-usage-statistics-trends-by-rapidvalue-solutions&usg=AFQjCNGutMIcaJ5ydQhxHE8AmI-2Zbz7xg (accessed Apr 1, 2017).

Regalix Research. 2016. https://www.regalix.com/insights/m-commerce-trends-in-india-2016/ (accessed May 3, 2017).

Sandhu, P. Mobile Commerce: Beyond E-commerce. *IJCST* **2012,** *3* (1), 59–63.

Sharma, A. M-commerce: A Revolution in India. *Res. J. Manag.* **2016,** *5* (10), 42–46 (E-ISSN 2319–1171).

Statista. 2017. https://www.statista.com/statistics/257048/smartphone-user-penetration-in-india/ (accessed Apr 10, 2017).

Tarasewich, P.; Warkentin, M. Information Everywhere. *Inf. Syst. Manag.* 2002, *19* (1), 8–13.

Turban, E.; King, D.; Lee, J.; Viehland, D. *Electronic Commerce 2002: A Managerial Perspective*; Prentice Hall, 2002. ISBN 0, 13 (975285), 4.

Vetter, R. The Wireless Web. *Commun. ACM.* **2001,** *44*, 60–61.
http://www.digit.in/telecom/mobile-commerce-in-india-ready-to-hit-mainstream (accessed Jan 1, 2017)

PART II

M-COMMERCE: HARBINGER OF "PHYGITAL" RETAIL

CHAPTER 7

LET'S GET PHYGITAL: M-COMMERCE AS A CRUSADER FOR "PHYGITAL" RETAIL

PRIYANKA SINGH[1*], GURSIMRANJIT SINGH[2], and MANINDER SINGH[3]

[1]*Indian Institute of Management Udaipur, Balicha, Udaipur 313001, Rajasthan, India*

[2]*Department of Management Studies, Amritsar College of Engineering and Technology, I. K. Gujral Punjab Technical University, Amritsar 143001, Punjab, India*

[3]*Department of Management Studies, Amritsar College of Engineering and Technology, Amritsar 143001, Punjab, India*

[*]*Corresponding author. E-mail: priyanka.iimu@gmail.com*

ABSTRACT

In the digital era and technological advancements, m-commerce is providing new paradigms to m-service providers and users. One such technological sophistication is "phygital" retail. The term "phygital" is the integration of digital technology (digital) with the personal communications with the consumers (physical), thereby bridging the gap between the two. With the emergence of technology and explosive growth of internet users, m-commerce is becoming increasingly important to various businesses nowadays and is widely proclaimed as the "new service frontier." The phygital experience is a key to enhance the physical experience with the digital experience. Smartphones being the key component of this strategy has resulted in the companies getting more engaged in physical tactics through the social media, thereby allowing the companies to leverage so that they can face the competition. The biggest challenge posed by phygital as a paradigm is developing an understanding of how the companies can implement this strategy. However, the barriers have witnessed a decline as people of all the ages are engaging in the digital experiences. The globalization of markets has resulted in the shift in the consumers' mindset for the retail behavior. They are more prone to observing and understanding the differences between offline and online world. The present consumers demand one experience that combines both the traditional and digital aspects. Thus, here the concept phygital comes into picture. Phygital marketing concept has the potential to provide a solution by combining the elements of a traditional sales promotion with the elements of digital brand activation in an integrated complex which is irresistible for consumer of the 21st century.

The concept phygital is not only associated with the tastes and preferences of the consumers but is also related to the psychological effects in the interaction of modern marketing communication, perception of brand products and values, individual interpretation, and other allied services. It has been posited that phygital components can provide more personalized experiences to the customers and would further result in increased in-store visits, enhanced loyalty, value cocreation, customer engagement, greater share of wallets, and wider customer-based brand equity. The "always on" approach would facilitate the customer engagement by mapping the customers' move and allow the companies to achieve the ultimate phygital success. The present chapter deals with the specific paradigms of phygital retail so as to get an insight into the issues and challenges for managing the cointegration of technology with the brick and mortar and also how this integration can facilitate leveraging the physical retail networks so as to offer better customer service. The proposed chapter envisages addressing comprehensively issues and challenges for implementing the phygital strategy. The chapter would address on the one hand, the relevance of phygital marketing in retail with respect to the current trends such that the effectiveness of the marketing firm can be enhanced. On the other hand, it also envisages that the global marketers should make every effort to overcome the challenges such as untrained in-store staff and lack of knowledge, which lead to broken mobile strategies, so as to endure increased linkage with their audiences.

7.1 INTRODUCTION

The phenomenon of m-commerce has gained wider popularity with the advent of technology and growth of mobile users (Hung and Chang, 2003; Yang, 2005) and is widely proclaimed as the "new service frontier." Present era is witnessing the increased use of smartphones among consumers in India (Chun et al., 2012) and has led to an enlarged popularity of smartphones day by day (Persaud and Azhar, 2012; Soch et al., 2014). Primarily, mobile phones were used for communication but now they have become a sign of personality and individuality (Persaud and Azhar, 2012). M-commerce as an era of advancement is providing new paradigms to m-service providers and users. Researchers in the past have examined the influence of constructs such as rich information, low cost, search engines, portals, payment trust, legality, computer self-efficacy, and anxiety in the adoption of m-commerce (Elsaadany and Kadry, 2014; Tang and Ann, 2015).

Chong (2010) have propounded that retail industry has witnessed tremendous changes in its business modes and strategies with the emergence of m-commerce. The age of e-business has resulted in the growth of mobile and wireless technologies and thus, it is seen as a business model which is going to have a significant impact on the industry and business community at large. In the present times, the number of mobile phone subscribers is overtaking the number of internet users in many countries and have paved the way for the evolution and growth of the digital era.

The retail environment has witnessed a tremendous change because of the technological innovations in the present era of digital innovations and technological advancements, m-commerce is providing new paradigms to m-service providers and users. One such technological sophistication is "phygital" retail. The term phygital is the integration of digital technology (digital) with the personal communications with the consumers (physical), thereby bridging the gap between the two. The phygital experience is a key to enhance the physical experience with the digital experience. Smartphones being the key component of this strategy has resulted in the companies getting more engaged in physical tactics through the social media, thereby allowing the companies to leverage so that they can stay competitive.

Spencer (2016) has described the concept of phygital retail in context of mobile commerce. According to him, phygital is "a term that describes the bridge between the narrowing gap between the physical world and the digital world, and really—the physical, digital and mobile experience."

The concept of phygital marketing emerged with the objective of developing a platform that integrates the digital and personal communications between the consumers and the retail brands so as to identify and address the needs and preferences of the consumers for enhancing the social and personal relationships. It is the application of technology in an interactive and agile manner so as to directly influence the overall experience of the users. This facilitates creation of a seamless world by setting up a platform for promoting digital interactions, thereby expanding the contact with user. It is a strategy to leverage physical retail networks to offer better customer service. This is done by developing each brand store as a point of distribution, thereby reducing the warehousing costs and substantially improving the delivery speeds.

7.2 EVOLUTION OF THE CONCEPT PHYGITAL

In late October, the concept of phygital retail landed on front pages with the announcement that Lowe's is testing two retail service robots in a San Jose store. With the constant rise on the use of internet, smartphones are accessed at an average rate of 180 times a day. This furnishes a point of departure from the traditional means of communicating with the customers and merging the technology with the physical contact so as to entice the clientele. It has been seen that for consumers engage more in physical experience than the digital. Despite this fact, phygital offers diverse platform through expansion of the internet.

The objective of the phygital concept is to take the area of marketing communication ecosystem to new heights and establish association between brands and consumer. Phygital concept uses the most modern knowledge and innovations while interacting with customers which are then implemented in the physical environment. Tools of this concept facilitate organizations to interact with their consumer audience, as marketing communication using phygital concept is much more interesting, dynamic, faster, and pushes the limits of human perception (Moraveikova and Kliestikova, 2017).

Phygital concept is often referred to as integrating the different methods of shopping available to consumers, it means using multiple channels for selling a product or service provided to customer who makes purchases online, via m-commerce or in store. Phygital experience not only adds value to products or services but it also strengths the combination of digital experience and physical improvements. Organizations can use the so-called marketing automation, which enables implementation processes more efficiently, as an integral part of customer relationship management (CRM). It follows that the concept phygital is a combination of online, digital, and social media marketing communications being implemented in the online environment. With the increasing digitalization, there has been a great change in the way the consumers engage with the brands. Moreover, it has enforced the companies to discover new tools to differentiate themselves from their competitors.

Gradually, firms are putting more emphasis on branding to stay competitive. This has led to the emergence of brand management as one of the most critical success factors for companies than ever before.

Armstrong and Rutter (2017) propounded that the key driver of phygital experiences is the evolution of digital marketing. The digital revolution has forced the marketers to look for the new and innovative tools in marketing so as to provide their customers with implausible experiences. This has resulted in the emergence of the phygital retail experiences. The phygital retail reflects the trends of purchases and consumption across different platforms.

As business environment has changed over the years due to revolution and platform provided by internet to contact a more extensive gathering of people, companies throughout the world are using social media both to interact with the consumer and even to influence their conversations (Amichai-Hamburger, 2009). With the help of social media, companies are building their brand equity by creating brand awareness and also understanding and promoting stakeholder's participation to build up healthy relationship (Bruhn et al., 2012) resulting in brand loyalty (Palmatier et al., 2007). Online communities have a very significant role in enhancing brand loyalty, positive word-of-mouth and generating interest in product (Barreda, 2004).

Presently, the goals are more important as social media plays a very crucial role in promoting interactions among consumers and brands (Park et al., 2007). Moreover, smartphones create interactive platforms for consumers and communities to share and modify user-generated content (Kietzmann et al., 2011). Such platforms are stimulants for branding activities on social media, as opposed to pairing different branded products to form a unique product, that is, composite brand (Park et al., 1996). All these activities provide new opportunity for marketers to develop their brand identity. Customers interact and share their experiences on social media resulting in creating shared meanings for brands (Hung et al. 2003, 2012; Islam et al., 2011; Muniz and O'Guinn, 2000). As an interactive platform, social media enables business to engage existing customers and attract new customers and generate more sales (Marshall, 2012; Gilfoil, 2012) by building brand awareness thereby enhancing brand loyalty (Palmatier et al., 2007) and brand image (Barreda, 2014; Bruhn et al., 2012). Branding with the help of customer using social media will lead to relationship marketing.

Horky (2009) emphasized that the increased usage of smartphone acts as a strategically placed quick response tool and has the potential to turn a physical experience to a digital one, for instance, the Facebook or Twitter can be used to set up a spontaneous flash mob. Ellis (2014) propounded that the increased competition between the online and offline retail paved the way for the introduction of the concept of phygital retail. Despite the growth of e-commerce at a pace of four to five times as compared to bricks and mortar retail, physical stores still dominate the market by accounting for 93.6% of the total retail sales. According to Horky (2009), in order to satisfy the customers in a cost-effective manner, internet marketing is associated with other digital channels; the businesses have transformed their objectives to marketing activities. This has resulted in the integration of traditional marketing with the internet marketing so as to generate the elements of effective marketing communication models.

Gregory (2012) asserted the role of social media in promoting and implementing the strategy of phygital retail. As social media is an affordable and effective marketing tool, it builds up on its own exponentially. The use of quick response codes, which initiate the mobile devices to the internet platforms is a good example of phygital marketing. They asserted that phygital retail is not merely concerned with the intention to purchase physical product only rather, it is the accumulated experience that a customer generates during the shopping trip.

Vel et al. (2015) reported that the phygital concept integrates the use of physical and digital platforms to create interactive experiences for the customers. The use of phygital marketing involves the synchronization of the physical and digital technology as against the multichannel marketing. They highlighted that majority of the companies all across the globe are adopting this concept and hence, there is a need to induce investment in phygital retail segment so as to create customer value in experiences and unveil the potential of the phygital marketing.

According to a study conducted by Google and A.T. Kearney, about 25% of the total organized retail sales by 2020 will be contributed by the online channel. The report titled "Digital Retail 2020" says that online shopping in India will reach a gross merchandising value (GMV) of $60 billion by the end of 2020. The report also predicts that the total number of online shoppers will grow to 175 million and about 33% customers will drive two-thirds of total shopping. According to Customer Growth's Johnson "In the mid-'80s, 45 percent of consumer spending went to goods, the remainder to services. Today those figures are 31 percent (goods) and 69 percent (services), respectively."

7.3 TOOLS OF CONCEPT PHYGITAL

The 21st century, has witnessed digital transformation in the way that the traditional marketing approach is enriched with the various elements of online marketing and digital technology, with the objective of providing the individuals more personalized experiences and opportunities to interact with the brand in an easier and effective way. The concept phygital also requires certain tools for working efficiently. These tools are discussed as follows:

1. QR codes: The system of QR, that is, quick response was developed by a division of Toyota in the year 1994. With the increased proliferation of mobile phone devices with other devices such as cameras, systems QRs were utilized by almost all the businesses. A QR code is a typical two-dimensional code that consists of specifically organized black and white squares. Figure 7.1 shows a QR code:

FIGURE 7.1 QR code.

Source: https://www.google.co.in/url?sa=i&rct=j&q=&esrc=s&source=images&cd=&cad=rja&uact=8&ved=0ahUKEwjN-Krz-dXVAhWJuI8KHcCrC2kQjB0IBg&url=http%3A%2F%2Fwww.qr-code-generator.com%2F&psig=AFQjCNH3RP0NQC-4Vy3khS9QiGesq3yG0A&ust=1502773461710036.

The typical application of the concept phygital is the QR system. It is the most versatile system using the phygital system for various platforms, vis-à-vis Apple iOS, Android OS, Blackberry OS, Java, Windows Phone 8, etc.

2. Augmented reality: As the name suggests, augmented reality is the synchronization of the real world with the virtual environment. It uses a total replacement principle for providing a real environment by integrating the different text information and graphic objects to the real picture and hence is a kind of virtual reality. Augmented reality technology works on the basis of mobile applications to real-time phenomenon and is inserted in the camera image generated content, for example, text, still graphics, video, 3D animation, and so on. That is wherever the camera lens is directed, mobile phone application detects through continuous evaluation of data obtained from the GPS chip (Fig. 7.2).

FIGURE 7.2 (See color insert.) Augmented reality example.
Source: https://leafcutter.com.au/blog/augmented-reality-enhancing-experience-and-engagement/.

One of the popular cases for the successful augmented reality application is the Brand Pokemon Go, which merged the virtual with the real-world experience.

3. Mapping technology: Based on the theme of virtual reality, mapping technology uses a variety of tools and technologies that can be controlled by the gestures by making use of virtual reality. It is a system of marketing communication that uses slipping and simulation of physical world. It is also a tool of phygital concept and is free from implementation problems. The two popular mapping technology practical tools are Google Tango and Vuforia. One of the widely used mobile platform services is the Vuforia to provide the customers a fun and interactive mobile experience that can diversify

the way the users can see the utility of a particular product or service. It is being used by many of the world's leading brands such as Mercedes Benz, Coca-Cola, Nissan, Honda, McDonalds, Hyundai, etc.

7.4 CURRENT TRENDS

7.4.1 PHYGITAL EXPERIENCE AND RETAIL BUSINESS

The roots of the buzzword "phygital" lie in the retail industry. Since the term bridges the gap between the physical and the digital worlds, the phygital experience fits in very well the way the consumers behave nowadays. In other words, it has become one of the widely adopted ways to engage the customers at the right time and place. It is the biggest reality that is embraced by the new businesses and it creates a new ecosystem for the consumers and the brand across the two worlds by blending digital and physical. The customers of today expect a seamless experience and hence, are not loyal to any specific channel, whether online or offline.

The key premise behind the success of this marketing mantra is the integration of the consumers for the information search, for rating and reviewing the products with respect to buying through brick and mortar or online platform and even at the times of accessing m-commerce or e-commerce.

Some of the real marketing applications of this concept are listed below:

1. With the Brooklyn-based community store, Nike aimed to creating a space that fosters the spirit of the community thereby promoting the unifying power of sport.
2. The increased usage of apps is attributable to the application of phygital marketing. It has been widely seen that people nowadays make use of various applications to replace the bureaucratic procedures. For instance, certain applications allow consumers to choose a restaurant in a shopping center, order the meal, and pay through the cell phone.
3. A retail chain in South Korea placed virtual shelves cloaked as a billboard in subway stations that allows its customers to shop using their phone and have their products delivered at home.
4. Bike-sharing systems via mobile apps, labels that locate the bags they are on, sellers using gamification apps to reach sales goals.

5. Fifteen brands are currently running the phygital pilot. Two of these are Tata brands—Chroma and Westside. The others include KKCL (Killer Jeans), Metro, Mochi, etc.
6. Titan Company Limited offers a unique retail experience to its customers through its online channel titan.co.in, thus promoting it as a lifestyle company and adding to its journey toward upgraded margins.
7. The opening up of brick and mortar stores by Amazon is another example of synergy between digital and physical retail.

The concept of phygital marketing has its significance for brand retailers also. They offer stimulating and engaging experiences to their customers and thus, signify and validate their brands. The use of digital interactions enhances the image of the physical store in both physical and digital environments. This further leads to an increased tendency on the part of the customers to make a purchase. According to the survey done by MJV Technology and Innovation (Brazil), "Approximately 50% of consumers shop in an almost phygital way, doing online search and then proceeding to the store. Stores that offer check-in through apps may send messages and promotions in-store, gaining the preference of at least 10% of customers and increasing sales up to 40% at department stores."

"Getting Retail to Go Phygital," an article in *The Economic Times* (September 9, 2015) lays emphasis on the transformation of the retail sector with the emergence of new-age technology. According to their findings, "The increasing use of smartphones, tablets, and internet broadband and 3G has led to developing a strong consumer base which is likely to increase further. This, combined with a larger number of homegrown retail companies with their innovative business models, has led to a robust retail market in India rearing to expand at high speed. True, several new retail models are also emerging, including physical, virtual and hybrid formats. Alongside, physical retailers are also rapidly extending into an omnichannel avatar, with the inclusion of internet-enabled commerce platforms and mobile commerce models."

According to Krish Iyer, President and CEO, Walmart India and Chairman of India Retail Forum (IRF) 2015, "In this important year for all of us in the retail industry, I see a great opportunity to build a period of Retail Renaissance, where with our collective wisdom and collaboration, we can grow the pie of modern Indian retail together. The 12th edition of this mega congregation of retail and consumer market intelligence will be driven by the theme 'Capturing the New Phygital Scenario,' with the objective of harnessing the future of retail in the Omnichannel era. IRF 2015 will attempt to build the roadmap for customer centric retail organizations through technological advances, new and innovative concepts, digital and social media." Amitabh Taneja, Chief Convenor, India Retail Forum also added to this concept, "Indian consumers are increasingly looking for experiences that wow them on several levels. Retailers and retail support companies should collaborate and create retail experiences that blend the best of all shopping channels."

As per the report titled "Forget e-commerce, m-commerce is where India's potential lies" by Kit Tang and Connie Tan Hui Ann, CNBC (Thursday, November 26, 2015), "Projections by Cisco put the number of Smartphone users in India at 651 million by 2019, a near five-fold jump from 140 million by end-2014. The study, released in February, noted a 54 percent surge in the number of Smartphone users in 2014 as the average price of handsets fell to around $150 last year and as Smartphone penetration increases in rural India. India has a huge opportunity for mobile commerce. This is the first time a majority of Indians are getting connected to the internet. They are discovering products at costs that are lower than they've never seen before, and they are getting products that were not available in their market before. So it's a huge opportunity. For one, India's focus on cash usage and security concerns about e-transactions is creating friction with the burgeoning online shopping market, analysts say. Large players are wary of the 'Cash on Delivery' system as it is manpower intensive, and requires time to collect the cash from the consumer's doorstep. In addition, India's poor logistics infrastructure creates a challenge for e-retailers to offer quick delivery services, while the lack of stable telecommunications infrastructure across the country could also limit the pace of growth."

7.4.2 BRAND BUILDING WITH THE CONCEPT PHYGITAL

The concept phygital is one of the best platforms for marketers throughout the globe for branding a product or service. Phygital concept works as a sales and marketing channel providing various opportunity to meet the needs and expectations of customers throughout the buying cycle. Physical communication channels comprise product packaging, brochures, advertising banners, whereas digital communication is more focused on social media, websites, or emails. Threats can arise from lack of communication between the channels, indicating that the brand lacks insight into its customers.

The main objective of the organization is to understand how the physical and the digital world of marketing communications cooperate in the process of branding. This emphasizes that it is not enough to focus their efforts in the field of physical communication mix, but also the digital and hence, raise the standards and reputation of the brand in both worlds. This concept changes the strategy of an organization. Era of internet has caused a further shift from action in the market to action in cyberspace. Organizations all over the globe expand activities not only in the market but also on websites on the internet.

The digital technology is used in number of ways such as buying, selling, recruiting, e-learning, published experience, blog discussions, etc. Moreover, phygital is the transition from marketing which uses one channel to marketing that uses multiple communication channels. Organizations, to be competitive, no longer rely only on one channel to reach the customers and provide them with their services. Customers also prefer different channels, enabling access to business products and services.

7.4.3 SHIFT FROM BRICK TO CLICK

The present retail environment has undergone a tremendous change. The customers are no longer dependent upon the physical stores for making a purchase. Rather they have opted for in-store, online e-commerce, and mobile and social commerce networks. This highlights the relative importance of merging the physical and online retail channels so as to gain a deeper understanding of their influence on each other.

Retail brands are emerging with the enhanced capability to serve their customers. This is mainly because the digital influences occurring at home and in the store have a significant impact on the customer experience of mobile natives, Millennial, more and more. This has led the retail stores to adopt a customer-centric approach and hence, become "clicks and mortar" from the traditional "bricks and mortar." The key concern at this point of time is to understand the phenomenon of Millennial centric in the present phygital era. Traditionally, the retailers adopted the branding strategies according to the needs of the audience. But with the advancement of technology and developments in the field of retailing, these traditional retailers have to adapt with the retail technologies so as to provide better customer experiences.

7.4.4 CONCEPTS FROM THE TOP BRANDS

In recent times, majority of stores and showrooms are designed in such a way that they appeal to customers who are digitally inclined. The main attraction of these stores is that they combine innovative technologies such as displays which act as a medium of interaction with the customers, e-commerce, sensors, m-commerce, and digital content engrossed in store environments that act as a different innovative digital experience for the customers.

The concept of phygital shopping is gaining new heights because of the convenience of e-commerce and the increase in use of smartphones all around the globe. As a result of the technological advancement and realizing the significance of digital retail stores, retailers have started to digitize their stores and even they are changing the customers' shopping experience. The new concept of phygital retail compliment technology by not only increasing consumer convenience but also by adding up an element of entertainment which ultimately increases the emotional timbre of the in-store shopping experience. Following are the examples of some of the major retail brands that use the concept of phygital retail.

Sephora, the global cosmetics retailer, proactively uses the concept of phygital so as to improve the customer's retail experience. They work with TIP concept (where TIP stands for teach, inspire, and play) which includes wonderful and innovative concept of Beauty Workshop stations equipped with touch screens that guide customers in makeup. Moreover, there are also Beauty Boards, screens featuring user-generated customer videos, which make an interactive platform for potential customers. The Fragrance IQ station allows customers to experience and try scents which are possible with the help of InstaScent technology. The Skincare IQ diagnostic station assists customers choose the best product for their skin type, condition, or for specific issues. The top management of Sephora believes that they adopted the concept of phygital retail as to completely redesign their store experience to reimaging it for their client today and into the future. Sephora also provides its customers access to the entire catalog via digital interfaces, with the help of which they can place orders for home delivery and pay once for both in-store and online purchases. Customers can choose to interact with real sales associates. Sephora is clearly at the forefront of experimentation when it comes to phygital retail.

The other major example of phygital retail is Sport Check which is the largest Canadian retailer of sporting clothing and sports equipment. In its Toronto flagship store, they provide opportunity of phygital experience to their customers through in-store touch screen kiosks, 3D holographic displays, sensors, golfing infinity screen, digital stride, and gait analysis equipment. The concept flawlessly inclines digital shopping with product information with add on flavor of entertainment. Moreover, the products have radio-frequency identification (RFID) tags that customers can use to get product information and for availability from interactive displays. The company revealed that by going phygital, the company increased store sales by 50%.

Zoom systems, retail automation company took a bold step when it worked with Nestlé to create Nespresso Cube, a completely autonomous, robotic store-in-store concept that provides a personalized experience to Nespresso club members. Nespresso Cube allows customers to place their orders on a touch screen and watch as robotic arms quickly prepare their orders. Nespresso Cube is a tremendous example of combination of social, mobile, and local technology and could be the future of retail commerce. More generally, to serve and attract a new generation of shoppers, retailers are experimenting with smaller-footprint stores, trading limited physical shelf space for the digital "infinite shelf."

Coca-Cola also took a phygital initiative which was launched in 2009 called the Coke Freestyle Machine. This initiative by coke provided users with a unique opportunity to create their own drink mix by using a touch screen that directs the machine to dispense a mixture of tastes from micro-cartridges. Moreover, the customers can share their physical experience with digital through mobile and social channels. With the help of this, Coca-Cola can leverage the data it collects through those machines to create new products from popular mixtures.

Brazilian fashion retailer C&A used a very innovative method to engage their customers by installing a real-time "like" counter on its clothes hangers with the help of C&A Facebook page, where items were displayed, allowing customers at the store to see shoppers' opinions and buyers' feedback. Likewise, in a different way, Nike created a phygital experience by generating feedback through its sneakers. For basketball players, Nike built their Hyperdunk shoes with sensors that track the wearer's every move during games and then syncs to iPhones, providing statistics and game feedback.

McDonald's created phygital experience by creating an interactive outdoor campaign in Stockholm, Sweden called Pick N Play, where consumers could interact with a billboard using their mobile phones to play games displayed on the billboard to win McDonald's treats and prizes.

7.5 ISSUES AND CHALLENGES FOR PHYGITAL MARKETING

Gilliland (2016) has outlined the emerging role of phygital marketing in retailing as brands nowadays are targeting the much popularized "phygital generation." Chris Sanderson (Future Laboratory) proclaims that "it combines the physical and digital worlds for an experience that is full of 'immediacy, immersion and interaction.'".

The most important concerns for the retailers in adapting to the strategy of phygital marketing are detailed as follows:

1. Rethinking the location: Although the emergence of technology has moved the consumers from bricks and mortar approach to clicks and mortar approach, the popularity of in-store shopping remains just as ever. This has posed the biggest challenge for the retailers to provide the customers with a store format which is relevant to generate a unique experience. The efforts have to be directed toward making the store a place that touches upon the emotions and sentiments of the customers, thus allowing them to shop nearby. The retailers need to personalize the retail experience of the customers so as to enhance their satisfaction and gain their trust, loyalty, and positive word-of-mouth.
2. Promoting point of view: The traditional approach that the consumer is always right is no longer followed by the present retailers. As the brands are focusing more on the core values and beliefs, many of them are making efforts to promote a unique point of view so as to generate loyalty among the customers. For instance, take the case of REI. Last year, the outdoor apparel company chose to boycott Black Friday with its bold #optoutside campaign. By refusing to slash prices, not only did REI emphasized its position as a voice of authority but it also managed to promote consumer choice (when every other retailer was pushing the inevitable sale) as well as conveniently highlight its core marketing message of living life outdoors.
3. Targeting emotions: The primary objective of the retailers in the current retail scenario is not merely to anticipate the purchase behavior of the consumers; rather it aims at predicting the emotions that drive the consumers to make a purchase decision. This concept has been popularized as "mood retail." Many brands all across the globe are widely adopting the construct of mood as a marketing tool. This is based on the fact that when the retailers use emotions as a driver of consumers' product choices, the users are able to obtain an immersive and intuitive shopping experience.
4. Miscellaneous: Besides these key considerations, there are also certain other factors that need to be taken into mind while opting for the phygital retailing as a marketing tool. These include the following:

a. Provide an educational journey and complementary services to the customers so as to allow them to have an edge over the competitors.
b. Advocate their brand story that will generate brand associations in the mind of the customers.
c. Create an emotional stimulus for the customers so that they can be engaged emotionally and this resonates with how the customers perceive their shopping experience.
d. Showcase the products in a better manner so as to reflect the changing customer trends in the novel designs of the products.
e. Utilize the customer analytics to learn about their customers and strive toward customer retention.
f. Make effective utilization of product analytics to gain a better understanding about the products and also facilitate product recommendation.
g. Integrating technologies such as internet of things (IoT), iBeacons, RFID tags, virtual reality (VR), augmented reality (AR), retail Androids, SMS push notifications, and more advanced personalization related to loyalty reward programs that will enable the brick and mortar stores to create a unified ambience that is entailed in phygital experience.

7.6 CONCLUSION

The phygital experience is a key to enhance the physical experience with the digital experience. Smartphones being the key component of this strategy, it has resulted in the companies getting more engaged in physical tactics through the social media, thereby allowing the companies to leverage so as to stay competitive. The biggest challenge posed by phygital as a paradigm is developing an understanding of how the companies can implement this strategy. However, the barriers have witnessed a reduction as people of all ages are engaging in digital experiences. While some authors have argued that e-stores will not overthrow physical shopping due to the fact that people enjoy the experience of physical shopping and may choose it over the convenience and affordability of e-stores, marketers must note that phygital marketing may provide a more enjoyable experience to the customer regardless of physical shopping experiences of any product. The concept is poised to transform the traditional marketing mix elements which focus on the product, price, promotions, and place, into one that creates an enjoyable experience by associating digital aspects of marketing with the physical experiences. So, before we do dismiss "phygital" as yet another buzzword, it is evident that consumers increasingly desire the best of both worlds. The challenge for brands is to know how to effectively combine the two.

KEYWORDS

- **m-commerce**
- **phygital retail**
- **branding strategy**
- **digital marketing**
- **customer engagement**
- **value cocreation**
- **social media**

REFERENCES

Amichai-Hamburger, Y. Ed. *Technology and Psychological Well-being*; Cambridge University Press, 2009.

Armstrong, K.; Rutter, C. Exploring the Enigma of the Happiness Construct in Phygital Fashion Experiences. In *Advanced Fashion Technology and Operations Management*; IGI Global: Philadelhia, USA, 2017; pp 220–233.

Barreda A. Creating Brand Equity When Using Travel-related Online Social Network Web Sites. *J. Vacat. Mark.* **2014,** *4*, 365–379.

Bruhn, M.; Schoenmueller, V.; Schäfer, D. B. Are Social Media Replacing Traditional Media in Terms of Brand Equity Creation? *Manag. Res. Rev.* **2012,** *35* (9), 770–790.

Chong, A. Y. L.; Ooi, K. B.; Darmawan, N.; Lee, V. H. Determinants of 3G Adoption in Malaysia: A Structural Analysis. *J. Comput. Inf. Syst.* **2010,** *51* (2), 71–80.

Chun, H.; Lee, H.; Kim, D. The Integrated Model of Smartphone Adoption: Hedonic and Utilitarian Value Perceptions of Smartphones Among Korean College Students. *Cyberpsychol. Behav. Soc. Netw.* **2012,** *15* (9), 473–479.

Ellis, R. Let's Get Phygital: How Content can Help Blend Physical and Digital Worlds. *J. Consum. Mark.* **2014,** *27* (4), 336–344. http://www.thismoment.com/content-marketing-blog/phygital-content/Fun (accessed Mar 25, 2017).

Elsaadany, O.; Kadry, M. Evaluation of the Factors that Affect the Mobile Commerce in Egypt and Its Impact on Customer Satisfaction. **2014**. http://link.springer.com/chapter/10.1007/978-3-319-05503-9_21 (accessed Mar 25, 2017).

Getting Retail to Go Phygital. *The Economic Times*, Sept 9, **2015**. http://epaperbeta.timesofindia.com/Article.aspx?eid=31818andarticlexml=GETTING-RETAIL-TO-GO-PHYGITAL-09092015023005 (accessed Mar 25, 2017)..

Gilfoil, D. M. Mapping Social Media Tools for Sell Vs Buy Activities into Emerging and Developed Markets. *Int. J. Manag. Inf. Syst.* **2012,** *16* (1), 69 (Online).

Gilliland, N. *Think Retail: How Brands Are Targeting the 'Phygital' Generation*; **2016.** https://econsultancy.com/blog/68023-think-retail-how-brands-are-targeting-the-phygital-generation/ (accessed Mar 25, 2017).

Gregory, Z. *Guerrilla Marketing Makes most of Little Cash*; **2012**. http://search.proquest.com.ezproxy.snhu.edu/docview/963779574 (accessed Mar 25, 2017).

Horky, Vit. *Types of Guerrilla Marketing*; **2009**.

Hung, S. Y.; Ku, C. Y.; Chang, C. M. Critical Factors of WAP Services Adoption: An Empirical Study. *Electron. Commer. Res. Appl.* **2003,** *2* (1), 42–60.

Hung, M. C; Yang, S. T; Hsieh, T. C. An Examination of Determinants of Mobile Shopping Continuance. *Int. J. Electron. Bus. Manag.* **2012,** *10* (1), 29.

Islam, A. Md. ; Khan, A. M.; Ramayah, T.; Hossain, M. Md. The Adoption of Mobile Commerce Service Among Service among Employed Mobile Phone Users in Bangladesh: Self Efficacy as a Moderator. *J. Int. Bus. Res.* **2011,** *4* (2), 80–86.

Kietzmann, J. H.; Hermkens, K.; McCarthy, I. P.; Silvestre, B. S. Social Media? Get Serious! Understanding the Functional Building Blocks of Social Media. *Bus. Horiz.* **2011,** *54* (3), 241–251.

Marshall, T. C. Facebook Surveillance of Former Romantic Partners: Associations with Post-breakup Recovery and Personal Growth. *Cyberpsychol. Behav. Soc. Netw.* **2012,** *15* (10), 521–526.

Moraveikova, D.; Kliestikova, J. Brand Building with Using Phygital Marketing Communication. *J. Econ. Bus. Manag.* **2017,** *5* (3), 148–153.

Muñiz, A. M.; O'Guinn, T. C. Marketing Communications in a World of Consumption and Brand Communities. In *Marketing Communication: New Approaches, Technologies, and Styles*; 2005; pp 63–85.

Palmatier, R. W.; Scheer, L. K.; Steenkamp, J. B. E. Customer Loyalty to Whom? Managing the Benefits and Risks of Salesperson-owned Loyalty. *J. Mark. Res.* **2007,** *44* (2), 185–199.

Park, C. W.; Jun, S. Y.; Shocker, A. D. Composite Branding Alliances: An Investigation of Extension and Feedback Effects. *J. Mark. Res.* **1996,** 453–466.

Park, J.; Yang, S.; Lehto, X. Adoption of Mobile Technologies for Chinese Consumers. *J. Electron. Commer. Res.* **2007,** *8* (3).

Persaud, A.; Azhar, I. Innovative Mobile Marketing Via Smartphones: Are Consumers' Ready? *Mark. Intell. Plann.* **2012**, *30* (4), 418–443.

Soch, H.; Kaur, K.; Gill, A.; Gill, H. Role of Corporate Image, Perceived Risk and Trust on the Usage Frequency of Smartphones in Emerging Markets. *Int. J. Adv. Res. Bus.* **2014,** *1* (1), 14–16.

Spencer. What's Phygital in Retail? **2016**. https://www.linkedin.com/pulse/whats-phygital-retail-michael-spencer.

Tang, K.; Ann, C. T. H. Forget E-commerce, M-commerce Is Where India's Potential Lies. *CNBC*, Nov 26, 2015. http://www.cnbc.com/2015/11/26/forget-e-commerce-m-commerce-is-where-indias-potential-lies.html (accessed Mar 25, 2017).

Vel, K. P.; Brobbey, C. A.; Salih, A.; Jaheer, H. *Data, Technology and Social Media: Their Invasive Role in Contemporary Marketing,* 2015, *14* (4). E-ISSN: 2177-5184.

Yang, K. C. Exploring Factors Affecting the Adoption of Mobile Commerce in Singapore. *Telemat. Inform.* **2005,** *22* (3), 257–277.

CHAPTER 8

M-COMMERCE: CRUSADER FOR "PHYGITAL" RETAIL

SHIVANI ARORA* and ANUKRITI VERMA

Department of Commerce, Shaheed Bhagat Singh College, University of Delhi, India

**Corresponding author. E-mail: dr.shivani.research@gmail.com*

ABSTRACT

India has a "curious case of online commerce." Why curious?—Since online commerce companies are reporting losses running into crores but the venture capitalists are still investing heavily in it. India is an interesting place; the uniqueness is in the diversity of population. India is second, only to China in terms of internet population; the negative is that the internet penetration in India is as low as 34.8% (Internet Live Stats, 2016). With the low prices of smartphones and the cheapest (sometimes free as in case of JIO) data plans, m-commerce is the way forward. The low internet penetration is the potential that e-commerce and the venture capitalists see, that is, once the internet penetration increases, the customer base will increase with it. The boom in the demand of smartphones, in recent years, accentuates the fact that people are becoming more reliant on the mobile phone. The number of smartphone and internet users is increasing at a fast pace. With more people using mobiles as the preferred device for online transactions (mostly shopping), it clearly shows that m-commerce has significant ramifications for e-commerce as well as bricks and mortar retailers. It is evident that m-commerce is expected to be a game changer when online shopping giants such as Flipkart, Amazon, and Myntra are reporting an increase in the m-commerce traffic. All these retailers have tried to go "app-only" at one point or the other. The impact of Digital India is huge for the m-commerce. Expansion of mobile infrastructure and integration of mobile payment gateways have made the consumer interaction possible in such a way that is blurring the line and creating an overlap between online and physical channel of retail and creating a phygital retail milieu.

This chapter would aim to help the readers understand the concept of phygital and m-commerce. To further the discussion of how m-commerce has the potential to save the cash-crunched and loss-spinning online companies, the chapter will provide detailed insight as to how m-commerce can further pave the way for phygital retail. The chapter would aid the reader in understanding how m-commerce is influencing and paving way for phygital retail, help in figuring out the relevance of phygital retail in the e-commerce space, aid in getting an insight into the applicability of omnichannel presence across different stages of customer purchase, and help find out the future prospects of phygital retail.

8.1 INTRODUCTION

The retailing landscape is dynamic. The way we shop is changing. With the tech-savvy customers occupying the market, technology is indeed making an impact on e-commerce which is stronger than ever (Agarwal, 2017). The penetration of smartphones in the retail space is the testimony to the fact that technology is indeed influencing the e-commerce, and to such extent that retailers are switching from e-commerce to mobile-commerce for business-to-consumer transactions. Considering the current trend, industry experts believe that m-commerce would contribute up to 70% (Press Trust of India, 2014) of their total revenues. Such estimates have caused many companies to "take the web out of e-commerce" and focus entirely on their apps (Sathe, 2015). The transition from the websites to the apps has opened up many avenues for the retailers, who can now reach out to the customers anytime and anywhere with the propagation of smartphones.

With smartphones accounting for 45.1% (Harty, 2016) of web shopping traffic (edging out computers at 45%) and taking over computers as top e-commerce traffic source, it is evident that m-commerce has gained relative significance in the past few years. It has greatly changed the way merchants and retailers interact with the consumer, as a consequence of which new opportunities for both the spheres are emerging.

M-commerce has undoubtedly served as a catalyst to revolutionize the retail milieu by introducing the concept of phygital retailing. A phenomenon has been observed and it is that in the recent years websites are behaving more like stores and stores are behaving more like websites.

Where multichannel (brick-and-mortar and online channel) was in vogue earlier in retailing, m-commerce has made it possible to take a mid-way out and employ the best of both worlds. This combination of physical and digital has originated a new term "phygital" in retailing. DigitasLBi coined the term "phygital" (Spector, 2014) as an apt acronym which denotes the coexistence of physical and the digital world. It is apt to mention that the digital world is as much a mobile phenomenon now, as it was a computer one.

Given the purchasing behavior of the digital natives, retailers have started picking up on omnichannel retailing, however, there are only as few as 12% (Khurana and Basu, 2015) retailers in the world who offer such omnichannel experience. Omnichannel approach is preferred by the customers but it is an expensive proposition for the retailer.

According to Kreuger (2015), success today means reaching the consumer everywhere irrespective of their location or the technology that they might be using. Nurturing a meaningful relationship with the consumers has become the key to success for the retailers. By presenting Amazon Go to the world, Amazon leapfrogged the competition and set up an amazing example of phygital retail. The concept of Amazon Go highlights the three essentials for retailers to observe if they do not want to lose out on potential buyers. *Better basics, convenience, and consumer experience* are the main elements which should be prioritized by the retailers—both the physical as well as the online retailers.

While retailers are making an effort to employ techniques, like QR codes for direct check outs, installing digital touch points such as tablets and mobiles to serve as information kiosks, and inculcating omnichannel presence across different stages of the customer purchase; customers on the other hand are creating their own omnichannel experience with the aid of integrated payment systems and shopping apps.

Phygital retail has dissolved the boundaries between the brick and click and brick and mortar stores, bridging the gap between "no touch and feel" by providing interactive websites.

Given the importance of the emergence of "phygital world," we believe an in depth discussion of this movement is warranted.

8.2 OBJECTIVES

The following are the objectives of this study:

- To understand how m-commerce is influencing and paving way for phygital retail

- To find out the relevance of phygital retail in the e-commerce space
- To get an insight into the applicability of omnichannel presence across different stages of customer purchase
- To find out the future prospects of phygital retail.

8.3 LITERATURE REVIEW

The retail landscape underwent immense and drastic changes after the dot-com implosion, with e-tailing gaining prominence. While e-commerce had been on the massive rise, Verhoef et al. (2015) note that the advent of new digital and specifically, mobile channels have resulted in another disruptive change in the retail environment. The ubiquitous presence of smartphones heralded m-commerce or mobile commerce which is now becoming a game changer in the retail ecosystem. Clarke (2001) define m-commerce as "M-commerce or mobile commerce is the ability to purchase anywhere through a wireless internet-enabled device. M-commerce refers to any transaction with monetary value that is conducted via a mobile network. It will allow users to purchase products over the internet without the use of personal computers."

The growing popularity of m-commerce has urged many online players to expand their business on mobile platforms by developing their apps and making it available on app store. Where some companies such as Amazon and Flipkart chose to exist both on online as well as mobile platform, other companies such as Myntra chose to use the "app-only" approach. In a press statement, Amazon India Senior Vice President Amit Agarwal has shared that computer traffic has taken a back seat to mobile internet traffic. M-commerce has upsurged due to the penetration of smartphones in India (Press Trust of India, 2014)

Kreuger (2015) has a very interesting take on the relationship between digital driving e-commerce as well as offline store shopping. The consumers, more precisely 71% of such shoppers, have claimed to use smartphones for in-store research. They have also agreed to the fact that in-store researching has become an important part of their purchase experiences. All the research points in the direction of upsurge in the usage of smartphones due to affordability and cheap data plans, which have become front runners for the phygital growth.

Within the literature, Verhoef et al. (2015) had observed the impacts of mobile channel on customer purchase behavior and found that customer touchpoints interaction have a stronger impact on brand preferences and liking than others. Stating "showrooming" and "webrooming," both being the forms of research shopping-as common practices among consumers, the authors had suggested that in order to provide a seamless purchase experience to their customer, the retailers-at some time will have to integrate the online and offline channels and switch to "omnichannel retailing." Adding to the same, a case study by IBM (2013) mentioned that those consumers, who possess smartphones, while strolling through brick and mortar shops, often check competitor pricing and products before making a purchase decision. Blurring the lines between the physical and virtual, this growing practice of showrooming is forcing traditional retailers to rethink marketing and merchandising strategies and counter with innovations to minimize the loss of sales to online competitors and leverage mobile to enhance the in-store buying experience (IBM Smarter Commerce, 2013).

Discussing the increase in demand of an omnichannel, the A.T. Kearney and Google study (2016) states that a time has come when retailers will have to embrace a retail mix which combines the benefits of brick and mortar as well as digital retail—appropriately termed as phygital retail. Furthermore, the study says that those retail players who do not espouse the idea of phygital retail have higher probability of losing out on 20–30% of potential customers in select categories (Google-A. T. Kearney, 2016).

While showrooming and webrooming are one of the most significant practices that showcase how m-commerce is a harbinger of phygital retail, Grewal et al. (2017) state that there is more to m-commerce than research shopping that is revolutionizing the current retail scenario. According to the authors (Grewal et al., 2017), from mobile apps to geofenced-targeted offers, to constant access to online environment, the advances in this realm have led to constantly changing consumer behaviors and enhanced ability of retailers to connect with consumers (Grewal et al. 2017). Citing the example of Amazon Go which allows customers to scan their smartphone as they enter the physical store, pick up the products they want, and leave; the authors (Grewal et al., 2017) say that such ideal consumer experience—which leverage new technologies—have set expectations of what shopping should be in future.

Brynjolfsson et al. (2013) had aptly argued: "In the past, brick-and-mortar retail stores were unique in allowing consumers to touch and feel merchandise and provide instant gratification; Internet retailers, meanwhile, tried to woo shoppers with wide product selection, low prices and content such as product reviews and ratings. As the retailing industry evolves toward a seamless omnichannel retailing experience, the distinctions between physical and online will vanish; turning the world into a showroom without walls."

It has been observed that customers have a liking to retailer touchpoints, in all forms and combinations (Sopadjieva et al., 2017). Khurana and Basu (2015) had made a similar observation and based on that they emphasized on responding to customer expectations for real-time, relevant, omnichannel, and contextual experiences. The concept of Code Halo is discussed which is the unique individual identity of the individual, created on the basis of every action performed online. Code Halo aids in targeting individuals with precision through the mobile apps or websites. Such customization—taken to another level—aids in beating the competition.

Arora and Budree (2016) have also discussed the scenario of e-commerce, wherein in spite of huge losses, the hype around e-retailing is on the high. The tussle goes on between the online retailers and the race is of the basket loyalty, rather than customer loyalty (Arora et al., 2016). In a time, when shopper's visit to traditional retail store is declining on one hand and big online players are reporting losses on the other, Sopadjieva et al. (2017) believe that omnichannel strategy is a panacea for such difficult environment (Sopadjieva et al., 2017). Not only does phygital retail give a competitive advantage to the traditional retailers but also benefit the e-commerce. In order to provide a frictionless purchase experience to its customers, it has been suggested that firms make the best usage of the research shopping attribute and provide tech-enabled services in the store where the customers can seek information and order them (Verhoef et al., 2015). By employing and integrating technologies such as cloud, computer vision, sensor fusion, internet of things (IoT) , radio-frequency identification (RFID) tags, retail Androids, virtual reality (VR), augmented reality (AR), etc., brick and mortars can create an environment that "phygital" entails (Spencer, 2016). On the other side, the online retail—by offering services like offline pick-ups and establishing physical presence for trials and after-sale services—can provide seamless experience to the consumers.

Phygital retailing has been categorized as a type of omnichannel retailing and though it presents a dramatic shift in the way we think about retail, it has many opportunities in the coming future (Kreuger, 2015). With digital natives occupying the market, retailers have a better prospect of providing rich purchase experience to such consumers by synchronizing the physical and digital worlds.

8.4 RESULTS AND ANALYSIS

A quantitative analysis was imperative to understand the consumer behavior and their purchasing pattern, and how these factors have paved the way for m-commerce, which according to the experts, is further laying down the foundation for phygital retail. A questionnaire was framed for the same. The sample size for the survey was 600 out of which 120 consumers responded to a list of questions for the study. Judgment sampling was undertaken in order to choose the respondents.

The respondents were categorized to identify their distribution across age and cities (Table 8.1).

TABLE 8.1 Summary of Respondents: Distribution Across Age and Cities.

Age group			Cities		
Age	**No. of respondents**	**%**	**Cities**	**No. of respondents**	**%**
0–25	92	76.7	Tier 1	71	59.2
25–50	28	23.3	Tier 2	49	40.8

Source: Compiled by authors.

The analysis of the result has been mentioned in following sections.

8.4.1 TO ANALYZE IF THE RESPONDENTS WERE AWARE OF THE TERM "M-COMMERCE"

The question was included in the questionnaire to check whether the consumers, despite transacting over the phone (be it for the purpose of payment, shopping, etc.), and using apps for the respective purpose, are aware of the specific term "m-commerce." While 81.7% respondents were aware what m-commerce is, the interesting aspect was that 18.3% respondents could not comprehend the term (Table 8.2).

TABLE 8.2 Awareness About M-commerce.

	No. of respondents	%
Yes	98	81.7
No	22	18.3
Total	120	100

Source: Compiled by authors.

8.4.2 TO ANALYZE THE PREFERRED MODE OF SHOPPING USED BY THE RESPONDENTS

Table 8.3 represents the mode of shopping that the respondents prefer to use the most. The analysis show sparkling results—on one hand, 53.3% respondents prefer to transact online, however on the other, there are 46.7% respondents who still prefer to shop in the traditional brick and mortar stores and retail outlets. While experts have time and again indicated the negative fate of physical stores compared to online stores, the response straightforward contradicts it. In fact, it shows that with more enhancements, brick and mortars may hold high chances against e-stores.

TABLE 8.3 Preferred Mode of Shopping.

	No. of Respondents	%
Online mode	64	53.3
Offline mode	56	46.7
Total	120	100

Source: Compiled by authors.

This particular aspect negates the myth that e-commerce will end the existence of offline stores. It reiterates that both would coexist and with equivalent vigor.

Another segment of the same analysis was to determine if respondents, while transacting online, prefer to use website or mobile apps for the same (Table 8.4). It was assumed that at some point or other (even if the respondent preferred offline shopping) the respondent must have performed an online transaction.

TABLE 8.4 Preferred Mode of Online Shopping.

	No. of respondents	%
Online website	32	26.7
Mobile apps	88	73.3

Source: Compiled by authors.

While transacting online, it was found that where 26.7% respondents preferred online websites for online transactions, a majority of 73.3% respondents preferred to transact online using the respective mobile apps. This is a good indicator for business enterprises and start-ups who already have an online presence but want to launch their own mobile app. Mark Zuckerberg, Founder and CEO of Facebook, said that if Facebook was launched a few years later, he would not have launched a website, but an app-only. Moreover, it gives a good prospect to enterprises planning to go "app-only."

8.4.3 TO ANALYZE THE MOST PREFERRED ONLINE SHOPPING APP OF THE RESPONDENTS

Assuming that most part of online transaction comprise shopping, the respondents were asked about their most used and preferred shopping app. The results were fascinating with Amazon coming out as a clear winner in the race. In total, 44.2% respondents stated that they had used Amazon for the purpose of shopping online. Flipkart grabbed the second spot among its competitors with 25% respondents voting for it, followed by Myntra with 20% respondents preferring it (Table 8.5).

TABLE 8.5 Most Preferred Shopping Apps.

Name of organization	No. of respondents	%
Amazon	53	44.2
Flipkart	30	25
Snapdeal	6	5
eBay	0	0
Myntra	24	20
Jabong	6	5
Others	1	0.8

Source: Compiled by authors.

The competition between Amazon and Flipkart has always been intense in the India e-commerce space. Things changed for Flipkart in India, since June 2013, when Amazon entered Indian online retail market. Innovation has been driving the two to stay in constant competition whether by offering memberships, loyalty programs, or exclusive deals. With Amazon introducing the concept of Amazon Go (though not implemented in India), Flipkart needs to buckle up.

Moreover, when it comes to specialized e-commerce stores (like stores dedicated solely for apparels or electronics, etc.), Myntra seems to be the top choice among respondents with 20% votes followed by Jabong. This shows that enterprises venturing to start their e-commerce stores may consider opening up specialized online stores.

8.4.4 TO ANALYZE THE SCOPE OF PHYGITAL STORES LIKE AMAZON GO IN INDIA

As mentioned earlier, Amazon has introduced the concept of Amazon Go—which combines the features of physical + digital stores. It requires customers to have the Amazon app in their smartphones and integrated online payment system. The idea has gotten positive views from industry experts; however, the result shows that this idea is still unheard of by many in India. Given the fact that the respondents belong to Tier 1 and Tier 2 cities, while 54.2% respondents are aware of Amazon Go, 45.2% respondents (which is a significant figure) still do not know about it (Table 8.6).

TABLE 8.6 Awareness About Phygital Stores (Amazon Go).

	No. of respondents	%
Yes	65	54.2
No	55	45.8

Source: Compiled by authors.

To further the analysis, the respondents were asked if they would like to avail of services or like to shop from a retail store similar to the lines of Amazon Go. The question was optional for the respondents, and so a total of 112 respondents out of 120 answered the question (Table 8.7).

TABLE 8.7 Shopping from Phygital Store (Amazon Go).[a]

	No. of respondents	%
Yes	55	67.1
No	27	32.9

[a]Optional for respondents.
Source: Compiled by authors.

Table 8.7 shows that 67.1% respondents would like to have a purchase experience similar to that offered by Amazon Go but 32.9% did not prefer to shop from such a store.

Assuming that the 45.2% who were not aware about Amazon Go represent the majority of Indian consumer mind-set; it still is a great option for e-commerce competitors to look into the omniretail channel and introduce the similar concept in India to get the first mover advantage. It is certain that once Indian consumers get familiar with the concept, phygital retail would become the store of the future and gain prominence over online stores and physical stores.

8.4.5 TO ANALYZE THE MODE OF PAYMENT PREFERRED DURING SHOPPING TRANSACTIONS

The respondents were asked to rank various modes of payments, which a normal vendor offers, according to their usage frequency (Tables 8.8 and 8.9).

TABLE 8.8 Preferred Mode of Payment During Shopping.

Mode of payment	Rank 1	Rank 2	Rank 3	Rank 4	Rank 5
Credit cards	29	11	10	18	52
Debit cards	38	30	27	16	9
Cash	36	34	20	11	19
Mobile wallets	7	29	43	30	11
e-banking	10	16	20	45	29

Source: Compiled by authors.

TABLE 8.9 Ranking Summary of Preferred Mode of Payment.

Mode of payment	Ranks
Debit cards	1
Cash	2
Mobile wallets	3
e-banking	4
Credit cards	5

Source: Compiled by authors.

Tables 8.8 and 8.9 show the ranking of various modes of payments where debit card came out as the winner followed by cash as the preferred payment mode (the ranking has been done on the basis of how many customers have given a specific rank to a particular mode of payment).

While cash is convenient for consumers, it involves high risk and high cost to the retailers. Therefore, the key concern for retailers would be how to reduce cash method of payment. Several approaches can be adopted as follows:

- Direct incentives, for example, coupons/vouchers for prepaid purchases, cash back schemes on mobile wallets (as has been implemented by Paytm) and plastic money cards
- Retail branded debit or credit cards for payment
- Electronic point-of-sales (PoS) devices at delivery (card on delivery).

It was assumed that post demonetization, the customers would move away from cash usage but the analysis indicates otherwise. Another segment to the same analysis was to learn if the situation of "digitalization" in India through the "Digital India" campaign persuaded consumers and whether they intended to use mobile wallets or e-banking in future. While majority of respondents claimed that they would prefer to use digital mode of payments in the future, 8.3% of the respondents showed their inclination toward traditional payment methods (Table 8.10).

TABLE 8.10 Preference of Digital Modes of Payment in Future.

	No. of respondents	%
Yes	110	91.7
No	10	8.3

Source: Compiled by authors.

The rise of Paytm can vouch for this result. Paytm claims that it registered 700% rise in overall traffic on the platform and 1000% growth in the value of currency added to the wallets, post demonetization.

To further the analysis, the respondents were asked if they preferred mobile wallets or e-banking. This was an optional question so as to have a comparative analysis between m-wallet or e-banking as the preferable mode of payment (both of which were relatively less used predemonetization). For the same reason, 112 respondents answered the question. The analysis showed that 72.3% respondents preferred mobile wallets such as Paytm,

MobiKwik, etc., to e-banking which was only preferred by 27.7% respondents (Table 8.11).

TABLE 8.11 Preferred Digital Payment: Mobile Wallets or E-banking?[a]

	No. of respondents	%
M-wallets	81	72.3
E-banking	31	27.7

[a]Optional for respondents.
Source: Compiled by authors.

If the Digital India campaign does succeed in propagating the reliability of digital mode of payments among consumers, then it may bring relief to online companies as digital wallets would be easy on their pocket and cash payment is something which e-commerce firm usually do not prefer.

8.4.6 TO ANALYZE THE CONSUMER BEHAVIOR ON PERSONALIZED OFFERS AND LOYALTY PROGRAMS

The respondents were asked if they liked to subscribe to loyalty programs or membership cards offered by retailers as a part of promotions. It was found that 60% respondents subscribed to such loyalty programs, whereas 40% reported that they were fine without subscribing to such programs (Table 8.12).

TABLE 8.12 Subscription to Loyalty Programs Initiated by Retailers.

	No. of respondents	%
Yes	72	60
No	48	40

Source: Compiled by authors.

The next part of the same analysis was whether the respondents availed of offers from such loyalty programs (Table 8.13).

TABLE 8.13 Availing of Benefits from Loyalty Programs.

	No. of respondents	%
Yes	73	60.8
No	47	39.2

Source: Compiled by authors.

Upon analysis it was found that approximately 60.8% (the same number of people who like to subscribe to such loyalty programs) like to avail of benefits from the same while shopping, whereas 39.2% respondents reported that they do not use membership points or any such promotional offers while shopping.

The respondents were also asked if they like to receive personalized offers or promotional notifications from the retailers. The analysis showed that 59.2% liked to receive notification regarding personalized offers and 40.8% did not prefer to have personalized offers (Table 8.14).

TABLE 8.14 Personalized Offers or Promotional Notification.

	No. of respondents	%
Yes	71	59.2
No	49	40.8

Source: Compiled by authors.

The potential here lies in the 40% respondents (approximate) who do not prefer to opt or subscribe to such promotional programs. Sometimes, it happens that due to constant notifications of similar sort, annoyed consumers tend to stop purchasing from a particular retailer. In such cases, the retailers are advised to provide geotargeted offers. Personalization is one of the major factors which comprise the core of phygital retail, and infrequent location-based offers would not only make the consumers aware but may also encourage the consumers to visit a retail outlet or website.

8.4.7 TO ANALYZE THE CROSS DEVICE AND ONLINE-TO-STORE SHOPPING BEHAVIOR

The respondents were asked if they used mobile to look up for better prices and offers while shopping in retail outlets/mobile shopping (the process of "showrooming" as mentioned by Verhoef et al., 2015). The results were that 55% respondents accepted that they compared the prices online over internet. On the other hand, 45% respondents claimed that they just bought from the outlets at the price catalogued in the tags without comparing it online (Table 8.15).

TABLE 8.15 Showrooming Behavior.

	No. of respondents	%
Yes	66	55
No	54	45

Source: Compiled by authors.

This analysis is an avenue for traditional retailers to entice 55% respondents (who have the tendency to move to online shopping if they get a competitive price over the internet) to their outlets by incorporating and installing interactive touchpoints at every phase of consumer purchase.

While the analysis endorses traditional retailers' logic of embracing an omnichannel strategy and using it as a differentiator to fight the competition from the pure-online players, omnichannel retailing has benefits for online retailers as well.

It is well known that one major defect of online shopping is that there is high probability of products differing from what the consumer has asked for, thereby adversely affecting the consumer expectation. Keeping that in light, we asked the respondents if they would want online retailers to set up physical outlets for trial and testing of products and after-sales service. The analysis showed that majority of respondent, that is, 91.7% respondents wanted such e-commerce firms to set up physical stores for the purpose of trial and after-sales service, whereas 8.3% were satisfied with just online presence of these e-commerce firms (Table 8.16).

TABLE 8.16 Physical Presence of Online Retailers.

	No. of respondents	%
Yes	110	91.7
No	10	8.3

Source: Compiled by authors.

Incorporating such an idea would benefit the e-commerce firms as it will enhance the credibility of their product and restore consumer's faith in the quality of their products, thereby attaining high conversion rate optimization.

8.5 CONCLUSION

M-commerce has without a doubt gained significant popularity among consumers. While e-commerce had been on the rise massively in recent years, it may not be the next true frontier for shopping as m-commerce continues to become more popular (Clarke et al., 2001) Shopping and payment through smartphones is expected to be a game changer and this in turn has impelled retailers—both online and traditional—to inculcate the elements of m-commerce in their retailing channel. The rise of m-commerce has indeed provided a food for thought to the retailer if they should opt for omnichannel or phygital retail, or if they should stick to the multichannel of retailing.

Coexistence: The analysis also concludes that respondents are using apps more than the website. But should it be taken as an indicator for the businesses to go app-only. The study of literature reveals that resistance to change is an inherent part of human nature. People take time to adjust with new technology; they did the same with e-commerce, when it was launched. Credit cards as a mode of online payments were resisted; digital wallets were also not easily accepted. But once people are given time for the same, they adjust. The pattern uptil now has been that the things coexist. Online and offline retail coexists, and so do all the modes of payment, namely, credit cards, debit cards, "cash on delivery," digital wallets, National Electronic Fund Transfer (NEFT), and hence the chances of thriving of both websites and apps are higher. The proportion may vary with time.

Also, the respondents around 91% want to have the best of both the worlds—online and offline stores.

Comparative shopping: Analysis shows that 55% respondents do compare the price over the internet while shopping in a physical store. That is where convenience comes in, as smartphones have become the new personal assistant for people. Now imagine a scenario where such tech-enabled interactive touch points are installed in a retail outlet. The purchase experience would be seamless, and thus it is suggested that retailers should leverage digital to engage and interact with consumers.

Customization: Adding to the same, findings from the survey show that customers are also avid users of retail touch points in the form of customized offers and coupons. This is where personalization enters the retail scenario. The figures show that approximately 60% of the respondents like to have personalization at some point during their purchase journey. With the advances in this realm which has enabled retailers to connect with consumers, the retailers are suggested to provide geofenced targeted offers which will serve the dual purpose of consumer awareness and increase the number of shoppers turning up at the retail outlet.

Boutique websites: The move from online retail types of Flipkart and Amazon, the shift may be toward the "boutique websites," that is, the websites which deal in specialized categories, for example, luxuryhotels.com, which is an industry mushroomed with retailing of online travel; this website caters to only customers looking for luxury stay.

Go digital: Integrating digital mode of payments—such as plastic cards and mobile wallets would be an added advantage to this perfect consumer purchase experience. The important thing here is that though mobile wallets were the third preferred mode of payment, at the same time, 91.7% respondents reported that they would prefer to use mobile wallets over traditional payment methods in the future. While mobile wallets already imbibe the element of convenience, with a touch of personalization in the form of customized coupons and cash backs being offered by various mobile wallet companies, prediction shows that mobile wallets are rising toward a soaring growth. The whole exercise of demonetization would be beneficial if digital payments take precedence over cash; the positive outcome from the research is that in future, the respondents plan to go digital.

Loyalty programs: The online retailers have been struggling to get their share of customer and the market. The analysis provides an indicator to that. The loyalty programs being offered by the companies are being appreciated by the customers and benefits being availed. The amount charged by the companies for the loyalty programs can go a long way in improving the cash crunch.

Convenience and personalization: Consumers are the king of the market, and convenience and personalization are the two favorite factors for the consumer to live by. Therefore, it becomes imperative for the retailers to consider omnichannel retailing in order to incorporate both the factors.

Another major inference that can be drawn is that both online and offline players have realized the significance of "experience" for the customers. One of the best instances is Pepperfry, an online furniture shopping store that has also established its showroom by the name of Pepperfry Studio. Through its showrooms, Pepperfry helps the customer to "touch and feel" the furniture around. However, the variety is more on the online store. They have also incorporated the facilities where the store representatives enable the customers to choose furniture online in the physical store and place the order. Even though this idea has not yet been widely applied in India, there is a big scope for it as upon analysis it was found that 91.7% respondents wanted such type of facilities for trial and testing and after-sales services.

The shopping experience of the ever-demanding customer needs to be improved and a blend of physical and digital is the way forward.

KEYWORDS

- **m-commerce**
- **phygital retail**
- **omnichannel retail**
- **online**
- **e-commerce**

REFERENCES

Agarwal, A. J. 5 Ways Technology Is Changing Ecommerce. *Entrepreneur,* 2017. https://www.entrepreneur.com/article/288149 (accessed Jan 31, 2017).

Arora, S.; Budree, A., 2016 Basket Loyalty Tussle Amongst Indian Online Retailers. *IEEE Xplore Digital Library*. DOI: 10.1109/INCITE.2016.7857580.

Brynjolfsson, E.; Hu, Y. J.; Rahman, M. S. Competing in the Age of Omni-channel Retailing. *MIT Sloan Manag. Rev.* **2013,** *54* (4), 23–29. DOI: 10.1108/IJRDM-09-2015-0140.

Clarke, I. Emerging Value Propositions for M-commerce. *J. Bus. Strat.* **2001,** *18* (2), 133–148.

Google-A. T. Kearney. Digital Retail in 2020: Rewriting the Rules. *The Economic Times,* 2016. http://retail.economictimes.indiatimes.com/etanalytics/reports/e-commerce/digital-retail-in-2020-rewriting-the-rules/400 (accessed June 12, 2016).

Grewal, D.; Roggeveen, A. L.; Nordfält, J. The Future of Retailing. *J. Retail.* **2017,** *93* (1), 1–6. DOI: 10.1016/j.jretai.2016.12.008.

Harty, D. Smartphones Overtake Computers as Top E-commerce Traffic Source. *Economic Times,* 2016. http://economictimes.indiatimes.com/magazines/panache/smartphones-over-take-computers-as-top-e-commerce-traffic-source/articleshow/53413455.cms (accessed Aug 14, 2016).

IBM Smarter Commerce. M-commerce Strategies for Delivering an Omni-channel Customer Experience. *IBM White Paper,* 2013. http://www-01.ibm.com/software/commerce/offers/pdfs/ibm-mobile-commerce-wp-final.pdf (accessed July 11, 2016).

Internet Live Stats. Internet Users by Country. *Internet Live Stats,* 2016. http://www.inter-netlivestats.com/internet-users-by-country/ (accessed Jan 12, 2017).

Khurana, K.; Basu, A. The Future of Retail is 'Here' and 'Now'. *Cognizant Whitepapers,* 2015. https://www.cognizant.com/whitepapers/the-future-of-retail-is-here-and-now-codex1053.pdf (accessed Jan 12, 2016).

Kreuger, J. Omni-channel Shoppers: An Emerging Retail Reality. *Think with Google,* 2015. https://www.thinkwithgoogle.com/marketing-resources/omnichannel/omni-channel-shop-pers-an-emerging-retail-reality/ (accessed Mar 31, 2016).

Press Trust of India. M-commerce to Contribute up to 70 Per Cent of Online Shopping: Experts. *NDTV Profit,* 2014. http://profit.ndtv.com/news/corporates/article-mobile-commerce-to-contribute-up-to-70-per-cent-of-online-shopping-experts-706283 (accessed Jan 31, 2015).

Sathe, G. Taking the Web Out of E-commerce—Why Indian Companies Are Ditching Their Websites. *NDTV Gadgets 360,* 2015. http://gadgets.ndtv.com/apps/features/taking-the-web-out-of-e-commerce-why-indian-companies-are-ditching-their-websites-678591 (accessed Aug 14, 2016).

Sopadjieva, E.; Dholakia, U. M.; Benjamin, B. A Study of 46,000 Shoppers Shows That Omni-channel Retailing Works. *Harvard Business Review,* 2017. https://hbr.org/2017/01/a-study-of-46000-shoppers-shows-that-omnichannel-retailing-works (accessed Jan 12, 2017).

Spector, N. Let's Get Phygital. *Geo Marketing,* 2014. http://www.geomarketing.com/lets-get-phygital (accessed Feb 12, 2016).

Spencer, M. What's Phygital in Retail? *LinkedIn Pulse*; 2016. https://www.linkedin.com/pulse/whats-phygital-retail-michael-spencer

Verhoef, P. C.; Kannan, P. K.; Inman J. J. From Multi-channel Retailing to Omni-channel Retailing. *J. Retail.* **2015,** *91* (2), 174–181. DOI:10.1016/j.jretail.2015.02.005.

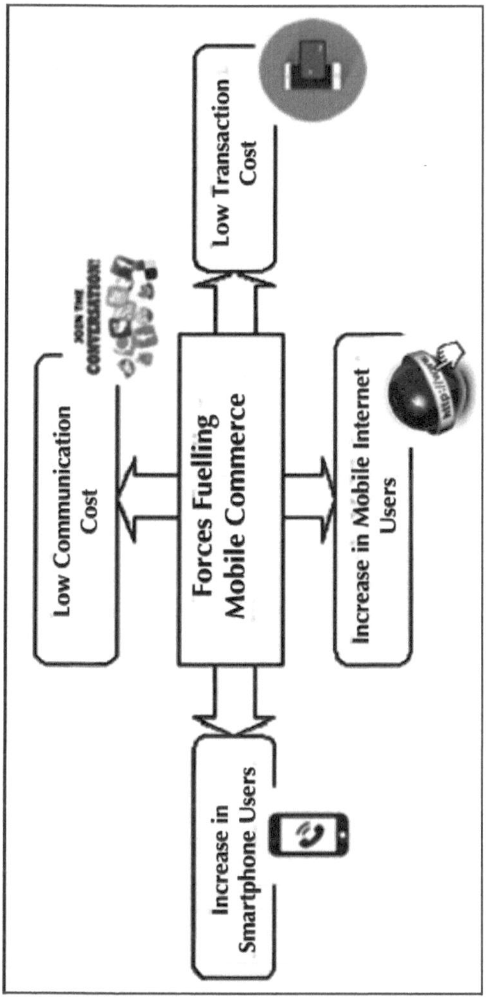

FIGURE 4.2 Forces fuelling mobile commerce.
Source: Compiled by authors.

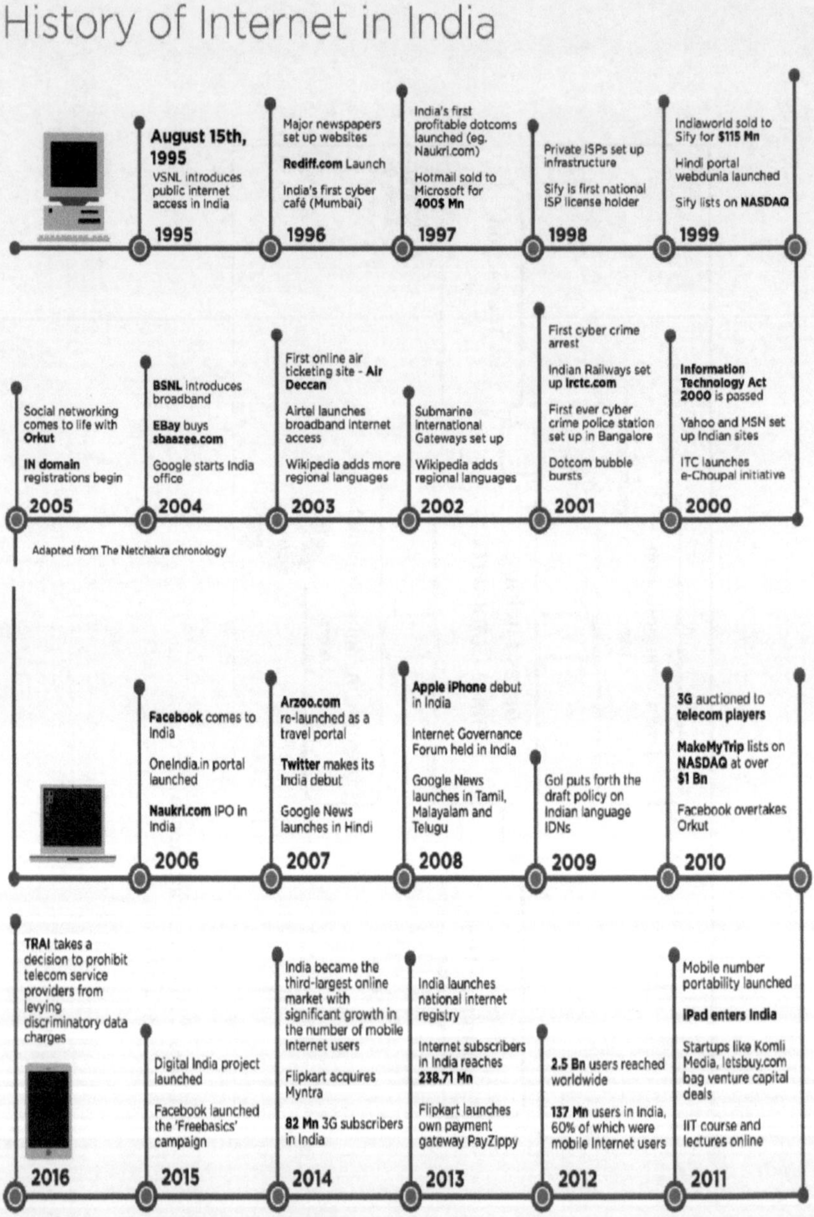

FIGURE 6.1 Internet history in India.

Source: Adapted from NASSCOM. The Future of Internet in India; 2016. http://www.nasscom.in/sites/default/files/Article_News/PR_NASSCOM_AKAMAI_Rise%20of%20Internet_17082016.pdf (accessed March 10, 2017).

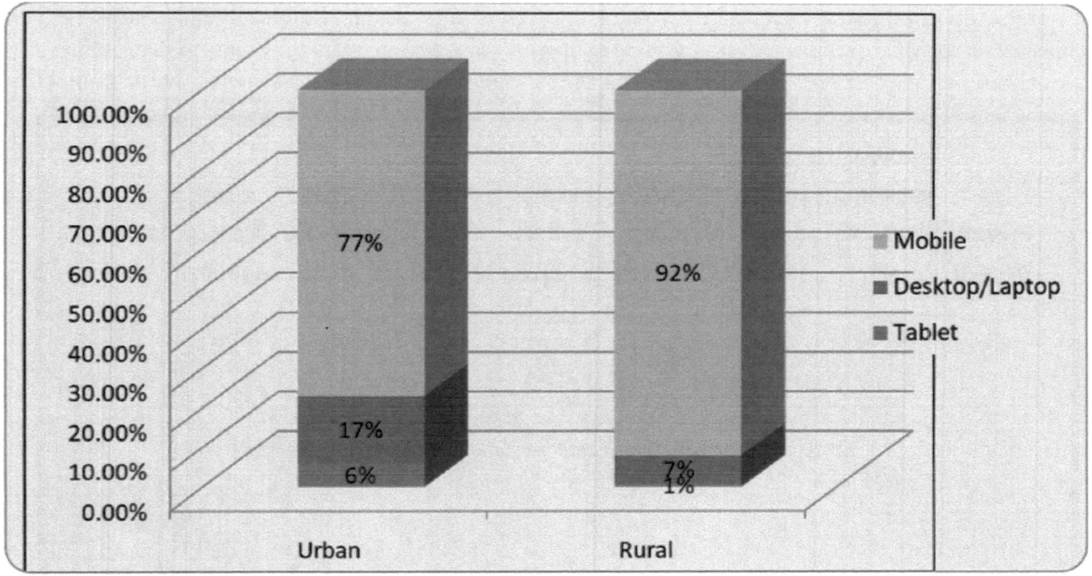

FIGURE 6.8 Device used for internet usage.

Source: Adapted from An IAMAI & KANTAR IMRB Report. *Internet in India 2016*; 2016. http://static.bestmediainfo.com/wp-content/uploads/2017/03/Internet-in-India-2016.pdf (accessed April 4, 2017).

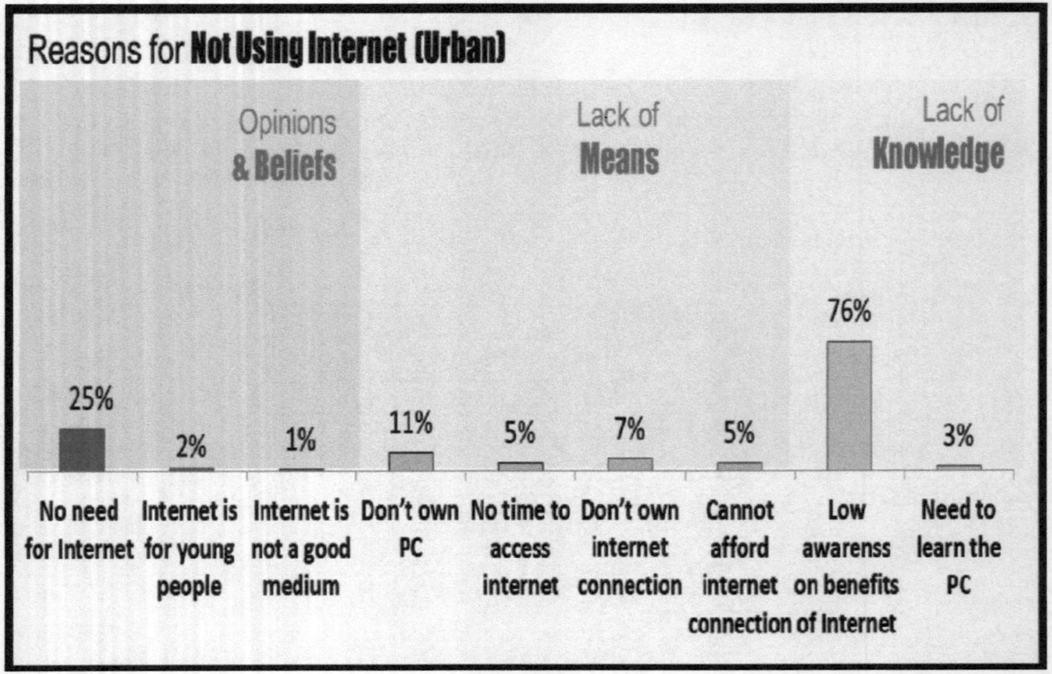

FIGURE 6.12 Noninternet users in urban India.
Source: Adapted from An IAMAI & KANTAR IMRB Report. *Internet in India 2016*; 2016. http://static.bestmediainfo.com/wp-content/uploads/2017/03/Internet-in-India-2016.pdf (accessed April 4, 2017).

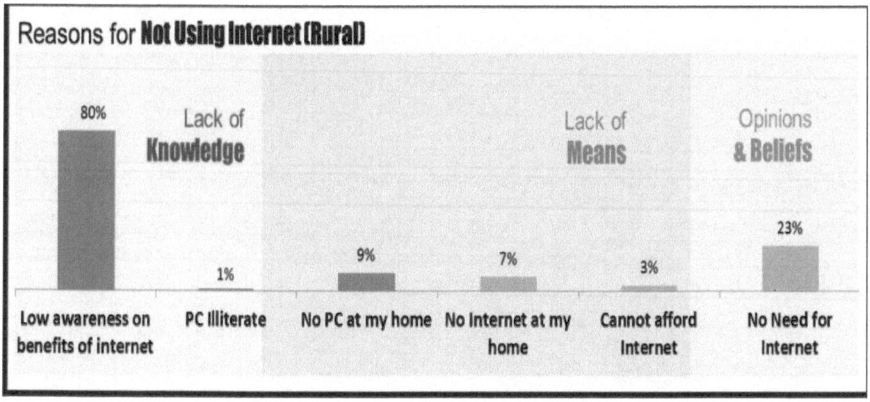

FIGURE 6.13 Noninternet users in rural India.
Source: Adapted from An IAMAI & KANTAR IMRB Report. *Internet in India 2016*; 2016. http://static.bestmediainfo.com/wp-content/uploads/2017/03/Internet-in-India-2016.pdf (accessed Apr 4, 2017).

FIGURE 7.2 Augmented reality example.
Source: https://leafcutter.com.au/blog/augmented-reality-enhancing-experience-and-engagement/.

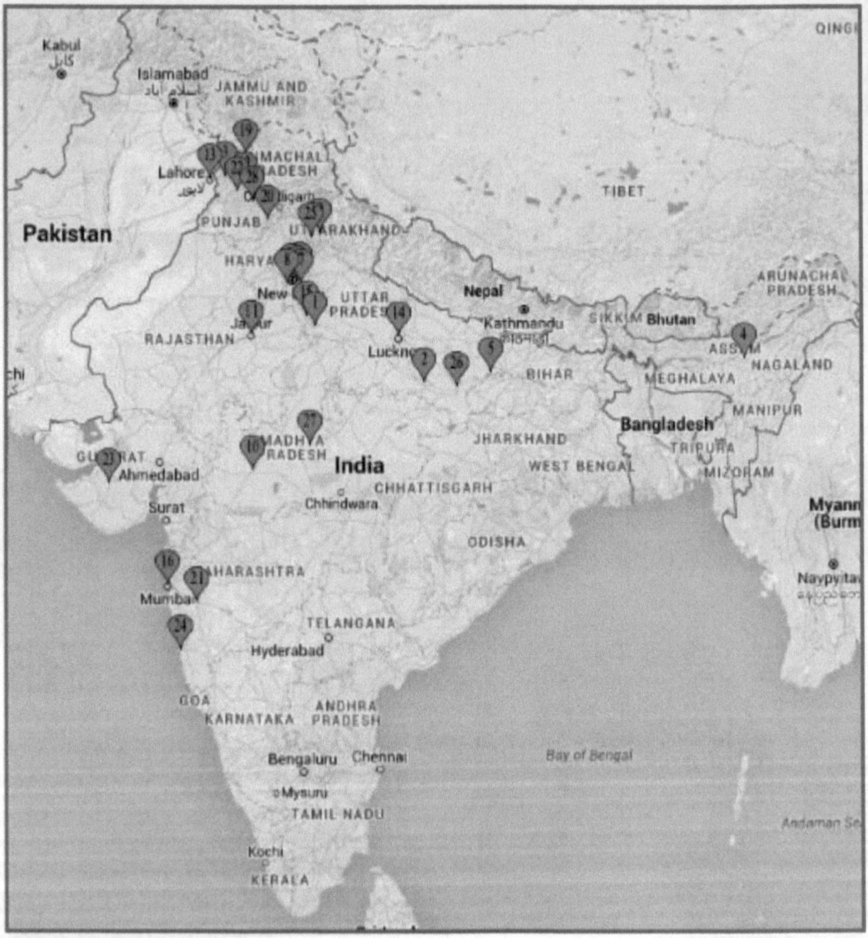

FIGURE 13.6 Cartogram representing location of respondents.
Source: Compiled by authors (created on https://www.mapcustomizer.com).

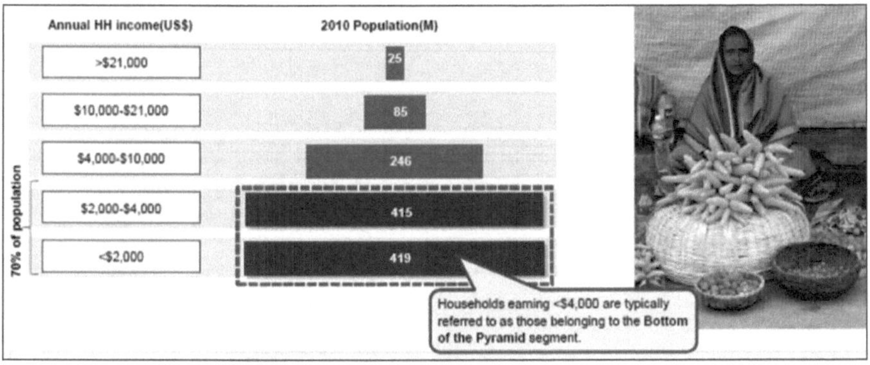

FIGURE 16.2 Bottom of pyramid in India.
Source: Modified from the Luce (2016).

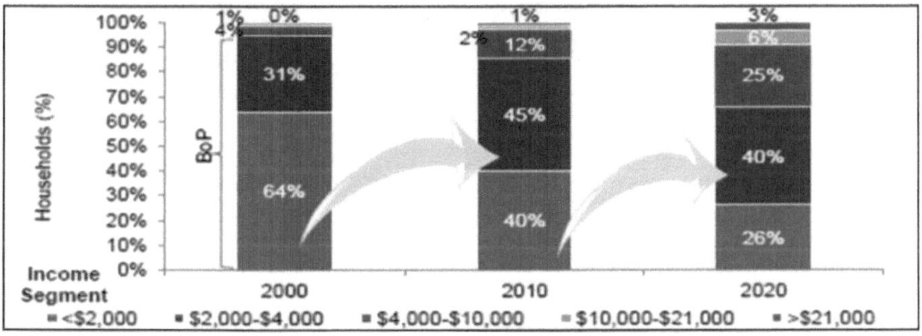

FIGURE 16.3 Households by income bracket (2000–2020).
Source: Modified from the Luce (2016).

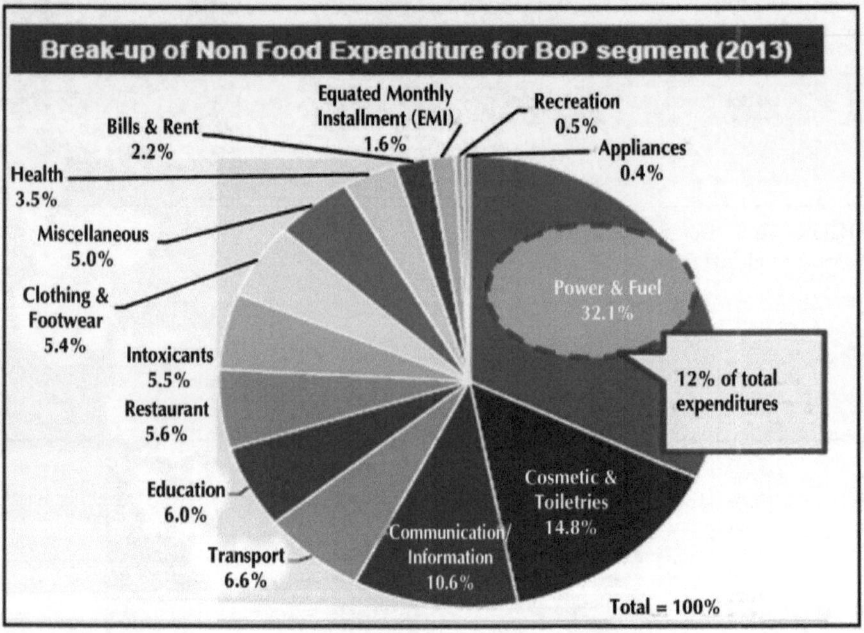

FIGURE 16.4 Expenditures by the BoP segment in India.
Source: Modified from Luce (2016).

PART III

FACETS OF M-COMMERCE AND CONSUMER PERCEPTIONS

CHAPTER 9

FACTORS DETERRING THE ADOPTION OF M-TOURISM IN INDIA

S. MEERA and A. VINODAN*

Indian Institute of Tourism and Travel Management (An Autonomous Organization Under Ministry of Tourism, Government of India), Nellore 524321, Andhra Pradesh, India

Corresponding author. E-mail: vinodan_tt@yahoo.co.in

ABSTRACT

M-commerce with distinct advantage over e-commerce has been gaining popularity in travel and tourism management. M-tourism is an extension of m-commerce into tourism, travel, and allied sectors. M-tourism applications are widely construed both from demand and supply dimensions. The present study tries to identify various factors deterring the adoption of m-tourism in demand paradigm. An exploratory sequential method consisting of in-depth interview and factor analysis was adopted for the study. The study shows that there are 7 identified dimensions, that is, performance expectation factor, perceived effort factor, socially influencing factors, facilitation factor, technology specific factor, personal centric factor, and country specific factor with 35 deterring variables prevailing in the context of m-tourism in India.

9.1 INTRODUCTION

Tourism is one of the emerging economic sectors of the world, contributing annually 10% of the world Gross Domestic Product (GDP), and creating one in eleven jobs globally (UNWTO, 2014). In terms of exports, tourism ranked fourth in 2013 after fuel, chemicals, food, and automobiles (UNWTO, 2014). Emerging and developing countries account for 45% of all international tourists arrivals and it is expected to reach 57% in 2030 (UNWTO,

2011). As far as domestic tourism is concerned, it occupies significant share of overall tourism markets, mostly representing more than 50% or more in many advanced countries. International tourism is considered as an important source of foreign exchange earnings for many emerging economies and several less developed countries and it represents 6% of overall exports of goods and services and 30% of the world's exports of services (UNWTO, 2015).

In India, leisure travel spending (inbound and domestic) generated 73.8% of direct travel and tourism GDP in 2011, as compared to business travel spending whose contribution is 26.2% (WTTC, 2012). The same are expected to rise by 7.6% per annum in 2022 for both leisure and business travel spending. Domestic travel spending has also generated almost five times more revenue than that of international tourism receipts. The breakdown of indirect contribution from travel and tourism industry is categorized into: supply chain (44.8%), investment (7.3%), and Government collective (1.5%) with a total collection of 53.7% of the total contribution (WTTC, 2012).

Tourism industry is growing, keeping pace with the technological changes in the society by adopting new technology and applications for transparency, value addition, accuracy, ease of use and so on. As far as applications are concerned, online booking is considered as one of the much sought after transaction with technology in tourism sector followed by purchase decision, evaluation, and after-sale service.

Internet and Mobile Association of India (IAMAI) and Indian market research study show that travel was the first industry to secure significant digital sales in India (www.emarketor.com). According to Juman (2012), the survey conducted by Nielson stated that about 40% of the Indians buy airline tickets and reservations online and 29% of the Indians opt to plan for online tours/hotel reservations. Subsequently, consumer confidence also increased and helped air and hotel suppliers strengthen their transactions (Juman, 2012).

Even though the growth of travelers using technology is increasing at a rapid pace, there are challenges in adopting technology in all tourism segments. This includes gap in tourism and technology industry, selection of right technology, tools, lacunae of strategy to utilize technology correctly, absence of fair or reliable information or education of technological applications, etc. Factors such as investment for the acquisition are not found to be a big challenge as the cost of hardware and connectivity has come down significantly over the years.

9.1.1 M-TOURISM

Like m-commerce, the advent of mobile technologies has opened up number of opportunities for service providers and destinations to facilitate and personalize experience in tourism today. Tourists are increasingly looking forward to take advantage of the portable internet connectivity offered by their smartphones or other portable devises to gather holiday information and execute travel plans. Generally, mobile technologies offer convenience, efficiency, and spontaneity along with a promise of authentic destination information in a unique, targeted, and engaging manner for better experiences.

Development of m-tourism apps provide information, linkages, travel write-ups and articles, photos, and details of unexplored destinations. Other intervention areas of m-tourism apps include walking tours, Global Positioning System (GPS) tracking, directions, destination search, hotels, and rental car reservations. Applications of m-commerce have found place in m-tourism market also. These can be listed as:

1. Mobile ticketing: Mobile ticketing is gaining popularity in tourism and allied services which include transportation, accommodation, food and beverages, creative industries, usage of tourism resources in destinations, booking of leisure and excursion services, and support services.
2. Mobile money transfer: Mobile money transfer is a payment gateway which facilitates the transfer of information between a payment portal (such as a website, mobile phone, or interactive voice response (IVR) service) and the front-end processor or acquiring bank. Thus, a Payment Gateway enables a website to accept payments from customers over internet.
3. Location-based services: Location-based services have wide potential in m-tourism market, as most of the consumer decisions are on the basis of local weather conditions, seasonality, and availability of destination resources. These include understanding of local weather condition, tracking and monitoring of people such as availability and services of guides, escorts, tour leaders, etc. Special offers as a part of "seasonality" sale strategies of various tourism services and attractions is one of the major m-tourism segment practices across the world to counteract seasonality issues of tourism.
4. Location-based games: Location-based games can be developed for customer engagement by highlighting and linking the specialties

of the destination to create customer loyalty, brand awareness, and enhance the user experience in tourism destinations.

5. Promotion: M-based promotion programs are common in tourism, travel, and allied sector especially by providing mobile voucher, coupons, and loyalty cards to reach customers fast and attract them. Destination promotion, promotion of travel packages, tickets, and special offers were found to be one of the widely adopted options under m-promotion.
6. Information services: This refers to an array of information services useful to travelers in the present as well as in future. Many GPS-based applications pertaining to map, eateries, art and culture, fairs and festivals, etc., have found place in m-tourism. For example (from their business profile), Mapmyindia is a leading map and navigation services provider which introduced iNav, a GPS-based application for mobile phones; it helps to locate a destination anywhere in India, also developed Delphi Nav 300 and 430 classic navigator models for Indian streets, and Delphi Nav 200 for cars (www.mapmyindia.com). Similarly, restaurants and hotels navigator Zomato (www.zomato.com), Map My Temple (www.mapmytemple.com) and Bharatwiki (www.bharatwiki.com) are the other examples.
7. M-purchase and payments: The rapid development of new technologies, especially those relating to mobile networks and mobile devices, are creating opportunities for electronic purchase and payments to serve new markets and merchant segments. For example, a mobile phone or app designed for travel service providers and intermediaries with a secure integrated adaptor, that accepts both magnetic stripe and chip cards, merges low-cost and widespread GSM/GPRS mobile phone technology with the capabilities of a point-of-sale terminal, bringing electronic payments to all locations. Mobile message-based purchase and payments are also expanding the choice to consumers and merchants of today.
8. M-marketing and advertising: With regard to reachability of advertisement and promotion packages of various tourism services, m-tourism app helps the customers particularly in availing those benefits on time. However, it requires special capability in mobile phone to identify and install such app.

9.1.2 REVIEW OF LITERATURE

A new platform for business transaction emerged in recent years with the adoption of mobile and wireless networks, known as mobile commerce (m-commerce). M-commerce connects wirelessly in a mobile environment using handheld mobile devices. M-commerce is based on the use of wireless technology, usually mobile internet and handheld mobile devices, for transaction processing, information retrieval, and user task performance (Varshney, 2003). Large number of studies has been done in the context of e-tourism services by touching various aspects of operations. These are international access (Nedelea and Balan, 2010), problems of e-travel markets (Oorni and Klein, 2003), meeting diversified needs of customers with value addition (Qirici et al., 2011), value-added service through tourism websites (Buhalis and O'Connor, 2005), internet purchase intensions (Urban, 2004), consumer trust (Gregori, 2009), and generating visitors interest through websites (Skadberg et al., 2005). According to Buhalis (2003), e-tourism reflects the digitization of all processes and value chains in the tourism, travel, and hospitality, and catering industries. In practice, it includes e-commerce and applies information and communication technology for maximizing the efficiency and effectiveness of the tourism products and enterprises.

However, Chen and Barnes (2007) investigated key obstacles to vendors succeeding on the internet medium. Study identified that perceived usefulness, perceived security, privacy, good reputation, and willingness to customize are the important obstacles.

M-commerce, in practice, extends wider flexibility in tourism industry, that is, travelers and suppliers (Lee and Mills, 2010). Mobile application is a software application designed to run on mobile devices (Wang et al., 2013). This application is able to replace the traditional marketplace in the tourism industry as the mobile apps help to connect users to internet. Through m-commerce, travelers can access the web, news updates, and conduct transactions using their mobile devices, whereas supplier's can have promotional and marketing means easier and faster as compared to the use of traditional media (Lee and Mills, 2010). Tourism industry considers m-commerce as an integral part of delivering better travel experience to travelers. Mobile devices such as smartphones, personal digital assistants (PDA), and tablets expedite communication and nonlocation-based information to tourists like a guide (Kenteris et al., 2009).

While examining the benefits of m-tourism, observations of Hu and Liu (2013) on m-tourism are found important. They have stated that m-tourism is beneficial to consumer in many ways, that is, convenience, ubiquity, positioning, destination-specific information, and instant access to relevant information. Personalized services, purchase decision-making based on relevant content/information, and management of user profiles are the other benefits of m-tourism (Gavalas and Kenteris, 2011).

According to UNCTAD (2005), linkage between tourism and information and communication technology (ICT) has opened numerous opportunities for developing nations. These include the enhanced number of internet users looking for travel experience, technology-enabled support to destination, and national tourism providers to develop, manage, and sell products worldwide, thereby creating brand image and development of new products thus making promotion easy. This, in turn, leads to enhancement of foreign exchange earnings and contribution to local development.

As indicated in Scott et al. (2014), mobile phones influenced travelers' behavior with regard to searching information, purchasing, post purchase evaluation, and travel aspects such as providing directions, public transportation navigation, and air travel along with other mobile tourist application such as AirAsia, MHmobile, Agoda, and Expedia.

Eriksson (2012) examines the user experience of arranging a trip of low–intermediate complexity over a mobile device using existing travel services. Study explored potential problems while making trip arrangements by using mobile devices. Factors such as type of services, mobile devices, and user skills affect self-arrangement experience on a mobile device. In another study, he has stated that m-tourism supports customers in meeting various needs such as spontaneous needs, time-critical arrangements, efficiency ambitions, mobility-related needs (location features), and entertainment needs, and cautioned that service should be customized to facilitate travelers' style of traveling (e.g., organized travel or independent travel) and should be easy to use, especially easy to take into use (access, install and learn) during a trip, keeping security and financial concerns of customers (Eriksson, 2013).

Study by Anckar and Walden (2002) presents expected consumer problems in internet travel bookings. Problems were excess time consumption, problems in price comparisons, limited industry knowledge, usability of websites, and practical difficulties such as location of websites of service providers, technical problems, and searching for availability of hotel rooms and flights. The other issue identified was increased assessment time to decide fair price of service offered.

While analyzing the factors affecting the adoption of m-tourism, the observations of Venkatesh et al. (2003) found performance expectancy as most relevant in the context of m-tourism. He explained performance expectancy as the extent to which a system could support a person in enhancing his job performance. They have further explored effort expectancy and social influence factors affecting the behavioral intention to use a technology.

Ericsson and Strandvik (2009) study tried to identify possible determinants of tourists' intended or actual use of mobile tourism services. Study identified three major barriers for the non usage of the trial services. These are linked to the value aspect of the packaged tour: price transparency and ease of use.

Previous studies highlight the importance of m-tourism and its resultant growth. Moreover, m-tourism practices are operationalized differently in different economic setup. The variability in acceptance also varies depending on cultural and social setup. In this direction, low level of adoption toward mobile tourism in India may affect country's digitalization initiatives and the economy in general. Hence, the present study investigates various deterring factors pertaining to the acceptance of m-tourism and its dimensions in the context of India.

9.1.3 OBJECTIVES

The objective of the study is to identify the deterring factors responsible for adopting m-tourism services in India. In this study, tourists' (customers') perceptions on various issues which hinder the acceptance and adoption of m-tourism services or factors that deter the digitalization of travel trade in India has been identified. The study revolves around these variables for domestic (customers) tourists planning to purchase or avail various tourism and related products. The study incorporates following specific objectives:

- To identify the deterring factors responsible for the adoption of m-tourism in India
- To examine the dimensional orientation of the deterring factors in adopting m-tourism

9.1.4 RESEARCH QUESTIONS

The study tried to investigate answers for the following questions: What are the different deterring factors prevailing in Indian society in adopting m-tourism transactions? And how are these factors dimensionally oriented?

9.1.5 HYPOTHESIS

- There exists significant factors deterring the adoption of m-tourism in India
- There exists dimensional orientation for the factors deterring the adoption of m-tourism in India

9.1.6 SCOPE OF THE STUDY

In the context of digitalization of Indian economy, this study assumes significance as it attempts to examine the deterring factors for the adoption of m-tourism among smartphone users and other mobile devices, revealing the bottlenecks on digitalization in the context of tourism. In this context, the term "adoption" is construed as "choice to mentally accept and acquire technology." Study enables the tourism service providers, government, and other intermediaries to gain better understanding about the issues and invite their attention to reduce the presence of these deterring factors to grab the upcoming opportunities.

9.2 METHODOLOGY

9.2.1 EXPLORATORY SEQUENTIAL METHOD

The present study followed a sequential mixed method because one stage was conducted after the previous stage was completed (Creswell et al., 2003). Furthermore, results of stage one of this study were used to develop the measurement instrument for stage two. The rationale for adopting sequential mixed method was justifiable on following grounds. The objective of the research was to identify certain factors which are unknown and its dimensional orientation that was not explored in prior studies. It is also observed that employing mixed methods could give more authenticity while identifying certain factors which are ever changing. Initial stage of the study adopted in-depth interview to collect qualitative data, and questionnaire-based survey was adopted for quantitative analysis.

9.2.1.1 IN-DEPTH INTERVIEW

Interview with experts among stakeholders of tourism was conducted as a part of this study to find out various indicators pertaining to the construct under study. Table 9.1 represents stakeholder-wise interviews held at different cities of South India. Expert interviews with stakeholders were organized during the months of January to March 2016. This method was introduced especially to examine more intensely certain topics that remained unexplored in the existing literature pertaining to m-tourism particularly in the context of India. Experts consisting of managers, not below the rank of Assistant Manager were interviewed among following stakeholder categories: (1) tour operators, (2) hospitality (hotels only), (3) tourist transport operators, and (4) tour guides. Table 9.1 indicates the spatial representation of data collected through interview method.

TABLE 9.1 Stakeholder-wise Interviews Held at Different Cities of South India.

Stakeholders	Cities			
	Hyderabad	Cochin	Chennai	Total
Tour operators	3	4	6	13
Hospitality enterprises	1	5	4	10
Tourist transport operators	4	9	12	25
Tourist guides	2	12	8	22
Total	10	30	30	70

Source: Compiled by authors.

9.2.1.2 INTERVIEW CRITERIA

As indicated in Table 9.1, Experts consisted of 13 assistant managers, having minimum 3 years of experience in the tour operations. Experts from hospitality sector consisted of five assistant managers, two marketing managers, and three entrepreneurs. Tourist transport operators were the major respondents of the study as they are widely exposed to the e-tourism/m-tourism transactions in recent years. They include nine assistant managers, nine entrepreneurs including partners of partnership firms, and seven branch managers. Tourist guides (22) who are working in the cities of Hyderabad, Cochin, and Chennai were also considered based on their experience in the field.

9.2.1.3 INTERVIEW RESULTS

Based on expert interview and the literature review, the study has produced an elaborative list of 41 indicators. The subsequent investigation by the researcher found that there were few redundancies in the identified indicators. In this regard, investigator approached two e-tourism consultants and two app developers. Based on their advice, 6 indicators were found to be vague and confusing, which were removed and 35 appropriate indicators were identified (Table 9.2) for measuring issues pertaining to deterrent factors of m-tourism adoption. Table 9.2 represents the indicators identified and explanations.

TABLE 9.2 Indicators and Its Explanations.

Usefulness: Mobile-based application is useless due to apprehensions in adoption	Motivation: M-tourism application does not usher any incentive	Productivity: M-tourism does not enhance productivity as it is no way better than the conventional system
Performance expectation: The actual performance may be lesser than expectations, as the existing system bestows optimum performance with human touch	Comparative advantage: M-tourism does not have any apparent comparative advantage over the existing mode of transactions prevalent in tourism as its application in all transactions is not known, or established	Accomplishments: Sense of accomplishment by adopting M-tourism is not considered significant
Time: Time taken for learning is very long	Ease of use: Handling or operational difficulties reduces the ease of use	Learning to use: Assimilation difficulties hamper the learning process
Network externalities: Nonavailability of uninterrupted network availability can lead to network externalities	Understanding: Cognitive difficulties are expected in understanding the application	Clarity: Clarity is lacking due to vagueness in application in the multifaceted tourism industry
Trialability: Scope of trial and error methods in application is negligible	Complexity: Complexity pertaining to modifiability backed by less reversibility	Installation: Issues in full-fledged installation of m-tourism application is complex and time consuming
Media influence: Influence of mass media on m-tourism is often confusing	Advisory: Advisory responses on m-tourism adoption are not encouraging	Peer influence: Negative peer influence on usage, comfort, and benefit

Knowledge level: Knowledge gap in adoption	Financial condition: Financial difficulties in embracing paid m-tourism applications	Compatibility: Less compatibility in installing mobile-based software in all mobile phones configuration
Assistance: Lack of timely assistance in operation and trouble shooting	Availability: Non-availability of required facilities for operating the downloaded application	Accessibility: Non-accessibility of support services like customer services centers exclusively for addressing the concerns of m-tourism applicants
Language: Familiarity with English or application language	Technical infrastructure: Availability of technical infrastructure like bandwidth, signal strength, network connectivity, etc.	Legal barriers: Lack of awareness on legal aspects of m-tourism application
Exposure: Limited exposure to tourism technology	Affordability: Less affordability in acquiring high-end smart phones	Diversity: Heterogeneous nature of tourism products gives less scope for uniform application
Technical inhibition: Inhibition of adopting technology in lieu of manual transaction with personal touch	Novelty: Continuing upgradation is required for m-tourism application	Self-efficacy: Effectiveness in terms of performance does not vary much from customer to customer
Trust: Distrust in wireless application	Education: Level of education is different among customers	

Source: Compiled by authors.

9.2.1.4 RELIABILITY AND VALIDITY

In order to assess the reliability and validity of the qualitative part of the present study, the Trochim (2006) criteria have been adopted. These are credibility, transferability, dependability, and conformability. In the present study, credibility has been established on the basis of criteria and informal conversations, transferability, as there is a possibility of replicating the same in the context of technology adoption, and related issues in similar or identical cases in other developing countries. Dependability is based on methodological consistency and conformability is established by selecting all respondents from related industry. Creditability seeks to ensure that the study measures or tests what is actually intended. Credibility and transferability corresponds to internal and external validity of the quantitative research. Lincoln and Guba (1985) argue that ensuring credibility is one of most important factors in establishing trustworthiness.

9.2.2 SCALE DEVELOPMENT FOR FACTORS DETERRING M-TOURISM

Developing a scale to measure various deterring factors of m-tourism was found cumbersome as every issue needs to be addressed in precision to have a holistic coverage of the topic. Since the word "deter" is often construed negatively, all questions indicating the construct were negatively worded to avoid confusion among respondents. The scale was developed based on Churchill's (1979) guidelines. A 5-point Likert agreement scale was designed to understand issues pertaining to m-tourism adoption.

9.2.3 SAMPLING DESIGN

The sampling design explains the definite plan for obtaining a sample from the population, that is, the entire group of people whom the researcher is interested to know about (Gupta and Gupta, 2013). In this study, the mobile users in South India were considered as the population for the study. The investigator identified four metros of South India, that is, Cochin, Hyderabad, Bangalore, and Chennai, as the technology adoption is growing not only in cities but also at small market locations (www.invasystems.com). These selections were made on the basis of purposive sampling. Purposive sampling is justified on the following grounds:

- It helps the researcher to use his discretion to select the respondents, so as to get the best samples to meet the purpose of the study.
- Purposive sampling is widely used in mixed method research (Maxwell and Loomis, 2002).
- In terms of sampling frame, it was basically judgmental in nature as the researcher's judgment has been combined with expert opinion which had been confirmed through the exploratory study.
- In this research, the sampling units are the citizens of metro cities of South India who had travel experience and, therefore, purposeful sampling is appropriate.

9.2.4 SAMPLING TECHNIQUE

In this study, the samples of respondents for the study were travelers. Selection of the respondents/customers was based on certain considerations, to match the objectives of the study.

- All respondents have completed at least one trip during the last 3 years and are willing to travel in near future with in India.
- All respondents have experience in using smart phones.

From the sample units selected on the basis of the criteria as discussed above, convenience sampling was used to select the sample units.

Convenience sampling refers to sampling by obtaining units or people who are most conveniently available. For example, it may be convenient and economical to select samples of customers who are approachable and reachable to get feedback on the subject matter. Moreover, convenience sampling is the only feasible way to proceed while attempting to learn about groups who are busy in their schedules.

9.3 DATA COLLECTION

The first stage involved convenience sampling method, to collect primary data using a structured questionnaire. As indicated in Table 9.3, 316 responses were finalized for the study. An item ratio of 1:9 (for 35 items 316 respondents) was maintained. Threshold range suggested by Flynn and Pearcy (2001) ranges from 1:4 to 1:10. Table 9.3 represents city-wise data used for the study.

TABLE 9.3 City-wise Data Used for the Study.

Locations	Respondents		
	Identified	Collected	Finalized
Cochin	135	121	105
Bangalore	100	87	65
Chennai	110	105	76
Hyderabad	100	84	70

Source: Compiled by authors.

All the sample units identified for the study on the basis of the criteria were not available at the time of data collection. As some of the questionnaires were incomplete, they could not be considered for the study. The total sample size is 316, comprising 105 from Cochin, 65 from Bengaluru, 76 from Chennai, and 70 from Hyderabad. Data were collected during the month of October to December, 2016.

9.4 ANALYSIS

9.4.1 *EXPLORATORY FACTOR ANALYSIS OF FACTORS DETERRING M-TOURISM*

The role of factor analysis is to identify the underlying structures derived from a set of variables (Hair et al., 1998). Exploratory factor analysis (EFA) was conducted for identifying the underlying factors deterring m-tourism adoption. All the indicator variables were subjected to factor analysis to get naturally occurring underlying variables (Rosen and Surprenant, 1998).

EFA with varimax rotation was performed to identify the number of factors with maximum explanations (Hair et al., 1998). In this study, items that load higher than 0.5 were retained. The result showed that the EFA identified seven latent constructs. The identified factors of all these constructs with an Eigen value greater than 1, together explained over 64.43 % of the variance, and hence, it was assumed that the model represents the data. There were no significant cross-loadings between items in this analysis. The Kaiser–Meyer–Olkin Measure of Sampling Adequacy was 0.884 for deterring factors. The Bartlett test of sphericity was significant ($p < 0.001$) with a Chi Square value of 5509 with 595 degrees of freedom which are considered to be appropriate for further analysis of factorization. Communalities between measured items loaded on the EFA model varied from 0.518 for N.6 item to 0.776 for N.13. All items loaded significantly were retained for further analysis. The rotated component matrix showed loadings of each measured item on each of the seven latent factors identified. Table 9.4 indicates factor loading for deterring factors for m-tourism adoption.

TABLE 9.4 Factor Loading for Deterring Factors for M-tourism Adoption.

Factor	Item no.	Item descriptions	Factor loading						
			1	2	3	4	5	6	7
Performance Expectation issues	N1	Usefulness	0.834						
	N2	Motivation	0.665						
	N3	Productivity	0.820						
	N4	Performance expectation	0.828						
	N5	Comparative advantage	0.732						
	N6	Accomplishments	0.658						
Perceived effort issues	N7	Time		0.823					
	N8	Ease of use		0.762					
	N9	Learning to use		0.765					
	N10	Clarity		0.724					
	N11	Understanding		0.716					
Technology-specific issues	N12	Network externalities			0.556				
	N13	Trialability			0.853				
	N14	Complexity			0.815				
	N15	Installation			0.824				
Facilitation issues	N16	Financial condition				0.741			
	N17	Knowledge level				0.739			
	N18	Compatibility				0.686			
	N19	Assistance				0.788			
	N20	Availability				0.677			
	N21	Accessibility				0.626			
Country-specific issues	N22	Language					0.691		
	N23	Technical infrastructure					0.767		
	N24	Legal barriers					0.690		
	N25	Diversity					0.735		
	N26	Exposure					0.719		
Personal centric issues	N27	Affordability						0.667	
	N28	Technology inhibition						0.789	
	N29	Novelty						0.723	
	N30	Self-efficacy						0.765	
	N31	Trust						0.737	
	N32	Education						0.744	
Social influence issues	N33	Peer influence							0.719
	N34	Advisory							0.813
	N35	Media influence							0.806

Extraction method: principal component analysis. Rotation method: varimax with Kaiser normalization.

Source: Compiled by authors.

EFA identified seven constructs. These are as follows:

1. Performance expectation issues (PEI): This factor has moderately explained variance accounting to 10.800%. The variables highly loaded in this factor are tabulated with their respective loadings in descending order. There are six indicators showing performance issues while adopting m-tourism operations; accordingly, the study conceptualized these issues as PEI.
2. Perceived effort issues (EEI): While adopting mobile-based travel arrangements or purchase, special efforts such as reduced time for learning, ease of use, etc., are also required. Indicators showing all such variables were formed under one construct called EEI. This factor accounts for the second highest proportion of variance, that is, 10.632% with five variables.
3. Technology-specific issues (TSI): All technology-related factors were loaded together, and then conceptualized as TSI. This factor had moderately explained variance accounting to 10.303%. The number of variables loaded in this factor was four.
4. Facilitation issues (FCI): There were factors which hinder the adoption of m-tourism; they were finalized as FCI. There are significant positive loadings in this factor also. All the facilitating variables are loaded in a single factor and also they are positively associated with each other.
5. Country-specific issues (CSI): All country-specific issues were clubbed as CSI. The number of variable loaded in this factor was five.
6. Personal centric issues (PCI): The number of variable loaded in this factor was six. Based on the loading pattern, these factors were conceptualized as PCI.
7. Social influence issues (SII): The numbers of variables related to general social factors were loaded together, accordingly conceptualized as SII.

So all those identified seven latent factors were retained, henceforth those identified constructs are called as the latent constructs. Table 9.5 indicates the identified latent construct and their corresponding indicators.

TABLE 9.5 List of Latent Constructs and Its Corresponding Variables.

Sl. no.	Constructs	No. of variables	Name of the variables	Individually explained
1	PEI	6	Usefulness, motivation, productivity, performance expectation, comparative advantage, and accomplishments	10.800
2	EEI	5	Time, ease of use, learning to use, clarity, and understanding	10.632
3	TSI	4	Network externalities, trialability, complexity, and installation	10.303
4	FCI	6	Financial condition, knowledge level, compatibility, assistance availability, and accessibility	9. 238
5	CSI	5	Language, technical infrastructure, legal barriers, diversity, and exposure	9.067
6	PCI	6	Affordability, technology inhibition, novelty, self-efficacy, trust, and education	8.615
7	SII	3	Peer influence, advisory, and media influence	5.771

CSI, country-specific issues; EEI, perceived effort issues; FCI, facilitation issues; PCI, personal centric issues; PEI, performance expectation issues; SII, social influence issues; TSI, technology-specific issues.

Source: Compiled by authors.

9.4.2 CONFIRMATORY FACTOR ANALYSIS OF FACTORS DETERRING M-TOURISM

In order to determine the ability of a predefined factor model to fit an observed set of data confirmatory factor analysis (CFA) was conducted. CFA provides estimates for each parameter of the measurement model. CFA is also useful to (1) test the relationship between two or more factor loadings, (2) test the significance of a specific factor loading, (3) assess the convergent and discriminant validity of a set of measures, and (4) test whether a set of factors are correlated or uncorrelated.

The statistical significance of the relationships among various deterring factors and its identified and extracted dimensions such as PEI, EEI, FCI, CSI, PCI, TSI, and SII were taken together as shown in Figure 9.1. Figure 9.1 represents the CFA of deterring factors.

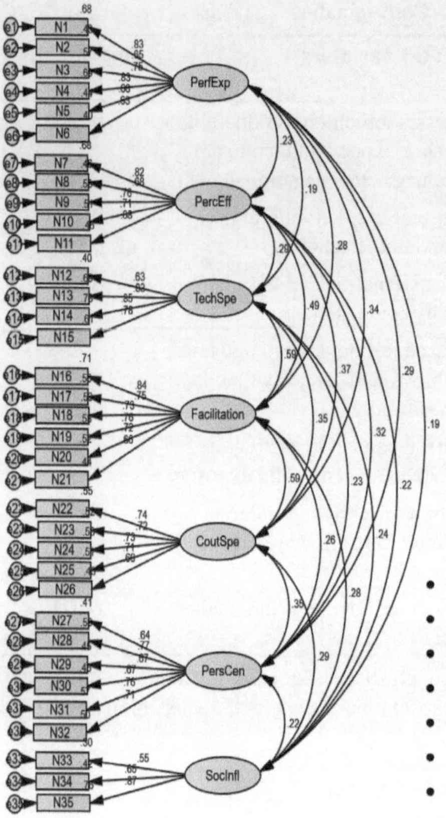

- PerfExp - Performance Expectation Issues
- PercEff - Perceived Effort Issues
- TechSpe - Technology Specific Issues
- Facilitation - Facilitation Issues
- CoutSpe - Country Specific Issues
- PersCen - Personal Centric Issues
- Socinfl - Social Influence Issues

FIGURE 9.1 Confirmatory factor analysis of deterring factors.
Source: Compiled by author.

The measurement model of constructs showed that the initial estimate of the 35 indicator variable model was found to be a valid fitting model. All the fit indices were within the permissible limits. All the paths shown in the model were significant as the critical ratios (CRs) were above 1.96. So the identified model is considered to be right fitting model with 35 indicators as illustrated in Figure 9.1. The model fit summary and estimates of deterring factors are given in Table 9.6.

TABLE 9.6 Fit Indices of Deterring Factors.

Fit measures	Indicators	Value obtained
Absolute fit measures	CMIN/DF	1.879
	RMSEA	0.053
	GFI	0.835
Incremental fit measures	NFI	0.823
	CFI	0.908
Parsimony fit measures	AGFI	0.807

AGFI, adjusted goodness of fit index; CFI, comparative fit index; GFI, goodness of fit index; NFI, nonnormal fit index.
Source: Compiled by authors.

There are two important considerations which are used to test the statistical significance using Amos output. Firstly, the CR, which represents the parameter estimate divided by its standard error, based on a probability level of 0.05, and the CRs are to be > ±1.96 for statistical significance Anderson and Gerbing (1988) At the same time, nonsignificant parameters, with the exception of error variances, can be considered unimportant to the model and hence they have to be removed from the model (Byrne, 2010). Secondly, the standard residual covariance should be less than the threshold limit of 2.58 to conclude statistically significant covariance between two variables (Byrne, 2010). In such cases, these observations cannot be considered for further analysis. In the present model, standard residual covariance variables were within the threshold limit, that is, 2.58. Accordingly, model can be considered as a good fitting model by considering empirical reasoning as well as the appropriateness of the model. Overall reliability scale for deterrent factors was 0.904.

9.5 RESULTS

Based on CFA, study identified 7 constructs with 35 indicators scale, developed for identifying deterring factors affecting the adoption of m-tourism in India. Table 9.7 shows constructs identified and finalized with its corresponding variables and reliability coefficients.

TABLE 9.7 Variables After Confirmatory Factor Analysis.

Constructs	Name of the variables	No. of variables	Cronbach's alpha
PEI	Usefulness, motivation, productivity, performance expectation, comparative advantage, and accomplishments	6	0.870
EEI	Time, ease of use, learning to use, clarity, and understanding	5	0.849
TSI	Network externalities, trialability, complexity, and installation	4	0.853
FCI	Financial condition, knowledge level, compatibility, assistance, availability, and accessibility	6	0.879
CSI	Language, technical infrastructure, legal barriers, diversity, and exposure	5	0.838
PCI	Affordability, technology inhibition, novelty, self-efficacy, trust, and education	6	0.842
SII	Peer influence, advisory, and media influence	3	0.727

CSI, country-specific issues; EEI, perceived effort issues; FCI, facilitation issues; PCI, personal centric issues; PEI, performance expectation issues; SII, social influence issues; TSI, technology-specific issues.
Source: Compiled by authors.

9.6 DISCUSSION

Findings of the study reiterate the presence of various factors which deter the adoption of m-tourism services among smartphone or other mobile users among domestic tourists of India. By considering the huge market potential of domestic tourism, it is imperative to have a glance on their concerns/issues while adopting m-tourism services in the context of cashless economy or economic digitalization. The study encompasses various aspects of concerns of Indian society by indicating 7 constructs consisting of 35 indicators. These constructs are: PEI, EEI, FCI, CSI, PCI, TSI, and SII. Understanding these issues gain paramount importance while initiating m-tourism services among all stakeholders. The multifaceted nature of tourism industry and vividness of service profile adds complexity to the tourism services to a greater extent. This reiterates the observations of Mupfiga (2015) to address the imbalances in digitalization for an enhanced user-driven travel planning. Moreover, frequency of use of various applications is also found to be less in tourism-related transactions. This invites the attention of policy-makers, stakeholders, as well as industry associations to gain understanding about such complexity of service delivery and its consistent applications. This is in tune with Oskam and Boswilk (2016) observation, that the network hospitality and tourism services are growing and also creating a platform for cocreated market in the digitalization of travel business. Identified variable are to be studied further jointly or severally to address specific or collective issues of certain applications in tourism, for instance, m-application of leisure booking, purchase of souvenirs, familiarization of cultural events, etc. If these factors are taken care of, by the stakeholders or trade association in tourism, implementation of m-tourism among Indian consumers (domestic tourists) can be strengthened and thereby tourism services can contribute toward digitalization of Indian economy. As Standing et al. (2014) pointed, paucity of investigation is evident both in demand as well as supply side environment; providers and consumers must opt for adoption of best practices for a hassle-free digitalized environment.

9.7 CONCLUSION

The study shows that all identified deterring factors or technology adoption aversion factors are found to be significant in the Indian domestic tourism market, which contributes a substantial part of tourism revenue with significant impact on economy in terms of employment and income. In general, digitalization of tourism services through m-tourism applications helps to balance both demand and supply side of tourism. If the policy-makers, trade associations, and stakeholders are able to pay attention to the factors identified in the study, it may bring considerable changes in the demand and supply side of tourism by providing instant and easy services delivery for tourists followed by significant benefits to tourism service providers, government, and other intermediaries.

KEYWORDS

- **m-tourism**
- **exploratory sequential method**
- **deterring factors**
- **performance**
- **expectation factor**
- **perceived effort factor**
- **socially influencing factors**
- **facilitation factor**

REFERENCES

Anckar, B.; Walden, P. Self Booking of High- and Low Complexity Travel Products: Exploratory Findings. *Inf. Technol. Tour.* **2002,** *4* (3/4), 151–166.

Anderson, J.; Gerbing, W. Structural Equation Modelling in Practice: A Review and Recommended Two Stage Approach. *Psychol. Bul.* **1988,** *27* (1), 5–24.

Bharatwiki. Business Profile, (n.d). www.bharatwiki.com (accessed Feb 19, 2017).

Buhalis, D. *E-tourism: Information Technology for Strategic Tourism Management*; Pearson: Cambridge, 2003.

Buhalis, D.; O'Connor, P. Information Communication Technology Revolutionizing Tourism. *Tour. Recreat. Res.* **2005,** *30* (3), 7–16.

Byrne, B. M. *Structural Equation Modelling with AMOS: Basic Concepts, Applications, and Programming*, 2nd ed.; Routledge: New York, 2010.

Chen, Y. H.; Barnes, S. Initial Trust and Online Buyer Behaviour. *Ind. Manag. Data Syst.* **2007,** *107* (1), 21–36.

Churchill, G. A. A Paradigm for Developing Better Measures of Marketing Constructs. *J. Mark. Res.* **1979,** *16* (2), 64–73.

Creswell, J. W.; Plano Clark, V. L.; Gutmann, M. L.; Hanson, W. E. Advanced Mixed Methods Research Designs. In *Handbook of Mixed Methods in Social and Behavioral Research*; Tashakkori, A., Teddlie, C., Eds.; Sage: Thousand Oaks, CA, 2003; pp 209–240.

Ericsson, N.; Strandvik, P. Possible Determinants Affecting the Use of Mobile Tourism Services. In *E-business and Telecommunications*; Filipe, J., Obaidat, M., Eds.; Springer: Arcada University of Applied Sciences. Heidelberg, 2009; pp 61–73.

Eriksson, N. User Experience of Trip Arrangements: A Comparison of Mobile Device and Computer Users. *Int. J. E-Services Mobile Appl. (IJESMA)* **2012,** *4* (2), 55–69.

Eriksson, N. Drivers and Barriers of Mobile Travel and Tourism Service Adoption A Study of Individual Perceptions and Business Model Development in a Travel and Tourism Context. Dissertation, 2013. https://www.doria.fi/bitstream/handle/10024/92187/eriksson_niklas.pdf?sequence=2 (accessed Jan 22, 2017).

Flynn, L. R.; Pearcy, D. Four Subtle Sins in Scale Development: Some Suggestions for Strengthening the Current Paradigm. *Int. J. Mark. Res.* **2001,** *43* (4), 409–423.

Gavalas, D.; Kenteris, M. A Web-based Pervasive Recommendation System for Mobile Tourist Guides. *Pers. Ubiquit. Comput.* **2011,** *15*, 759–770

Gregori, A. Optimised Mobile Marketing as Part of Integrated Marketing Campaigns Building Quality Leads, 2009. http://mobilemarketingwinners.com/resources/Presentations/08%20Alexan der%20Gregori.pdf (accessed Feb 22, 2016).

Gupta, M.; Gupta, D. *Research Methodology*; PHI: New Delhi, 2013.

Hair Jr., J. F.; Anderson, R. E.; Tatham, R. L.; Black, W. C. *Multivariate Data Analysis*, 5th ed.; Prentice Hall: New Jersey, 1998.

Hu, J.; Liu, X. An Influence Study on Adopting Will of Tourism Mobile E-commerce by the Perceived Risk and Trust. *iBusiness* **2013,** *5* (1), 39–42.

Juman, D. Online Travel Bounces with Consumer Confidence, 2012. http://www.phocuswright.com/research_updates/online-travel-bounces-withconsumerconfidence (accessed Feb 22, 2017).

Kenteris, M.; Gavalas, D.; Economou, D. An Innovative Mobile Electronic Tourist Guide Application. *Pers. Ubiquit. Comput.* **2009,** *13* (2), 103–118.

Lee, J. K.; Mills, J. E. Exploring Tourist Satisfaction with Mobile Experience Technology. *Int. Manag. Rev.* **2010,** *6* (1), 91–93.

Lincoln, Y. S.; Guba, E. G. *Naturalistic Inquiry*; Sage: Beverly Hills, 1985.

Mapmyindia. Business Profile, (n.d). www.mapmyindia.com (accessed Feb 14, 2017).

Maxwell, J.; Loomis, D. Mixed Methods Design: An Alternative Approach. In *Handbook of Mixed Methods in Social and Behavioral Research*; Tashakkori, A., Teddlie, C. Eds.; Sage: Thousand Oaks, CA, 2002; pp 241–271.

Mupfiga, P. S. Adoption of ICT in the Tourism and Hospitality Sector in Zimbabwe. *Int. J. Eng. Sci.* **2015,** *4* (12), 72–78.

Nedelea, A.; Bălan, A. E-tourism and Tourism Services Consumer Protection. *Amfiteatru Econ.* **2010,** *12* (28), 492–503 (Online).

Online Travel Continues to Dominate India Ecommerce. www.emarketor.com/m/Article/Online-Travel- Continues-Dominate-India-Ecommerce/1009964 (accessed Feb 19, 2017).

Oorni, A.; Klein, S. Electronic Travel Markets: Elusive Effects on Consumer Behavior. In *Information and Communication Technologies in Tourism*; Frew, A. J., Hitz, M., O'Connor, P., Eds.; Springer Wien: New York, 2003; pp 29–38.

Oskam, J.; Boswijk, A. Airbnb: The Future of Networked Hospitality Business. *J. Tour. Futures* **2016,** *2* (1), 22–42. http://doi.org/10:1108/JTF-11-2015=0048 (accessed Feb 19, 2017).

Qirici, E.; Theodhori, O.; Elmazi, L. E-marketing and ICT-supported Tourist Destination Management: Implications for Tourism Industry in Global Recession. *Int. J. Manag. Cases* **2011,** *13* (3), 152–158.

Rosen, D. E.; Suprenant, C. Evaluating Relationships: Are Satisfaction and Quality Enough? *Int. J. Serv. Ind. Manage.* **1998,** 9(2), 103–125

Scott, A. C; Prayag, G.; Moital, M. Consumer Behavior in Tourism: Concepts, Influences and Opportunities. *Curr. Issues Tour.* **2014,** *17* (10), 872–909.

Skadberg, Y. X.; Skadberg, A. N.; Kimmel, J. R. Flow Experience and Its Impact on the Effectiveness of a Tourism Website. *Inf. Technol. Tour.* **2005,** *3* (4), 147–156.

SMB's in India are Gearing up to Use and Deploy Smart Technologies in Their Operations. http://www.invasystems.com/90-of-smbs-in-india-are-adopting-smart-technologies/ (accessed Feb 17, 2017).

Standing, C.; Tang-Taye, J.; Boyer, M. The Impact of Internet in Travel and Tourism: A Research Review 2001–2010. *J. Travel Tour. Mark.* **2014,** *31* (1), 82–113.

Trochim, W. M. K. Introduction to Validity. Social Research Methods, 2006. www.socialresearchmethods. net/kb/ introval.php (accessed Feb 12, 2016).

UNCTAD. *E-Commerce and Development, 2005*; Report by the UNCTAD Secretariat, UNCTAD/SDTE/ECB/2005/1, UNCTAD Science Technology and ICT Branch: Geneva, 2005.

UNWTO. *Exploring the Full Impact of Tourism for Policy Making. Extending the Use of Tourism Satellite Account Through Macroeconomic Analysis Tools,* 2011; UNWTO: Madrid, 2011.

UNWTO. Tourism Highlights, 2014. http://mkt.unwto.org/publication/unwto-tourism-highlights-2014-edition (accessed Feb 11, 2017).

UNWTO. Tourism Highlights, 2015. http://mkt.unwto.org/publication/unwto-tourism-highlights-2015-edition (accessed Feb 23, 2017).

Urban, G. L. *Digital Marketing Strategy: Text and Cases*; Pearson Prentice Hall: New Jersey, 2004.

Varshney, U. Issues, Requirements and Support for Location-intensive Mobile Commerce Applications. *Int. J. Mobile Commun.* **2003,** *1* (3), 247–263.

Venkatesh, V.; Morris, M. G.; Davis, G. B.; Davis, F. D. User Acceptance of Information Technology: Toward a Unified View. *MIS Q.* **2003,** *27* (3), 425–478.

Wang, H.; Liao, C.; Yang, L. What Affects Mobile Application Use? The Roles of Consumption Values. *Int. J. Mark. Stud.* **2013,** *5* (2), 11–22.

WTTC. *Travel and Tourism Economic Impact: India*, 2012 (Online). www.wttc.org/site_media/uploads/downloads/sub_saharan_ africa2012.pdf (accessed Feb 17, 2017).

Zomato (n.d). Business Profile. http://www.zomato.com (accessed Feb 11, 2017).

CHAPTER 10

ADOPTION OF M-COMMERCE AMONG WOMEN: AN EMPIRICAL STUDY IN INDIA

ANUSHRUTI VAGRANI*, DILPREET KAUR, and
P. VIGNESWARA ILAVARASAN

Department of Management Studies, Indian Institute of Technology Delhi, New Delhi, India

**Corresponding author. E-mail: smz148410@dms.iitd.ac.in*

ABSTRACT

Using a two-step methodology, qualitative interviews and quantitative surveys, the paper explores the mobile commerce adoption among women in India. M-commerce is revolutionizing various business sectors and creating an economic shift. This shift is visible in various business sectors from entertainment to banking to health and education; and it is affecting people across the spectrum of demographics. With increasing number of women using mobile phones, use of m-commerce among women is on rise. While the sales influenced by women are on rise, market is also coming up with women-specific segments and new ideas in various sectors including m-commerce. However, the understanding of use of m-commerce among women, particularly in developing countries is still underdeveloped. Using literature review and qualitative interviews, factors that motivate mobile commerce usage among women were identified and incorporated into the questionnaire. The findings reveal role played by the personal and technical motivators in adopting m-commerce usage. It is found that perceived ease of use and perceived purchasing power among women significantly influence the m-commerce adoption, along with technical factors such as internet connectivity and security.

10.1 INTRODUCTION

With technology gaining the status of one of the structural constituents of life, it has become crucial for stakeholders to develop a deeper understanding of acceptance of technologies among various different segments of users. Vast number of studies in the field of adoption of technology strengthens this point further (Boudreau, 2013; Maity and Dass, 2014). Gender-based segmentation in terms of adoption of technology captured focus of researchers a bit late; even later in case of developing countries (Morris et al., 2005; Slyke et al., 2010; Faqih and Jaradat, 2015). These days worldwide, the purchases influenced by women are on rise (Partners, 2014). Though gender-based studies in the field have given mixed results initially (Hasan, 2010); studies also indicate increased likelihood of women participating in e-commerce activities. Potential in this area calls for further understanding of the factors that affect the intention to use, and usage pattern of technologies such as e-commerce and m-commerce, in women. Earlier attempts in this direction indicate toward several factors such as social norms, ease of use, attitude, etc. defining gender as a differentiating construct that affects adoption of e-commerce activities (Topi et al., 2010; Hwang, 2010; Kim et al., 2011). However, most of the literature available discusses the difference in attitude of buying/selling/usage behavior of men and women on an online e-commerce platform. Literature regarding gender difference in terms of consumer behavior in m-commerce is inadequate so far.

Buying/selling of information, products, and services via computer networks is called e-commerce (Stafford and Gillenson, 2003). With the advancement in technology and availability of internet with an all-time access, the buying and selling of goods and services on the go are on rise. Users are embracing mobile internet increasingly. It is done with the use of "mobile" devices such as smartphones, tablets, and other handheld devices using applications and/or personal digital assistants. All economic transactions with these devices can broadly be termed as mobile commerce or m-commerce (Niranjanamurthy et al., 2013). Applications designed for m-commerce provide powerful, advanced, and innovative features such as delivery tracking, payment portals, customized and personalized services, one-to-one interactions, and other location- and context-based features (Boudreau, 2013). All these appealing features positively alter the perceived value of m-commerce, and that can lead to usage transfer from web- to mobile-based applications (Cao et al., 2014; Faqih and Jaradat, 2015). On these platforms, m-commerce can be identified as the next technology to e-commerce (Liang et al., 2007).

As the penetration of smart mobile phones is increasing, use of mobile internet and possible m-commerce are gaining momentum. Increasing number of mobile devices with internet access leads to its stronger influence on consumer shopping patterns as well (Ha and Im, 2014). M-commerce is revolutionizing various business sectors and creating an economic shift. This shift is visible in various sectors including telecom, banking, tourism, transport, health and education, entertainment, retail, and public sector. The shift is affecting people across the spectrum of demographics. Some of the earlier studies have focused on understanding the demographics and profiles of the users, and their usage of m-commerce. With this study, we aim to identify the factors that lead to m-commerce usage amongst women in India. The outcomes of the study are relevant with increasing number of women using mobile phones, sales influenced by women being on rise, and with market responding with women specific categories such as, lingerie, jewelry, maternal items, etc. They are expected to help the firms that are exploring women consumers who are using internet. It also adds to the existing literature of information systems, especially, adoption of m-commerce. It can help market players to plan better approaches for targeting women for usage of m-commerce.

10.2 REVIEW OF LITERATURE

10.2.1 M-COMMERCE

M-commerce is defined in many different ways in literature; the basic understanding from many studies suggests that m-commerce is "e-commerce with handheld or mobile devices"; it refers to the ability to do a monitory transaction anywhere through internet-enabled mobile devices (Clarke, 2008). Tiwari and Buse (2007) present m-commerce as a subset of e-commerce and e-business; they take into account the nonmonetary transactions as well. Other researches refuse to go with this basic notion and define m-commerce as a new and innovative technology which is time critical and unique in the way of interactions (Feng et al., 2006; Chong, 2013). Feng et al. (2006) use the following definition: "any transaction, involving the transfer of ownership or rights to use goods and services, which is initiated and/or completed by using mobile access to computer-mediated networks with the help of mobile devices."

From being a luxury to being a necessity, the evolution of m-commerce has been comparatively rapid as well as substantial; and it holds potential for further growth (Ngai and Gunasekaran, 2007). The field has grown along with continuous technological advancements in the field of wireless data transfer standards and mobile devices. Starting with the first-generation (1G) analogue phones in 1980s which were based on frequency division multiple access (FDMA), the generation of mobile technology went on to grow as 2G circuit-switched phones in 1991, 2.5G in 2001, 3G in 2004, 3.5G in 2008, and then 4G in 2012 (Chaffey, 2007). Fifth generation (5G) has been proposed to handle internet of things as well. Kourouthanassis and Giaglis (2012) discuss the history of m-commerce through three different eras and describe the journey in reference to close and open-endedness of the platforms. First era was that of m-portals which were closed in nature and could not survive in the market for long. Next, was the time of m-internet model where technologically advanced mobile devices were the means to access open internet, and e-commerce providers now aimed to design their websites so as to better fit mobile operation. Recent has been the era of m-applications. These platforms are comparatively less open in nature yet the choice of selection of app, using open internet, remains with the user and it becomes increasingly significant with the increasing number of mobile apps coming up in the market.

While generally accepted as a form of e-commerce (Kourouthanassis and Giaglis, 2012), m-commerce is different at fundamental levels. Some of the distinct features that associate with m-commerce include ubiquity, immediacy, convenience, instant connectivity, proactive functionality, personalization, localization, and simple authentication process (Liang et al., 2007; Tiwari and Buse, 2007). These features change the perceived value and in turn its usage (Cao et al., 2014; Faqih and Jaradat, 2015). M-commerce also has its own set of challenges including security, screen size (that can limit details), ease of use and connectivity among others (Clarke, 2008). All these leverages and challenges affect the behavioral patterns of the users and call for unique market strategies; as it is established that m-commerce with its features is different from e-commerce (Boudreau, 2013).

When comparing e-commerce and m-commerce, user's preference of using one over other platform is studied with two conflicting models "task-media fit" and "cognitive cost model" (Maity and Dass, 2014). The cognitive cost model defines that each individual has a finite cognitive capacity and their decisions are based upon the effort it takes to complete an action. More media availability will lead to lesser effort to complete an action, whereas, the task-media fit model describes media availability as a hindrance to task completion. It explains more media as a distraction for the completion of a task and suggests that consumer usually like to take up m-commerce when they want to perform simple tasks, and media richness is not preferred there. The perceived value that the user derives from the technology is altered with all these features based on their perception regarding convenience, personalization, risk, and security among others, which leads to usage transfer (Cao et al., 2014).

Existing research explores the growth, challenges, and acceptance of technology for m-commerce (Wen-Jang and Su-Fang, 2003; Liang et al., 2007; Clarke, 2008). Other studies explore adoption of m-commerce in different fields such as advertising (Merisavo et al., 2007; Boudreau, 2013), banking (Tiwari and Buse, 2007), etc. Going through these studies, it is observed that role of gender in acceptance of m-commerce is scarcely studied with the exception of few studies that we will discuss in the next section. Secondly, studies mostly focus on user's intention to use m-commerce, whereas the usage of m-commerce remains underexplored (Chong, 2013). With m-commerce characterized with its broad reach (Ngai and Gunasekaran, 2007), it becomes important to understand the usage part well. Research also identifies various barriers to the use of m-commerce, including perceived risk, security (of data), perceived cost, and illiteracy in developing countries (Coursaris and Hassanein, 2002; Rahman, 2013). Other factors causing barriers for shift from e-commerce to m-commerce include lack of rich and detailed information, hard to compare prices, lack of comfort in typing, and smaller screen.

10.2.2 TECHNOLOGY ADOPTION AND ROLE OF GENDER

Role of gender has been vastly studied in information system literature. Outside and inside of the domain of information systems, gender along with social influence, has great impact on behavior in various respects. It has been established that the decision-making process in men and women is different in terms of technology acceptance and use as well (Venkatesh and Morris, 2000; Hwang, 2010). The same has been studied in reference to mobile internet in few studies (e.g., Okazaki and Hirose, 2009; Okazaki and Mendez, 2013). These differences in behavior are understood to be the result of inbuilt value system and attitude rooted in the behavior since childhood (Hasan, 2010; Okazaki and Mendez, 2013). Riquelme and Rios (2010) find that female users highly relate ease of use to usability and social norms also play higher important role for them. Sangwan et al. (2009) establish differences in the online shopping motivations for female users. Contradictory studies also are there that show behavior is similar in male and female in terms of online shopping (Bigne et al., 2005; Faqih and Jaradat, 2015). In contrast, Hasan (2010) brings out that the cognitive attitude toward online shopping is lower in women and awareness is needed in terms of risks and security. However, in terms of mobile internet, Okazaki and Hirose (2009) establish that women are more willing to use and spend time to get detailed information. Okazaki and Mendez (2013) also finds that women perceive mobile devices as easy to use; though in case of online transactions, the study finds reluctance in women. These mixed and contrasting results in e-commerce and m-commerce practices suggest need of detailed context specific studies in case of m-commerce.

10.2.3 THEORIES OF TECHNOLOGY ADOPTION

In terms of technology adoption, literature contains several theoretical models addressing different determinants of technology acceptance (Davis, 1989; Venkatesh and Davis, 2000; Davis et al., 1992; Venkatesh et al., 2003). Later studies in the field tend to accept combined and context-relevant approach toward technology acceptance (Oliveira et al., 2014). The aim here is to understand the factors that lead to use of m-commerce in women, while having the understanding of unique sociocultural construct around the gender. Hwang (2010) on the basis of "Self-determination theory" (SDT) argues that differences in human behavior can be explained as self-determined, controlled, or motivated. "Self-determination theory" given by Deci and Ryan (1985) suggests that there are two different categories of motivation, namely, intrinsic motivation which gives inherent satisfaction and is enjoyable, and extrinsic motivation that exist because there is an outcome of that certain action (Ryan and Deci, 2000). Cognitive evaluation theory of Ryan and Deci (2000) further explains intrinsic motivators as variable under different social contexts with factors such as "self efficacy" along with "autonomy" as these can induce the "feeling of competence." "Perceived enjoyment" is discussed as an intrinsic motivator in the discussions of technology adoption, along with the extrinsic motivators "perceived ease of use" and "perceived usefulness" that are defined in detail in "Technology Acceptance Model" (TAM) (Davis et al., 1992; Davis, 1989; Teo and Noyes, 2011).

TAM has been the base of numerous studies aimed at understanding the motivation to use information technology. There are numerous empirical evidences in literature through several studies. It suggests that the willingness to use any new technology is majorly determined by the "perceived ease of use" and "perceived usefulness" of that technology. Here, perceived ease of use refers to "the degree to which a person believes that using a particular system would be free of effort," whereas perceived usefulness is "the degree to which a person believes that using a particular system would enhance his or her job performance" (Davis, 1989). Gender construct was not a part of TAM theory and it was later with studies like Venkatesh and Morris (2000) that implementation of TAM was studied in a gender-based context through incorporating psychological research. The study suggests that women are more driven by perceived ease of use and social norms.

The theoretical underpinning for TAM comes from the "Theory of Reasoned Action" (TRA) (Davis, 1989; Venkatesh and Davis, 2000). TRA has rather general application to human behavior and is used in several areas inside and out of the domain of technology research. In other studies, attempts have been made to explain technology adoption with TRA. TRA attempts to identify person's intention to perform/not perform a behavior based on their behavior intention. Behavior intention is affected by factors such as attitude toward the behavior and facilitating conditions. Attitude in turn reflects how a person responses to a situation; it is learned through experience in long term (Madden et al., 1992; Hasan, 2010; Mishra et al., 2014; Zhang et al., 2014). Social norms and resultant emotional and social experiences affect the intention of use of a technology. Unawareness regarding e-commerce results in lower cognitive attitude toward online shopping thus lower intention to use (Hasan, 2010).

In TAM2 (Venkatesh and Davis, 2000), which is a theoretical extension of TAM, Venketesh and Davis addressed other important constructs in line with TRA. "Subjective norms" explain how a person might not be positive about a behavior or its outcome, yet be motivated to perform it, if she perceives that people important to her think she should show a behavior. It also depends on the social image in that if all the members of a social circle are perceived to adopt a technology, it can motivate the person to adopt as well. Intention of use can also vary if the person perceives that as mandatory. These reflect the "social influence process." In order to gain influence and status in their work group, people also adopt technology that is perceived to improve job performance.

In an attempt toward a comprehensive theory Venkatesh et al. (2003) understand usage as a dependent variable and give "united theory of acceptance and use of technology" (UTAUT). This theory puts forward four constructs defining the behavioral intention and in turn behavior. Out of these four, three constructs are shown as gender relevant: "Performance expectancy" which defines how the person perceives the technology to help in job performance; "effort expectancy" which is the degree of ease related to the technology; and "social influence" which is the perception of individual how people important to her would think she should or should not use the technology.

10.3 SCENARIO IN INDIA

Telecom and internet penetration in India has been on rise (Government of India, 2017a; Government of India, 2017b). Decreasing mobile internet tariffs and availability of smartphone at prices as low as INR 5000 add to this growth. According to Telecom Regulatory Authority of India (TRAI), as on December 31, 2016, there were 1127.37 million wireless connections in India with a monthly growth of 2.53%. According to a recent news article in January 2017, the number of smartphone users in India has gone above 300 million (Firstpost, 2017). Due to mobile internet penetration, the primary shopping instincts of the consumer has been modified as instant information access is available to the consumer. Kalia (2016) describes the present m-commerce scenario and digital India as the "third wave" which is causing shift in the way businesses operate, and increase in the entry of foreign players and funding. E-wallets and m-commerce are the latest addition to the spree of digital India. M-commerce also provides opportunities for deeper rural penetration as it has higher penetration than internet.

The vision of "Digital India program" is to transform India into a digitally empowered society and knowledge economy (Government of India, 2017c). The "Digital India" program aims to bring inclusion and move toward improved, transparent, and secure services. The efforts include banking, education, healthcare, e-governance, etc., among others. Mobile devices provide easy and cost-effective penetration in different fields such as mobile banking (Deb et al., 2017). Application ranges from entertainment, shopping, travel to even traffic control. Use of mobile wallets is on exponential rise since the demonetization in country under which larger denomination notes (INR 500 and INR 1000) were withdrawn from the market, forcing the online transactions, payment, and business to come into focus.

While we look at all the shining aspects of growth in mobile, internet, and m-commerce, the skewed gender ratio of the country cannot be ignored which reflects in these sectors as well. Moreover, with the sociocultural construct of the country, this gap widens further. Studies show huge gender gap in terms of internet usage in India. It is said that from the pool of internet users in India only 29% are women. This gap is expected to narrow down in the years to come; however, in order to do so, women need to be motivated toward use of internet and technology. Internet can be an empowering tool with all the knowledge it can provide; yet with the huge gap existing as of now, it creates a threat of gender digital divide. Therefore with this study, we aim to identify the factors that lead to mobile commerce usage amongst women and for what purposes do the women use the mobile commerce in India.

Based on the literature, we will discuss two different types of motivators. One, "personal motivators" are those that motivate the adoption of technology, for example, perceived purchasing power, ease of use, enjoyment, and influence of social norms. Second, "technical motivators" are those that motivate switch from other traditional means to m-commerce and motivate its repeated usage, for example, connectivity, security, personalized offers and discounts, and different application features.

10.4 RESEARCH METHOD

From the literature, we see that the understanding of the adoption of e-commerce and m-commerce has been taken up by many studies which led to the identification of factors for intended usage of these technologies. A number of studies have also been conducted to identify the difference in consumer statistics and behavior of men and women while using these technologies. The behavior of men and women toward technology adoption is different which indicates toward the need to consider gender as an important construct while studying m-commerce as well. The number of studies conducted to understand usage of m-commerce is relatively low; very few of them consider gender construct, and even fewer of those are conducted in context to developing countries like India. Thus, the study that we conduct will lead to greater understanding of what motivates women to use m-commerce and what are the inhibitors for the same in India. This need to understand the factors leading to usage of m-commerce by women in India is backed up with the fact that the contribution of women in m-commerce as compared to men is very less. This study will fill the gap in literature with the understanding of the women behavior and m-commerce; it can be used by the m-commerce entrepreneurs toward a better approach for better reach to female consumers as well.

This study applies a two-step methodological approach. First step is a qualitative survey comprising semistructured interviews. Responses from the interviews lead to identification of factors influencing usage and inhibitors for the usage of m-commerce by women. This helps in synthesizing research questions and formulating hypotheses. In the second step, a detailed survey is conducted with a research instrument. The responses from which are used to validate the model constructed in the first step.

10.5 QUALITATIVE SURVEY AND MODEL DEVELOPMENT

Adequately backed up with the literature review; a semistructured interview was conducted with six women. With these interviews, we tried to understand the motivators on a broader level. The respondents belonged to different backgrounds as two homemakers, two students, and two working women. Out of the six respondents, four respondents had an exposure to m-commerce or e-commerce. Out of the two respondents who did not have exposure to m-commerce, one was a homemaker and other was a student. Both of them stated that the reason to not use m-commerce was no access to smartphone and use of basic mobile phones. The lack of access in case of the student was due to restriction of parents over their children to use mobile phones; in case of homemaker lady, the reason was found to be the individual's inability to afford a smartphone and an ongoing internet access on the smartphone. This highlights the importance of social norms and the perceived purchasing power of an individual.

Analysis of all the responses from the interviews gave us motivators and inhibitors to m-commerce. Factors leading to the use of mobile commerce included instant and uninterrupted connectivity, comparatively more discount available on mobile applications, security of payments, easy to use application, uncluttered application, comparing option, delivery tracking and personalized offers in terms of discounts availability by apps for being a repeated customer, etc., whereas factors that inhibit the usage of m-commerce were security issues of online payments and excessive advertisements on applications. Preference for e-commerce amongst the respondents was due to availability of detailed information and a customizable website, that is, availability of filter. We found that the respondents' buying behavior was influenced by various social norms. Going into details, it is seen that while media only makes them aware of the application, family and peers have a greater influence on the buying behavior. Respondent said they feel more comfortable while using m-commerce if their friends are also using the same applications. Family influence is reflected in their control over expanses.

Guided by literature, we analyze all the responses and classify all the above found factors into "personal motivators," and "technical motivators."

10.5.1 PERSONAL MOTIVATORS

Personal motivators include intrinsic and extrinsic motivators that motivate the adoption of technology. These include perceived ease of use, perceived enjoyment, individual's purchasing power, and social norms that affect their behavior. It was found that out of six, four women's purchasing behavior was influenced by the perception of their family and friends.

10.5.2 TECHNICAL MOTIVATORS

Technical motivators are focused around the features of the mobile application that attract female users for repeated usage. They motivate switch from other traditional means to use of m-commerce. These include high-speed internet connectivity, security, discounts, and freebies, accurate and complete product information, app customization, delivery tracking, and personalized offers.

10.5.3 USAGE ACTIVITIES

Usage activities are the different ways that mobile applications are used by women. The major categories that came out from the interviews were: (1) buying and selling of stuff, (2) ordering food and taxis, (3) newspaper subscription services, (4) reading books, (5) m-banking applications, (6) virtual games, (7) entertainment, and (8) mobile wallets.

Based on the above identified indicators, we suggest the following research model that explains the scope of the study. Based on this model, we will develop the hypotheses, and the model would be validated in next step of the research (Fig. 10.1).

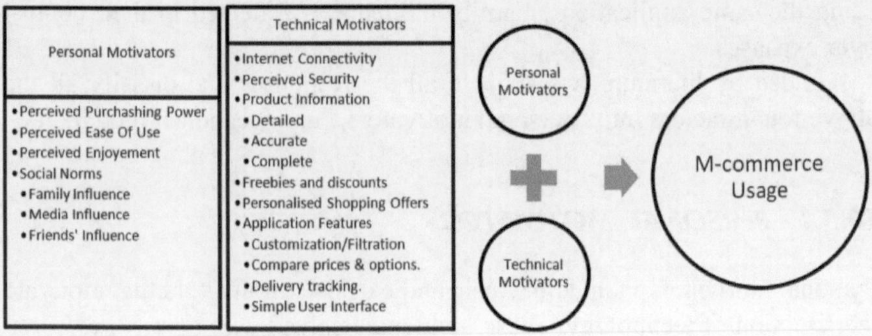

FIGURE 10.1 Research model.
Source: Compiled by authors.

10.5.4 RESEARCH HYPOTHESIS

Based upon the research model, we develop the following hypotheses that we would test in order to identify the motivational factors among women to use m-commerce.

The qualitative survey with the women led to the identification of "purchasing power" of an individual as a motivation to use m-commerce. An individuals' perception of her spending power would decide how flexible she can be in her spending, giving her a sense of autonomy. The "cognitive evaluation theory" (Ryan and Deci, 2000) discusses the "feeling of competence" as an intrinsic motivator that is induced by "self-efficacy" and "autonomy."

H1: Purchasing Power Perception of Women Positively Affects the Usage of M-Commerce

The initial survey with the women indicated toward a positive relation in the usage of m-commerce with the simplicity and ease of use of a mobile application. "Perceived ease of use" is one important components of TAM (Davis, 1989). Further gender-based studies indicate that women are more driven by perceived ease of use (Venkatesh and Davis, 2000; Riquelme and Rios, 2010). Hence, we hypothesize that ease of use as perceived by women has a positive effect on the motivation to use mobile applications for m-commerce.

H2: Ease of Use as Perceived by Women Positively Affects the Usage of M-Commerce

Purchases via mobile phones can be perceived as exciting as the fact that the access to internet is instant and can be used at anytime and anywhere. The qualitative survey shows use of m-commerce in terms of shopping, entertainment, books, games, etc.; all fulfilling experiences related to the enjoyment show positive influence on buying behavior of individuals. Previous studies discuss perceived enjoyment as an intrinsic motivator for technology adoption (Davis et al., 1992; Teo and Noyes, 2011).

H3: Perceived Enjoyment Positively Affects the Usage of M-Commerce

The purchasing pattern of a consumer is seen influenced by these social norms. The qualitative survey outlined the effect of these norms. In literature, TAM2 discusses social influence process in terms of subjective norms and social image (Venkatesh and Davis, 2000). Social image or social inclusion plays the driving role here. A number of studies suggest that women are more affected by social norms in technology adoption (Venkatesh et al., 2000; Riquelme and Rios, 2010; Hasan, 2010).

H4: Social Norms Positively Affect the Usage of M-Commerce

Among technical motivators in the model identified one of the major factors is the availability of uninterrupted and high-speed connectivity to internet. The factor of mobility, convenience, and anytime-anywhere functionality are all factors that make m-commerce exciting, and all of these are dependent on the availability of high-speed uninterrupted connection. Several of the studies discuss this as important component for m-commerce (Liang et al., 2007; Tiwari and Buse, 2007). It plays a great role in motivating the use of mobile applications for m-commerce.

H5: High-speed Internet Connectivity Positively Affects the Usage of M-Commerce

Statistics of Indian population in the adoption of mobile commerce enlists security of payments as a barrier to acceptance of mobile commerce. This fact was further realized during the interviews with the women on mobile adoption. Security of online payments and securing the details of credit and debit cards is of paramount importance. Security aspect is often studied as a challenge in literature concerning m-commerce (Clarke, 2008). Mobile applications provide simpler authentication process leading to lower chances of error and failure during transactions. We propose to establish that higher the sense of security, higher will be the behavioral intention to use m-commerce.

H6: Higher Perceived Security Positively Affects the Usage of M-Commerce

Security aspect is also connected to the availability of detailed and accurate information of the product. Providing detailed and accurate information on mobile applications can be challenging due to small screen size (Clarke, 2008). We find that consumer associate detailed and accurate information with lower risk and tend to opt for the mode that provides better information.

H7: Availability of Detailed Product Information Positively Affects the Usage of M-Commerce

H8: Availability of Accurate Product Information Positively Affects the Usage of M-Commerce

H9: Availability of Complete Product Information Positively Affects the Usage of M-Commerce

The strategy being followed by organizations to attract more and more customers to their mobile applications is to provide more freebies and discounts as compared to the e-commerce. The more discounts available will attract more users to the application. The rationale behind this strategy is that people are attracted toward the extra items available at the same or lesser price. An individual brain considers the extra item as a gift and feels in an advantageous position that leads to motivation or increased intention to use mobile commerce.

H10: Availability of More Freebies and Discounts Positively Affects the Usage of M-Commerce

Another factor that leads to higher intention to use mobile commerce is the availability of personalized shopping offers available to the consumers. The women surveyed in the qualitative survey conducted have mentioned that being a repeated customer, the expectation to get higher discounts and personalized offers would lead to higher customer loyalty and greater motivation to use m-commerce.

H11: Availability of Personalized Shopping Offers Positively Affects the Usage of M-Commerce

The mobile is associated with the simplicity or stripping down of the features available in the web 2.0 version of the same. This sometimes leads to availability of less functionality in the mobile applications which directly and indirectly affects the motivation to use mobile applications for m-commerce. The qualitative survey with women identified that the simpler user interface with features availability such as filtration, delivery tracking, comparing option, etc., would increase the motivation of usage. Hence in a nut shell, we can say that the features affect the buying pattern in women. Literature supports that availability of proactive functionality, personalization, localization, etc., build the perceived value for m-commerce.

H12: Availability of Features and Functionality Positively Affects the Usage of M-Commerce

10.6 QUANTITATIVE SURVEY AND DATA VALIDATION

10.6.1 DATA COLLECTION

Based on the qualitative survey, a research instrument was designed that was backed up by the literature as well. This questionnaire was majorly divided in four sections. First section collected demographic details and details about mobile and internet access. Second section asked details about usage activities in women. Third section tried to capture the behavioral intention to m-commerce use and related factors for the women. Last section captures the features that women find are important while using m-commerce. Apart from the demographic details section, rest three sections measure respondent's views on Likert scale. Using the questionnaire, data was collected from women on Facebook using convenient sampling. The questionnaire was shared over Facebook with around 200 women and we received responses from 79 women[1] majorly from the northern part of India. Participation in the study was voluntary. Participants were assured that the survey is anonymous and individual responses could not be identified. The survey was floated to women with educational level starting from primary school to postgraduation and occupational status of student, salaried individual, homemaker, and self-employed.

10.6.2 DESCRIPTIVE STATISTICS

All the respondents from the sample have a smartphone and internet connectivity. It was found that the major respondents were either students or earning women. The average age of participants is 24.59 years with a standard deviation of 1.53 years. Maximum respondents were the age of 24 and it was an equal mix of graduates and postgraduates, with salaried individuals having the greater portion of the pie as compared to students, that is, salaried individuals formed 65.8% of the respondents (Table 10.1). It was noted that none of the respondents were homemakers or self-employed individuals although the survey was shared with a fair mix of women from the given occupation status. Also, there were no respondents with an educational level at primary or secondary school. This could be an indication of the fact that the women with higher education level are more aware and able to shop online via use of handheld devices.

[1]Though an attempt was made to have a larger representative sample, we were able to get only 79 responses due to time constraints. The small sample size may be treated as a limitation of the study.

TABLE 10.1 Descriptive Statistics of Sample.

	Frequency	Percent
Age		
21	2	2.5
22	2	2.5
23	10	12.7
24	33	41.8
25	12	15.2
26	9	11.4
27	7	8.9
28	4	5.1
Education		
Graduate	40	50.6
Postgraduate	39	49.4
Occupation		
Salaried	52	65.8
Student	27	34.2

Source: Compiled by authors.

10.6.3 DATA VALIDATION

In the questionnaire, the data was collected on Likert type responses, Always to Never, Very High to Very Low and Strongly Agree to Strongly Disagree using a Likert scale on 5 points. SPSS 19 was used to conduct the statistics tests on the data collected to test the hypothesis. Factor analysis was carried out including Kaiser–Meyer–Olkin (KMO) and Bartlett's test, reliability correlation with Cronbach's alpha test, and ANOVA test using linear regression test. During validation, usage of m-commerce (dependent variable) was considered as a function of all the available factors (independent variables).

Based upon the value of Cronbach's alpha test as indicated in Table 10.2, the reliability of the overall scale used is above the accepted levels of Cronbach's alpha, 0.94.

TABLE 10.2 Reliability Statistics.

Cronbach's alpha	Cronbach's alpha based on standardized items	No. of items
0.940	0.923	14

Source: Compiled by authors.

To check whether the sample taken from population was adequate for data analysis, KMO and Bartlett's test were conducted. Kaiser–Meyer–Olkin measure (KMO test) defines whether the sample taken is adequate enough (Field, 2009; Rathore and Ilavarasan, 2014). A value between 0 and 1 is expected, whereas value higher than 0.8 is considered good. Thus, a value of 0.846 indicated that the sample collected with the survey is adequate for data analysis (Table 10.3). The Bartlett's test indicates whether the data collected is homogenous in terms of the variance of the sample. With the Bartlett's test, a value less than 0.05 are expected for the sample to have homogenous variance. Since the value of Bartlett's test is 0.00, it is inferred that the factor analysis is good for the data collected (Table 10.3).

TABLE 10.3 KMO and Bartlet's Test.

Kaiser-Mayer-Olkin measure of sampling adequacy.		0.846
Bartlett's test of sphericity	Approx. Chi-square	1784.456
	Df	91
	Significance	0.000

Source: Compiled by authors.

The following table of model summary explains two values, that is, the value of R square and Durbin–Watson statistics. The value of R square or adjusted R square gives us the percentage of variance by the predictors (independent variables). The model predicted 63 (R square) and 57(adjusted R square) percent which is considered to be good. The Durbin–Watson statistic indicates the independent error assumption (Field, 2009). The value for Durbin–Watson statistics should lie between 0 and 4. A value closer to 2 is better. This model has a data value of 2.527 for Durbin–Watson statistic as shown below (Table 10.4).

TABLE 10.4 Model Summary.[a]

Model	R	R square	Adjusted R square	Std. error of the estimate	Durbin–Watson
1	0.797[b]	0.636	0.570	0.50,804	2.527

[a]Dependent variable: Usage.
[b]Predictors: (Constant), TM Complete info, SN (Social Norms), PM Personalized Offer, PP (Purchasing Power), PM Enjoyability, TM Detailed Info, TM Discounts, PMEase, TM High Speed, Func, TM Security, TM Accurate Info (TM, Technical Motivators; PM, Personal Motivators).
Source: Compiled by authors.

10.6.4 DATA ANALYSIS AND RESULTS

Regression analysis was used to test if the indicators significantly predict participants' usage of m-commerce. Usage is identified as the independent variable and the dependent variables for all the hypotheses are listed below given in the list along with corresponding beta value and *t*-test from regression analysis. All the results given are calculated at 95% confidence interval (Table 10.5).

TABLE 10.5 Significance Value for Regression Test.

Hypothesis	Independent variable	B	Significance value	Status
H1	Perceived purchasing power	0.60	0.000	Accepted
H2	Ease of use perception	0.46	0.000	Accepted
H3	Perceived enjoyment	0.21	0.056	Rejected
H4	Social norms	−0.20	0.076	Rejected
H5	High-speed internet connectivity	0.66	0.000	Accepted
H6	Security	0.61	0.000	Accepted
H7	Detailed information	0.46	0.000	Accepted
H8	Accurate information	0.48	0.000	Accepted
H9	Complete information	0.42	0.000	Accepted
H10	Discounts	0.53	0.000	Accepted
H11	Personalized offers	0.34	0.020	Accepted
H12	Features	0.45	0.000	Accepted

Source: Compiled by authors.

For a hypothesis to be accepted, the significance value should be less than 0.05 then only the corresponding factor is considered a significant contribution to the model. Table 10.5 gives the summary of the regression analysis done for the stated hypotheses. From the table, it can be inferred that the usage motivation of m-commerce is affected by the perceived purchasing power ($\beta = 0.60$, $p < 0.000$), ease of use ($\beta = 0.46$, $p < .000$), high-speed internet connectivity ($\beta = 0.66$, $p < 0.000$), and security of payments ($\beta = 0.61$, $p < 0.000$), detailed information ($\beta = 0.46$, $p < 0.000$), accurate information ($\beta = 0.48$, $p < 0.000$), complete information ($\beta = 0.42$, $p < 0.000$), personalized offers ($\beta = 0.34$, $p < 0.000$) and discounts ($\beta = 0.53$, $p < 0.000$), and features availability ($\beta = 0.45$, $p < 0.000$).

The data analysis resulted in the rejection of perceived enjoyment ($\beta = 0.21$, $p < 0.056$) and social norms ($\beta = -0.20$, $p < 0.076$) as the motivators for the usage of m-commerce activities among women. This analysis negated the results found by the research done on the effects of gender on e-commerce adoption which stated that women are affected by the social influences in the adoption of online shopping via e-commerce (Hwang, 2010; Riquelme and Rios, 2010).

10.7 FINDINGS AND DISCUSSION

The study puts forward the understanding of the adoption on m-commerce in women in India in terms of "personal motivators" and "technical motivators." "Personal motivators" are those that motivate the adoption of technology, whereas "technical motivators" motivate switch from other traditional means to m-commerce and motivate its repeated usage. Together personal and technical motivators lead to usage of m-commerce. While testing the model, it was found that, whereas all the factors considered as technical motivators were found to be significant, not all the personal motivators were that effective in motivating the consumer for m-commerce. Perceived purchasing power of individual was a new factor found as a motivator to use m-commerce during the qualitative interview with the women in the initial stage of the study. This factor contributing significantly to the model was accepted as a motivator. This finding is supported by the "cognitive evaluation theory" (Ryan and Deci, 2000). The theory suggests the "feeling of competence" as an intrinsic motivator that is induced by "self-efficacy" and "autonomy." Higher purchasing power of women gives her autonomy; she can afford a smartphone and uninterrupted internet connection. It thus acts as a motivator for m-commerce adoption.

The data analysis revealed that along with perceived purchasing power of an individual; technical motivators such as discounts and personalized shopping offers have a significant positive influence on the intention to adopt m-commerce. This suggests that women are more inclined to shop using the mobile devices if they have enough money to spend, and at the same time, they like to spend less amount on the same item with the help of discounts and personalized shopping offers. The personalized shopping offers bring in the element of intimacy with the business. Discounts would lead to attracting new consumers while personalized shopping offers would result in the customer loyalty to the business. The theory of self-determination (Deci and Ryan, 1985) is realized by the study in terms of decision by the women to use m-commerce when discounts and personalized offers are provided to them.

The output from detailed survey found that the effect of perceived enjoyment and societal norms over the use of m-commerce were not significant. These results were in contrast with the qualitative survey where students and homemakers from the sample had identified with the impact of various social norms. The result is also in contrast to the findings of earlier studies on the online shopping behavior of males and females (Hasan, 2010; Teo and Noyes, 2011). While Hasan (2010) in his study of adoption of online shopping in women found that the women are more attracted toward acceptance of new technology if the social norms are in favor of the technology; our study found those not being so effective with regard to m-commerce usage. Similarly, "perceived enjoyment" discussed as an intrinsic motivator in the discussions of technology adoption (Teo and Noyes, 2011) was found less effective, rather ease of use came out as a significant factor. As the sample for our study consisted majorly of salaried self-dependent women, we perceive that independent decision based on ease of use play significant role with m-commerce usage, rather than social norms. As "cognitive evolution theory" (Ryan and Deci, 2000) suggests that the role of intrinsic motivators stands variable under various social constructs, role of personal enjoyment can vary across various segments of users. These results also indicate how decision-making process for adoption of technology and its usage can be different. Also the decision-making process in women varies with self-efficacy (Hackett and Betz, 1981).

The study also intended to determine the effects of technical motivators on the usage of mobile for the online shopping. The technical barriers of m-commerce identified by qualitative interviews and secondary research were security concerns, lack of information availability, and lack of features availability in the applications. The study found that the high-speed internet connectivity and high security of payment is required for the user to accept the m-commerce usage. While other factors identified to motivate the usage of m-commerce were information availability and the application to have advanced features such as delivery tracking, comparing option, and customization of app along with simpler user interface. This indicates that the consumer is more inclined toward getting the web 2.0 features with the portability of a handheld device. Therefore, the businesses should focus on providing the features available on website version on the application version with a simple user interface.

The study regarding women's motivation to use m-commerce puts forward important findings particularly the role of "perceived purchasing power" and technical motivators. The findings from the study are found aligned with "Theory of Reasoned Actions," TRA (Madden et al., 1992); "Self Determination Theory," SDT (Deci and Ryan, 1985); and "Technological Acceptance Model," TAM (Davis, 1989), while social inclusion theory is not supported by the study findings. With the understanding of technical motivators, the study would also be useful for m-commerce market players as it can help them plan better approaches for targeting women for usage of m-commerce. The study suggests that in terms of designing of mobile application, developing less complicated and easy to use user interface that do not compromise on providing information can help tap this market segment effectively. Technological advancements increasingly support multiple features in mobile devices without increasing the complexity. In terms of marketing strategy, spending less to buy more is preferred by women thus discounts along with personalized offers can be an effective market strategy. These personalized offers and discounts can be communicated through various mediums including social media and television.

10.8 LIMITATIONS AND FUTURE SCOPE

The current study has its own limitations. The choice of survey sampling via convenience sampling resulted in the educational background of the participants to be confined to be only graduation and postgraduation. Moreover, the occupational background came out to be only students and salaried. Although, the participants targeted were of fair mix, the voluntary participation led to the responses being skewed in favor of students and salaried individuals with an educational background of graduation and post graduation. In light of these limitations, we suggest future research with a more varied sample to understand further interplay between various social norms and self-efficacy in women under various social constructs.

KEYWORDS

- **m-commerce**
- **technology adoption**
- **usage**
- **women**
- **motivators**

REFERENCES

Bigne, E.; Ruiz, C.; Sanz, S. The Impact of Internet User Shopping Patterns and Demographics on Consumer Mobile Buying Behaviour. *J. Electron. Commer. Res.* **2005,** *6* (3), 193.

Boudreau, K. E. Mobile Advertising and Its Acceptance by American Consumers. *Management*, Paper 2, 2013 (Browser Download This Paper) http://docs.rwu.edu/management_theses/2 (accessed Mar, 2017).

Cao, Y.; Lu, Y.; Gupta, S.; Yang, S. The Effects of Differences Between E-commerce and M-commerce on the Consumers' Usage Transfer from Online to Mobile Channel. *Int. J. Mobile Commun.* **2014,** *13* (1), 51–70.

Chaffey, D. *E-business and E-commerce Management: Strategy, Implementation and Practice*; Pearson Education: England, 2007.

Chong, A. Y. L. Mobile Commerce Usage Activities: The Roles of Demographic and Motivation Variables. *Technol. Forecast. Soc. Change* **2013,** *80* (7), 1350–1359.

Clarke III, I. Emerging Value Propositions for M-commerce. *J. Bus. Strateg.* **2008,** *25* (2), 41.

Coursaris, C.; Hassanein, K. Understanding M-commerce: A Consumer-centric Model. *Q. J. Electron. Commer.* **2002,** *3*, 247–272.

Davis, F. D. Perceived Usefulness, Perceived Ease of Use, and User Acceptance of Information Technology. *MIS Q.* **1989,** *13*, 319–340.

Davis, F. D.; Bagozzi, R. P.; Warshaw, P. R. Extrinsic and Intrinsic Motivation to Use Computers in the Workplace 1. *J. Appl. Soc. Psychol.* **1992,** *22* (14), 1111–1132.

Deb, M.; Deb, M.; Agrawal, A.; Agrawal, A. Factors Impacting the Adoption of M-banking: Understanding Brand India's Potential for Financial Inclusion. *J. Asia Bus. Stud.* **2017,** *11* (1), 22–40.

Deci, E. L.; Ryan, R. M. The General Causality Orientations Scale: Self-determination in Personality. *J. Res. Personal.* 1985, *19* (2), 109–134.

Faqih, K. M.; Jaradat, M. I. R. M. Assessing the Moderating Effect of Gender Differences and Individualism–Collectivism at Individual-level on the Adoption of Mobile Commerce Technology: TAM3 Perspective. *J. Retail. Consum. Servic.* **2015,** *22*, 37–52.

Feng, H.; Hoegler, T.; Stucky, W. In *Exploring the Critical Success Factors for Mobile Commerce*, International Conference on Mobile Business (ICMB'06), Copenhagen, June 26–27, 2006; IEEE, 2006; pp 40–40.

Field, A. *Discovering Statistics Using SPSS*; Sage Publications: London, 2009.

Firstpost Article. Title: Number of Smartphone Users Crosses 300 Million in India as Shipments Grew 18 Percent. http://tech.firstpost.com/news-analysis/number-of-smartphone-users-crosses-300-million-in-india-as-shipments-grew-18-percent-359075.html (Jan 25, 2017, 09:56).

Government of India. Ministry of Electronics & Information Technology, 2017a. http://www.digitalindia.gov.in/content/vision-and-vision-areas (accessed March 2017).

Government of India. Telecom Regulatory Authority of India, 2017b. http://www.trai.gov.in/sites/default/files/PR_No.9of2017.pdf (accessed March 2017).

Government of India. Telecom Regulatory Authority of India, 2017c. http://www.trai.gov.in/sites/default/files/Press_Release_11_17_Feb_2017_Eng_0.pdf (accessed March 2017).

Ha, Y.; Im, H. Determinants of Mobile Coupon Service Adoption: Assessment of Gender Difference. *Int. J. Retail Distrib. Manag.* **2014,** *42* (5), 441–459.

Hackett, G.; Betz, N. E. A Self-efficacy Approach to the Career Development of Women. *J. Vocat. Behav.* **1981,** *18* (3), 326–339.

Hasan, B. Exploring Gender Differences in Online Shopping Attitude. *Comput. Hum. Behav.* **2010,** *26* (4), 597–601.

Hwang, Y. The Moderating Effects of Gender on E-commerce Systems Adoption Factors: An Empirical Investigation. *Comput. Hum. Behav.* **2010,** *26* (6), 1753–1760.

Kalia, P. Tsunamic E-Commerce in India: The Third Wave. *The Global Analyst*, July 28, 2016; Vol. 5, pp 47–49. (Browser Download This Paper.)

Kim, E. E. K.; Mattila, A. S.; Baloglu, S. Effects of Gender and Expertise on Consumers' Motivation to Read Online Hotel Reviews. *Cornell Hosp. Q.* **2011,** *52* (4), 399–406.

Kourouthanassis, P. E.; Giaglis, G. M. Introduction to the Special Issue Mobile Commerce: The Past, Present, and Future of Mobile Commerce Research. *Int. J. Electron. Commer.* **2012,** *16* (4), 5–18.

Liang, T. P.; Huang, C. W.; Yeh, Y. H.; Lin, B. Adoption of Mobile Technology in Business: A Fit-viability Model. *Ind. Manag. Data Syst.* **2007,** *107* (8), 1154–1169.

Madden, T. J.; Ellen, P. S.; Ajzen, I. A Comparison of the Theory of Planned Behavior and the Theory of Reasoned Action. *Pers. Soc. Psychol. Bull.* **1992**, *18*(1), 3–9.

Maity, M.; Dass, M. Consumer Decision-making Across Modern and Traditional Channels: E-commerce, M-commerce, In-store. *Decis. Support Syst.* **2014,** *61*, 34–46.

Merisavo, M.; Kajalo, S.; Karjaluoto, H.; Virtanen, V.; Salmenkivi, S.; Raulas, M.; Leppäniemi, M. An Empirical Study of the Drivers of Consumer Acceptance of Mobile Advertising. *J. Interact. Advert.* **2007,** *7* (2), 41–50.

Mishra, D.; Akman, I.; Mishra, A. Theory of Reasoned Action Application for Green Information Technology Acceptance. *Comput. Hum. Behav.* **2014,** *36*, 29–40.

Morris, M. G.; Venkatesh, V.; Ackerman, P. L. Gender and Age Differences in Employee Decisions about New Technology: An Extension to the Theory of Planned Behavior. *IEEE Trans. Eng. Manag.* **2005,** *52* (1), 69–84.

Ngai, E. W.; Gunasekaran, A. A Review for Mobile Commerce Research and Applications. *Decis. Support Syst.* **2007,** *43* (1), 3–15.

Niranjanamurthy, M.; Kavyashree, N.; Jagannath, S.; Chahar, D. Analysis of E-commerce and M-commerce: Advantages, Limitations and Security Issues. *Int. J. Adv. Res. Comput. Commun. Eng.* **2013,** *2* (6), 2360–2370.

Okazaki, S.; Hirose, M. Does Gender Affect Media Choice in Travel Information Search? On the Use of Mobile Internet. *Tour. Manag.* **2009,** *30* (6), 794–804.

Okazaki, S.; Mendez, F. Exploring Convenience in Mobile Commerce: Moderating Effects of Gender. *Comput. Hum. Behav.* **2013,** *29* (3), 1234–1242.

Oliveira, T.; Faria, M.; Thomas, M. A.; Popovič, A. Extending the Understanding of Mobile Banking Adoption: When UTAUT Meets TTF and ITM. *Int. J. Inf. Manag.* **2014,** *34* (5), 689–703.

Partners, A. *India E-commerce.* Accel Partners: New Delhi, 2014.

Rahman, M. M. Barriers to M-commerce Adoption in Developing Countries—A Qualitative Study Among the Stakeholders of Bangladesh. *Int. Technol. Manag. Rev.* **2013,** *3*(2), 80-91.

Rathore, A. K.; Ilavarasan, P. V. In *Mobile Adoption in Collaborating Supply Chains: A Study of Indian Auto SMEs*, Proceedings of the 2014 International Conference on Information and Communication Technology for Competitive Strategies, Udaipur; Nov 14–16, 2014; ACM: New York, 2014; p 56.

Riquelme, H. E.; Rios, R. E. The Moderating Effect of Gender in the Adoption of Mobile Banking. *Int. J. Bank Market.* **2010,** *28* (5), 328–341.

Ryan, R. M.; Deci, E. L. Intrinsic and Extrinsic Motivations: Classic Definitions and New Directions. *Contemp. Educ. Psychol.* **2000,** *25* (1), 54–67.

Sangwan, S.; Siguaw, J. A.; Guan, C. A Comparative Study of Motivational Differences for Online Shopping. *ACM SIGMIS Database* **2009,** *40* (4), 28–42.

Slyke, C. V.; Bélanger, F.; Johnson, R. D.; Hightower, R. Gender-based Differences in Consumer E-commerce Adoption. *Commun. Assoc. Inf. Syst.* **2010,** *26* (1), 2. (17–34.)

Stafford, T. F.; Gillenson, M. L. Mobile Commerce: What It Is and What It Could Be. *Commun. ACM* **2003,** *46* (12), 33–34.

Teo, T.; Noyes, J. An Assessment of the Influence of Perceived Enjoyment and Attitude on the Intention to Use Technology Among Pre-service Teachers: A Structural Equation Modeling Approach. *Comput. Educ.* **2011,** *57* (2), 1645–1653.

Topi, H.; Valacich, J. S.; Wright, R. T.; Kaiser, K.; Nunamaker Jr., J. F.; Sipior, J. C.; de Vreede, G. J. IS 2010: Curriculum Guidelines for Undergraduate Degree Programs in Information Systems. *Commun. Assoc. Inf. Syst.* **2010,** *26* (1), 18. (359–428.)

Tiwari, R.; Buse, S. *The Mobile Commerce Prospects: A Strategic Analysis of Opportunities in the Banking Sector*; Hamburg University Press: Hamburg, 2007.

Venkatesh, V.; Davis, F. D. A Theoretical Extension of the Technology Acceptance Model: Four Longitudinal Field Studies. *Manag. Sci.* **2000,** *46* (2), 186–204.

Venkatesh, V.; Morris, M. G. Why Don't Men Ever Stop to Ask for Directions? Gender, Social Influence, and Their Role in Technology Acceptance and Usage Behavior. *MIS Q.* 2000, *24* (1), 115–139.

Venkatesh, V.; Morris, M. G.; Davis, G. B.; Davis, F. D. User Acceptance of Information Technology: Toward a Unified View. *MIS Q.* **2003,** *27* (3), 425–478.

Wen-Jang, K. J.; Su-Fang, L. An Exploratory Analysis of Relationships Between Cellular Phone Uses' Shopping Motivators and Lifestyle Indicators. *J. Comput. Inf. Syst.* **2003,** *44* (2), 65.

Zhang, X.; Guo, X.; Lai, K. H.; Guo, F.; Li, C. Understanding Gender Differences in M-health Adoption: A Modified Theory of Reasoned Action Model. *Telemed. e-Health* **2014,** *20* (1), 39–46.

CHAPTER 11

MOBILE LEARNING: FACTORS AFFECTING STUDENTS' ADOPTION IN HIGHER EDUCATION

AMANDEEP KAUR*, NIDHI SHARMA, RUHEE MITTAL, and ANSHU LOCHAB

[1]*Department of Management Studies, Rukmini Devi Institute of Advanced Studies, New Delhi, India*

Corresponding author. E-mail: amandeep092@gmail.com

ABSTRACT

Mobile learning represents a permanent solution to many of the problems in education. Education in India has reached a critical stage. It is necessary for future leaders to develop their skills and contribute to the nation's growth and prosperity, which is a much desired and serious reform. The education style and pedagogies that has been developed were suitable for our agrarian and industrial era but education in 21st century needs an advanced and more interactive pedagogy. In this postindustrial world, education delivered through devices such as smartphones and tablets enable the students to gain access to digitized information and personalized assessment. Mobile learning today is a mania in research community. But the successful integration of mobile learning in our education scenarios is possible though user's acceptance. Thus, the purpose of this paper is to identify factors that affect students' adoption of mobile learning in higher education using Unified Theory of Acceptance and Use of Technology (UTAUT) (Venkatesh et al., 2003) model. The study adopted correlation and regression for analysis on data collected from 285 respondents. The findings obtained from the study indicated that performance expectancy, effort expectancy, lecturers' influence, quality of services, personal innovativeness were all significant factors that affect the behavioral intention of the students toward m-learning.

The discussions and results in the study provide apt information regarding designing and features of m-learning tools for the students.

11.1 INTRODUCTION

M-learning is an advanced tool that helps students and teachers to navigate the options available in the distance-learning world. This includes learning by mobile phones and computer handheld devices. Now tablets, handheld electronic devices, web-enabled phones, smartphones, digital readers, personal digital assistant devices and other mobile devices act as an educational tool using which students are able to download and access books, notes, and other study material. Moreover, people tapping away at smartphones makes the m-learning process easy and seems to make perfect sense. Learners can get material and resources that they want at any point of time using m-learning. In M-learning, "M" stands for mobile, so this can be known as mobile learning (Gikas and Grant, 2013).

The unique feature, and indeed the obvious benefit, of m-learning, is that it is not limited to the specific vicinity, including sitting in a lecture room or perched in the front of a personal computer inside a fixed space. It can take place anywhere and each time, responding to the needs of an always-on society along with giving the learner flexibility. Technology-rich tasks can lead to high level of student engagement as compared to the less technology-related activities.

Institutions and educators need to bring the feasibility of harnessing mobile platforms and employ them for the student's effective learning. This will increase the flexibility of teachers and instructors, who may also create learning substances immediately in response to specific desires, or offer instant guidance. In this respect, digital technologies facilitate unlimited access to educational sources and flexibility to learners to take charge of the direction and depth of learning. Basically, the focus of learning is everywhere but this digital platform provides the best chance to achieve much of this in education. Thus, m-learning is frequently viewed as an element of a learning program, something that helps the learning technique with an add-on tool, instead of being the essential studying technique. According to this perspective, there may be a connection between m-learning and the concept of blended learning, in which the knowledge is gained by both traditional coaching methods and electronic- or internet-based resources.

The features of mobile learning contemplated by Huang et al. (2008) are:

1. enhancing availability and accessibility of information networks,
2. engaging students in learning-related activities in diverse physical locations,
3. supporting of project-based group work,
4. improving of communication and collaborative learning in the classroom, and
5. enabling quick content delivery.

There are certain benefits of m-learning in the higher education for the institutions. M-learning presents an environment that allows access to peers, content, portfolio artifacts, professionals, etc. This level of constant accessibility can be from a laptop, handheld electronic devices, net-enabled phones, digital readers, personal digital assistants (PDAs), and different cellular gadgets. As the cloud platform drives mobility, it gives accessibility to study substances such as all records resources; hence, opening doors to inaccessible kinds of collaboration and revision (Rao et al., 2012). M-learning allows you to have a customized studying experience. The time provided is enough or adaptable to every person. Mobile learning addresses the entire idea of inclusive education (digital interplay) which includes being attentive to a podcast and learning from movies and other visual displays. Another feature of m-learning is transparency and that is the by-product of mobility, collaboration, and connectivity. Through social media platforms consisting of Facebook, Twitter, and LinkedIn, it is possible to reach and target audience in nearby and worldwide communities. That is all possible because of performance, reflection, assessment, questioning, and planning virtually and in real time by constantly converting the traditional and physical studying environments into the more advanced learning setup as a result of recent challenges, thoughts, and opportunities. The learning for new users to this exposure is continuous and direction accessibility via gadgets allows such beginners to keep abreast of genuine and useful information (Jacob and Issac, 2008).

M-learning though of great significance, have certain challenges to be implemented in India. A mobile device with limited capability might not be appropriate for m-learning. For this reason, a learner has no choice, however, to get one with the specified degree of functionality so that one can get access to online courses. This mobile tool can be expensive. Additionally, the cell device generally has a small display size where the scholar is compelled to take strain on eyes. The records supplied may also be incomplete because of the constrained length of the tool. Also, reminiscence storage that is to be had in mobile gadgets might not be enough to save numerous offline content and frequent downloaded materials. Moreover, most gadgets batteries discharge fast because of number of running apps within the gadget. Thus, there will be a chief hassle in delivering content for a lengthy duration of time. Mobile learning may be tough if there is no tool standardization. Instructors can also face the trouble of providing same learning to all students at same level because smartphones and tablets have different OS and hardware. The students may not be able to interact in online discussions or even access the coursework (Parsons and Ryu, 2006).

M-learning has proved to be an effective tool for day to day personal work, training, and commercial enterprises as well. Mobile learning works as an imperative tool to facilitate better understanding on different areas of concern. Mobile learning helps training across contexts and life transitions, which poses tremendous problems for assessment. Sharples et al. (2009) has mentioned that there is no method that would give consistent results in evaluation, the learning may also unfold across places and times, the curriculum cannot be prespecified, if someone would take it as a hobby, it would be for personal, institutional, and public technologies, it may be disrupted with different activities, and there would be many challenges in monitoring the students' behavior outside the classroom learning. In a summary of researchers' reflections on present-day mobile learning ways, Kukulska-Hulme (2009) proposed four key concepts for destiny studies: "it need to be in track with new thinking about learning; it must not ignore the effect of context; it ought to consider one of a kind forms of information and analysis; and it should involve beginners as codesigners or coresearchers." The conclusions being drawn from the research on cellular-based learning need to be dealt with warning. Within the lecture room, handheld reaction systems can be mainly effective if used to show a numeric response to open questions. Mobile interactive and collaborative learning is an extension of online collaborative learning by incorporating mobility as key that permits a shift from individual responses to significant research projects having the formal and informal learning settings facilitating self-directed learning system. This system works against the private ownership of gadgets which is disruptive to collaborative learning such that colleges, schools, and universities will no longer be able to justify expenditure on laptop and computer systems when college students have their very own powerful tools. Another option in mobile learning is of its applicability in connecting in formal and informal settings such that the queries in the classrooms or labs may be persevered in practical settings, or that personal casual mastering can be used as a resource for formal schooling.

Despite the increased usage of internet among the young populations and the development of the learning technology and infrastructure, m-learning as a part of education system is not yet diffused in developing countries like India. The debate on virtual learning adoption and the reason for students to adopt has always been in light and one of the well-known models that explain these reasons is Unified Theory of Acceptance and Use of Technology (UTAUT). The e-learning system in Indian education represents a web-based platform that allows students to learn from different tools. The major challenges that influence the adoption of m-learning relates to use of mobile phones in classrooms; it may disturb the classrooms and instructors and students may or may not adopt this technology. The institutions may avoid it because of lots of effort that will have to be put in for its effective implementation. Nevertheless, students' perceptions toward m-learning need to be researched at the first step of its execution in higher education (Cheon et al., 2012). Therefore, it is necessary to conduct research that identifies the factors college students consider important in the acceptance of m-learning. Few researchers have investigated the acceptance of m-learning using Technology Acceptance Models (TAM) (Wang et al., 2009; Liu et al., 2010; Park et al., 2011). However, very few researchers have studied the influence of quality of services (QoS), lecture's impact, and innovativeness on students' behavioral intention (BI) to adopt m-learning. Based on this fact, the study aims to achieve following objectives:

- To investigate the UTAUT factors influencing college students' acceptance of m-learning.
- To understand the influence of m-learning on students' behavior in Delhi.

11.2 LITERATURE REVIEW: MODEL OF TECHNOLOGY ACCEPTANCE

The widely used and accepted models related to technology acceptance and use is the TAM, originally proposed by Davis in 1986. TAM has been the perfect theoretical model in helping to explain and predict user behavior of information technology (Legris et al., 2003). TAM is considered an influential extension of Theory of Reasoned Action (TRA), according to Ajzen and Fishben in (1980). Davis (1989) and Davis et al. (1989) proposed TAM to explain why a user accepts or rejects information technology by adapting TRA. TAM provides a basis with which one traces how external variables influence belief, attitude, and intention to use. Two cognitive beliefs are posited by TAM: perceived usefulness and perceived ease of use.

Venkatesh and Davis (2000) extended the original TAM model and proposed TAM2. They explain perceived usefulness and usage intentions in terms of social influence process and cognitive instrumental processes.

The next popular and recent model in information technology acceptance is the UTAUT. This theory was proposed by Venkatesh et al. (2003) making attempts to integrate and evidently compare dimensions from different TAM in technology acceptance.

The UTAUT contains four determinants of IT user behavior and four moderators that are found to moderate the effect of the four determinants on the behavior intention and user behavior. UTAUT theorizes that performance expectancy (PE), effort expectancy (EE), social influence, and facilitating conditions are direct factors of behavior intention or user behavior. This gives recommendable improvement to the explanatory power of the model. Also, the moderating variables (gender, age, experience, and voluntariness of use) are very important in understanding the characteristics of different user groups.

There have been several studies that have used UTAUT for explaining the m-learning acceptance. The results have indicated that all the factors are important in designing a successful m-learning environment.

Liu and Chen (2008) identified other factors such as self-efficacy, perceived enjoyment, and self-management of learning along with basic factors of the model.

Lowenthal (2010) investigated the two factors of UTAUT model, PE and EE, significantly influences BI. Self-management of learning was not found to be significant determinant and demographics, age, and gender were also not having any mediating effects.

Baran (2014) explained the growing interest of students in mobile learning and a lack of fusions in the framework of coach training. Findings also include the methods as well as policies for applying mobile learning and itinerant gears in different teacher education environments. The author also mentioned that numerous research papers evolved in this area has augmented the concept of mobile learning and offered significant learning. It was also found that there was a limited report of hypothetical and abstract outlooks. Review revealed that insights, attitudes, and usage patterns among teachers varied. Mobile learning was reported as mainly beneficial in teacher education contexts.

Jairak et al. (2009) in their study using UTAUT model based on TAM presented results that indicated that PE, EE, and social influence, significantly affect the attitude and other basic factors such as facilitating conditions have significant positive impact on behavior intention.

Bellaaj et al. (2015) carried out research in Saudi Arabia and revealed the results of virtual learning environment. The researcher reported that stronger internet experience impacts the PE and leads to decrease in EE. In addition to this, the impact of social influence on user's intention was reported to be moderated through gender where women responded favorably toward mobile learning having a societal influence.

11.3 RESEARCH MODEL

After considering the factors that might affect user's acceptance of m-learning, the constructs have been modified to be applicable for the young Indian users. The final constructs in UTAUT to investigate the factors that might affect student's acceptance of m-learning are: PE, EE, lecturers' influence (LI), QoS, and personal innovativeness (PI) (other social influence and facilitating conditions were not included in the study) (Fig. 11.1).

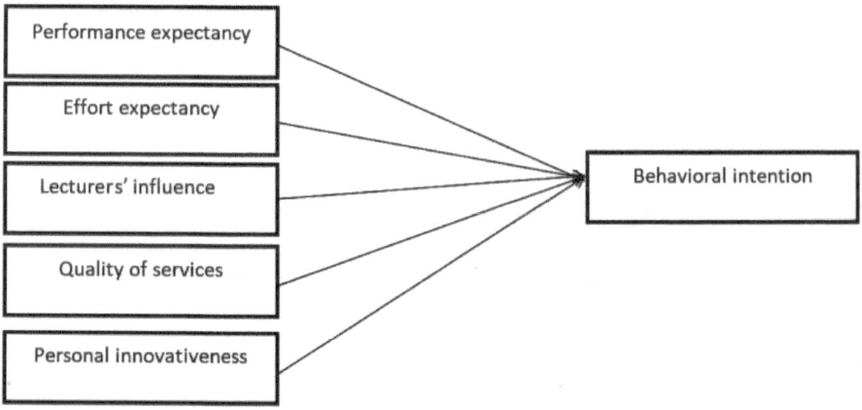

FIGURE 11.1 Research model.

Source: Compiled by authors.

11.3.1 RESEARCH DIMENSIONS AND HYPOTHESIS

11.3.1.1 PERFORMANCE EXPECTANCY

PE as defined by Venkatesh et al. (2003) is "the degree to which an individual believes that using the system will help the user to attain gains in job performance." However, in mobile learning, it would represent the degree to which students can enhance their performance and learn more in better way using mobile learning (Wang et al., 2009). Focusing on this belief will increase the student's intention to use this m-learning. Another similar study conducted in Ghana region and the United Kingdom using UTAUT found PE as a strong predictor (Adedoja et al., 2013, Abu-al-Aish and Love, 2013). This led to test the following hypothesis:

H1: PE will have a positive effect on BI to use m-learning.

11.3.1.2 EFFORT EXPECTANCY

Venkatesh et al. (2003) defined EE as the "degree of ease associated with the use of system." In developing countries, the technology is on accelerating stage but the convenience depends on the experience of using, thus in India, people have less than a decade of exposure to information systems (Ssekakubo et al., 2011), this construct is an important determinant of mobile learning acceptance. This variable is drawn from the TAM factor "perceived ease of use." Therefore, the proposition is derived as follows:

H2: EE will have a positive effect on BI to use m-learning

11.3.1.3 LECTURERS' INFLUENCE

LI is derived from social influence where the user attaches importance to the other person's (lecturer's) belief that the user should use m-learning (Venkatesh et al., 2003). Previous studies have indicated that social influence affects the BI of the students (Abu-al-Aish and Love, 2013; McMullan et al., 2003). The social influence is divided into two: superior influence and peer influence. This study incorporates the LI (superior's influence) on BI, which is defined as the degree to which the faculty or instructor encourage or motivate students to use m-learning. This led to test the following hypothesis:

H3: LI will have a positive effect on BI to use m-learning

11.3.1.4 QUALITY OF SERVICE

QoS has always determined the customer's expectation and the perceived service performance after the service delivery. Parasuraman et al. (1988) defined consumer expectation of QoS as what the consumer would like to be offered rather than what organizations are trying to sell to them. Zeithaml (1988) defined QoS as users' assessment of the overall superiority of the service. The excellence of services being provided to users can affect the level of acceptance of new technology (Xin, 2004). Lee (2010) indicated that students' perception of quality online service in education might be considered as an important factor that affects their BI toward the acceptance of e-learning. Thus, this study tested the following hypothesis:

H4: QoS will have a positive effect on BI to use m-learning

11.3.1.5 PERSONAL INNOVATIVENESS

Agarwal and Prasad (1998) explain it as the individual's willingness to try out any new information technology. Abu-al-Aish and Love (2013) tested its influence on BI and found out that it significantly influence the BI of the students and can be used as an effective strategy. Several other studies also investigated the effect of PI on BI of users ((Hung and Chang, 2005; Lu et al., 2005; Lian and Lin, 2008). The study also reflected that students with high innovativeness will take more risk and have positive intention toward m-learning adoption. The following hypothesis thus was tested:

H5: PI will have a positive effect on BI to use m-learning

11.4 RESEARCH METHODOLOGY

The questionnaire has been structured on the basis of the study of Venkatesh et al. (2003). The UTAUT model is used to study the factors influencing the adoption of m-learning among students pursuing higher education. The scale has been used from Abd-al-Aish and Love (2013) study conducted in London. In the present study, the data collection is done in two phases. In first phase, a pilot study was conducted in which five to six group sessions were conducted with five to seven students each pursuing higher education and using e-learning in their studies. The prestudy survey was used to gather students' perceptions toward mobile learning and these sessions helped in redesigning and finalization of the questionnaire. After this, the final questionnaire with 36 questions was used for the purpose of data collection based on 5-point Likert scale. In the second phase, an online questionnaire was made which was circulated among 300 college students pursuing under graduation and post graduation in affiliated college of central and state universities in Delhi region. The sampling method used was nonprobability convenient sampling. A total of 300 questionnaires were administered and 285 were returned, a response rate of 95%. The characteristic of participants are shown in Table 11.1.

TABLE 11.1 Respondent's Profile and Background Information.

Respondents' profile	Classification	Frequency	%
Gender	Female	120	42.10
	Male	165	57.90
Kind of mobile device	Mobile phone for calls and text	56	19.60
	Smartphone with advanced features	258	90.50
	PDA	2	0.70
	Tablet PC	19	6.70
	Other	2	0.70
Constant accessibility to the internet	Yes	256	89.80
	No	29	10.20
Use the internet from your mobile device	Everyday	276	96.80
	Every week	4	1.40
	Monthly	2	0.70
	Rarely	3	1.10
Pay to access the internet	Yes	251	88.10
	No	34	11.90
Opinion on the price of accessing the internet	High price	111	38.90
	Normal price	144	50.50
	Low price	30	10.50
Heard about mobile learning (m-learning)	Yes	227	79.60
	No	58	20.40
Your opinion of m-learning	Good idea and I would like to use it	228	80
	Good idea but I would not like to use it	38	13.30
	I do not think it is a good idea	12	4.20
	Other	7	2.50

Source: Compiled by authors.

Reliability analysis—The data indicate that the measures are robust in terms of their internal consistency reliability as indexed by composite reliability. The reliability of the collected data in this study was assessed by the Statistical Package for Social Science (SPSS). The composite reliabilities ranged from 0.81 to 0.95, which exceed the recommended threshold value of 0.70. Reliability results are given in Table 11.2.

TABLE 11.2 Reliability Analysis.

Reliability of research variable	Cronbach's alpha
PE	0.837
EE	0.824
LI	0.784
QoS	0.854
PI	0.866
BI	0.912

BI, behavioral intention; EE, effort expectancy; LI, lecturers' influence; PE, performance expectancy; PI, personal innovativeness; QoS, quality of services.
Source: Compiled by authors.

Correlation analysis—Convergent validity and discriminant validity are assessed by Pearson correlation analysis. Guidelines suggest that factor loadings be greater than 0.50 (Hair et al., 1998) or, under a stricter criterion, greater than 0.70 (Fornell, 1982). All of the factor results of items in this research model are higher than 0.50; most of them are above 0.70. Every item is loaded significantly ($p < 0.01$ in all cases) on its constructs. Therefore, all constructs in the model have adequate reliability and convergent validity. Correlation results are shown in Table 11.3.

TABLE 11.3 Correlation Matrix.

	PE	EE	LI	QoS	PInn	BI
PE	1.000					
EE	0.789*	1.000				
LI	0.694*	0.686*	1.000			
QoS	0.761*	0.794*	0.895*	1.000		
PInn	0.500*	0.522*	0.527*	0.662*	1.000	
BI	0.715*	0.716*	0.725*	0.845*	0.752*	1.000

BI, behavioral intention; EE, effort expectancy; LI, lecturer's influence; PE, performance expectancy; PInn, personal innovativeness; QoS, quality of services.
*Correlation is significant at the 0.01 level (two-tailed).
Source: Compiled by authors.

Table 11.4 illustrates the descriptive statistics for the UTAUT constructs. The mean values for PE, EE, LI, *QoS*, *PI*, and *BI* indicators are between 2 and 3, which imply that most of the students' responses were either agree or strongly agree. The descriptive statistics also suggest that most of the respondents agree with the statements in the questionnaire as observed in Table 11.2.

TABLE 11.4 Data for ICT Adoption Using UTAUT Indicators.

Descriptive statistics			
Indicator	**Mean**	**Std. deviation**	**N**
PE			
PE1. I find m-learning useful for my studies	2.320,423	1.118,179	285
PE2. Using m-learning would enable me to achieve learning tasks more quickly	2.292,254	1.084,329	285
PE3. Using m-learning in my studying would increase my learning productivity	2.334,507	1.123,507	285
PE4. Mobile learning could improve my collaboration with classmates	2.366,197	1.143,294	285
PE5. Using m-learning would improve my performance in my studies	2.359,155	1.085,344	285
EE			
EE1. I would find an m-learning system flexible and easy to use	2.253,521	1.079,436	285
EE2. Learning to operate an m-learning system is easy for me	2.165,493	1.139,124	285
EE3. It would be easy for me to become skillful at using an m-learning system	2.306,338	1.090,188	285
LI			285
LI1. I would use m-learning if it was recommended to me by my lecturers	2.309,859	1.110,077	285
LI2. I would like to use m-learning if my lecturers' supported the use of it	2.28,169	1.108,057	285
LI3. Lecturers in my course have not been helpful in the use of m-learning systems	2.429,577	1.101,864	285
QoS			
QoS1. It is important for institute's m-learning services to increase the quality of learning	2.207,746	1.04,787	285
QoS2. I would prefer m-learning services to be accurate and reliable	2.190,141	1.066,238	285

QoS3. It is important for m-learning services to be secure to use	2.207,746	1.119,602	285
QoS4. It is important for m-learning to focus on the speed of browsing the internet and obtaining information quickly	2.137,324	1.136,675	285
QoS5. Communication and feedback between lecturers and students would be easy using m-learning systems	2.288,732	1.09,983	285
QoS6. It is preferable that m-learning services are easy to navigate and download	2.274,648	1.034,011	285
PInn			
PInn1. I like to experiment with new information technologies	2.204,225	1.089,875	285
PInn2. When I hear about a new information technology I look forward to examine it	2.334,507	1.132,903	285
PInn3. Among my peer group, I am usually the first to try out a new innovation in technology	2.573,944	1.111,214	285
BI			285
BI1. I plan to use m-learning in my studies.	2.383,803	1.11,711	285
BI2. I predict that I will use m-learning frequently	2.383,803	1.0,553	285
BI3. I intend to increase my use of mobile services in the future	2.323,944	1.09,318	285
BI4. I will enjoy using m-learning systems	2.376,761	1.080,974	285
BI5. I would recommend others to use m-learning systems	2.28,169	1.123,889	285

BI, behavioral intention; EE, effort expectancy; LI, lecturers' influence; PE, performance expectancy; PInn, personal innovativeness; QoS, quality of services; UTAUT, unified theory of acceptance and use of technology.

Source: Compiled by authors.

Regression analysis—The regression analysis was used to investigate the influence of PE, EE, and social influence on intention to use. The results show that PE, EE, LI, QoS, and PI significantly affect BI. The results are presented in Table 11.5.

TABLE 11.5 Regression of Adoption Factors on Behavioral Intention.

	Beta
PE	0.224
EE	0.252
LI	0.353
QoS	0.162
PInn	0.552
R^2	0.655
Adjusted R^2	0.687

EE, effort expectancy; LI, lecturers' influence; PE, performance expectancy; PInn, personal innovativeness; QoS, quality of services.
Source: Compiled by authors.

The results showed that PE positively affects users' intentions to use m-learning ($\beta = 0.224$, $p < 0.001$). Therefore, H1 is supported. This implies that when students expect that using m-learning material would increase their performance, they increase their intentions to use it.

The results showed that effect expectancy positively affects users' intentions to use m-learning

($\beta = 0.252$, $p < 0.05$). Therefore, H2 is supported. This means that when students expect an m-learning website to be easy to use, they increase their intentions to use it.

The results showed that LI positively affects users' intentions to use m-learning ($\beta = 0.353$, $p < 0.001$). Therefore, H3 is supported. This means that when teachers and peers suggest them to use m-learning websites, they strongly develop intentions to do so.

The results showed that QoS positively affects users' intentions to use m-learning ($\beta = 0.162$, $p < 0.001$). Therefore, H4 is supported. This means that when websites offer quality services, they increase their intentions to use them.

The results showed PI positively affects users' intentions to use m-learning ($\beta = 0.552$, $p < 0.001$). Therefore, H5 is supported. This means that when user does want to innovate their learning, they increase their intentions to use them.

11.5 DISCUSSION AND CONCLUSION

Mobile learning in higher education is still in the beginning stages of implementation. The concepts and instructional issues surrounding mobile learning are still evolving (Kukulska-Hulme, 2007). The findings from this research add to existing technology acceptance literature and propose a framework for understanding, explaining, and predicting factors influencing individual acceptance of mobile learning. The study provides valuable baseline data for future studies on student acceptance and intention to use mobile devices for learning. The research model also establishes a foundational framework that administrators and educators can use to evaluate success factors for implementing mobile learning. Overall, the results from this study indicated that together the UTAUT predictors (PE, EE, LI, QoS, and PI) account for 0.687% of the variance in BI to use mobile learning. This is higher than 58% that was found in the research by Wang et al. (2009). Against the results of study by Venkatesh et al. (2003), PI and LI had a significant, positive influence on BI to use mobile learning. PI, EE, and QoS were also found to be a significant positive predictor of BI.

By understanding the determinants of student's mobile learning acceptance, these stakeholders are able to incorporate these factors into the design and implementation phases of a mobile learning initiative, particularly in Indian education system. Institutional preparation requires careful planning in infrastructure and strategy development necessary for implementing a mobile learning initiative to benefit college students. The results of this study identify factors that favor college students' mobile learning.

The findings of this research will help educators and administrators to promote mobile learning and provide insights into future research on mobile technology acceptance. These findings present implications for different approaches to support mobile learning. Understanding the determinants of students' acceptance and use of mobile technology for learning is essential for the successful delivery of academic, organizational, library, and instructional information. Before investing in development of mobile services and content, an institution must anticipate factors that influence students' technology acceptance. If students fail to accept mobile technology offered, then they will not use it to seek and exchange information. The outcome will be wasted budgetary expenses.

One of the impediments in adopting m-learning in higher education is the alignment and modification of the curriculum to accommodate mobile learning. Moreover, attitude and fear of change is one of the reasons that act as a barrier in effective implementation of m-learning. Another challenge in implementing m-learning in higher education is the excessive usage and reliance of students on mobile devices, that is, students may depend more on their mobile device than attending classes when they know that they can access learning materials through their mobiles and they miss out on important discussions or any other useful activities organized by their lecturers in the classroom; also, this may have a negative impact on the attendance of the students.

Given the integration of mobile devices into students' daily lives, faculty and instructional design staff can support mobile learning by identifying ways in which mobile devices can be utilized to support both classroom and remote learning.

11.6 SCOPE FOR FURTHER STUDY

Research in the area of mobile learning using technology acceptance theory is relatively new and more research continues to be needed to further develop our understanding of the determinants of mobile learning. Several opportunities are available to extend this research. Future theoretical technology acceptance research could examine self-management alone to understand its importance as a predictor in mobile learning and future research can examine EE using individuals with varying levels of experience with mobile devices and a wider age range of users.

To enhance understanding of student use of mobile learning in colleges in Delhi and the constructs tested, further research could help to determine whether or not there is longitudinal evidence supporting the findings of this paper. To increase generalizability, future research could use the same survey instrument and randomly sample university college students throughout the Delhi. For example, researchers could administer the survey at the beginning and end of a term. This would allow student response comparisons after training and access to mobile learning.

KEYWORDS

- **effort expectancy**
- **lecturers' influence**
- **performance expectancy**
- **personal innovativeness**
- **quality of services**

REFERENCES

Abu-Al-Aish, A.; Love, S. Factors Influencing Students' Acceptance of M-Learning: An Investigation in Higher Education. *Int. Rev. Res. Open Distr. Learn.* **2013,** *14* (5).

Adedoja, G.; Adelore, O.; Egbokhare, F.; Oluleye, A. Learners' Acceptance of the Use of Mobile Phones to Deliver Tutorials in a Distance Learning Context: A Case Study at the University of Ibadan. *Afr. J. Inf. Syst.* **2013,** *5* (3), 3.

Agarwal, R.; Prasad, J. A Conceptual and Operational Definition of Personal Innovativeness in the Domain of Information Technology. *Inf. Syst. Res.* **1998,** *9* (2), 204–215.

Ajzen, I.; Fishbein, M. Understanding Attitudes and Predicting Social Behavior; Englewood Cliffs: New Jersey, 1980.

Baran, E. A Review of Research on Mobile Learning in Teacher Education. *Educ. Technol. Soc.* **2014,** *17* (4), 17–33.

Bellaaj, M.; Zekri, I.; Albugami, M. The Continued Use of E-Learning System: An Empirical Investigation Using Utaut Model at the University of Tabuk. *J. Theor. Appl. Inf. Technol.* **2015,** *72* (3).

Cheon, J.; Lee, S.; Crooks, S. M.; Song, J. An Investigation of Mobile Learning Readiness in Higher Education Based on the Theory of Planned Behavior. *Comput. Educ.* **2012,** *59* (3), 1054–1064.

Davis, F. D. Perceived Usefulness, Perceived Ease of Use, and User Acceptance of Information Technology. *Manag. Inf. Syst. Q.* **1989,** *13* (3), 319–340.

Davis, F. D.; Bagozzi, R. P.; Warshaw, P. R. User Acceptance of Computer Technology: A Comparison of Two Theoretical Models. *Manag. Sci.* **1989,** *35* (8), 982–1003.

Fornell, C.; Bookstein, F. L. Two Structural Equation Models: LISREL and PLS Applied to Consumer Exit Voice Theory. *J. Mark. Res.* **1982,** *19*, 440–452.

Gartner. Gartner Says Worldwide Traditional PC, Tablet, Ultramobile and Mobile Phone Shipments to Grow 4.2 Percent in 2014, 2014.

Gikas, J.; Grant, M. M. Mobile Computing Devices in Higher Education: Student Perspectives on Learning with Cellphones, Smartphones & Social Media. *Internet High. Educ.* **2013,** *19*, 18–26.

GSMA. *Mobile Learning Policy Handbook*; GSMA: London, 2014.

Hair, J. F.; Anderson, R. E.; Tatham, R. L.; Black, C. B. *Multivariate Data Analysis*, 5th ed.; Prentice Hall Inc.: Upper Saddle River, 1998.

http://www.gartner.com/newsroom/id/2791017 (accessed Feb 10, 2016).

http://www.internetlivestats.com/internet-users/adresindenalındıon (accessed Feb 20, 2016).

http://www.marketingcharts.com/online/smartphone-penetration-rising-in-all-age-and-income-demos-hits-75-of-us-mobile-market-51585/ adresindenalındı (accessed Feb 25, 2016).

Huang, Y.-M.; Huang, T.-C.; Hsieh, M.-Y. Using Annotation Services in a Ubiquitous Jigsaw Cooperative Learning. *Educ. Technol. Soc.* **2008,** *11* (2), 3–15.

Hung, S. Y.; Chang, C. M. User Acceptance of Wap Services: Test of Competing Theories. Comput. Stand. Interfaces **2005,** *28*, 359–370.

Jacob, S. M.; Issac, B. Mobile Technologies and Its Impact: An Analysis in Higher Education Context. *Int. J. Interact. Mobile Technol.* **2008,** *2* (1).

Jairak, K.; Praneetpolgrang, P.; Mekhabunchakij, K. An Acceptance of Mobile Learning for Higher Education Students in Thailand. 2009.

Kukulska-Hulme, A. Mobile Usability in Educational Contexts: What Have We Learnt? *Int. Rev. Res. Open Distance Learn.* **2007,** *8* (2), 1–16.

Kukulska-Hulme, A. Will Mobile Learning Change Language Learning? *ReCALL* **2009,** *21* (2), 157–165.

Lee, J. W. Online Support Service Quality, Online Learning Acceptance, and Student Satisfaction. *Internet High. Educ.* **2010,** *13* (4), 277–283.

Legris, P.; Ingham, J.; Collerette, P. Why do People Use Information Technology? A Critical Review of the Technology Acceptance Model. *Inf. Manag.* **2003,** *40* (3), 191–204.Lian, J.; Lin, T. Effects of Consumer Characteristics on Their Acceptance of Online Shopping: Comparisons Among Different Product Types. *Comput. Human Behav. 2008, 24* (1), 48–65.

Liu, Y.; Chen, N. S. In *An Adoption Model for Mobile Learning.* Proceeding for the IADIS International Conference E-Commerce, 2008; pp 251–256.

Liu, Y.; Li, H.; Carlsson, C. Factors Driving the Adoption of M-learning: An Empirical Study. *Comput. Educ.* **2010,** *55* (3), 1211–1219.

Low, L.; O'Connell, M. In *Learner-centric Design of Digital Mobile Learning.* Proceedings of the OLT Conference, Sept 2006; pp 71–82.

Lowenthal, J. N. Using Mobile Learning: Determinates Impacting Behavioral Intention. *Am. J. Dist Educ* **2010,** *24* (4), 195–206.

Lu, J.; Yao, J. E.; Yu, C. S. Personal Innovativeness, Social Influences and Adoption of Wireless Internet Services Via Mobile Technology. *J. Strat. Inf. Syst.* **2005,** *14* (3), 245–268.

McMullan, M.; Endacott, R.; Gray, M. A.; Jasper, M.; Miller, C. M. L.; Scholes, J.; Webb, C. Portfolios and Assessment of Competence: A Review of the Literature. *J. Adv. Nurs.* **2003,** *41* (3), 283–294.

Parasuraman, A.; Zeithaml, V. A.; Berry, L. L. SERVQUAL: A Multiple-item Scale for Measuring Consumer Perceptions of Service Quality. *J. Retail.* 1988, 64, 12–37 (Spring).

Park, S. Y.; Nam, M.; Cha, S. University Students' Behavioral Intention to Use Mobile Learning: Evaluating the Technology Acceptance Model. *Br. J. Educ. Technol.* **2011,** *43* (4), 592–605.

Parsons, D.; Ryu, H. In *A Framework for Assessing the Quality of Mobile Learning*, Proceedings of the International Conference for Process Improvement, Research and Education, April 2006; pp 17–27.

Rao, N. M.; Sasidhar, C.; Kumar, V. S. Cloud Computing Through Mobile-learning, 2012. *arXiv Preprint arXiv: 1204.1594.*

Sharples, M.; Arnedillo-Sánchez, I.; Milrad, M.; Vavoula, G. Mobile Learning. In *Technology-Enhanced Learning*; (). Springer: Dordrecht, 2009; pp 233–249.

Ssekakubo, G., Suleman, H., & Marsden, G. In *Issues of Adoption: Have E-Learning Management Systems Fulfilled Their Potential in Developing Countries*? Proceedings of the South African Institute of Computer Scientists and Information Technologists Conference on Knowledge, Innovation and Leadership in a Diverse, Multidisciplinary Environment; ACM, 2011; pp 231–238.

Venkatesh, V.; Davis, F. D. A Theoretical Extension of the Technology Acceptance Model: Four Longitudinal Field Studies. *Manag. Sci.* **2000,** *45* (2), 186–204.

Venkatesh, V.; Morris, M. G.; Davis, G. B.; Davis, F. D. User Acceptance of Information Technology: Toward a Unified View. *MIS Q.* **2003,** *27* (3), 425–478.

Wang, Y.; Wu, M.; Wang, H. Investigating the Determinants and Age and Gender Differences in the Acceptance of Mobile Learning. *Br. J. Educ. Technol.* **2009,** *40* (1), 92–118.

Xin, X. In A Model of 3G Adoption, AMCIS 2004 Proceedings Paper 320, 2004.

Zeithaml, V. A. Consumer Perceptions of Price, Quality, and Value: A Means End Model and Synthesis of Evidence. *J. Mark.* **1988,** *52*, 2–22.

CHAPTER 12

AN EMPIRICAL INVESTIGATION OF ADOPTION AND SATISFACTION OF M-COMMERCE USERS IN THE SELECTED CITY OF GUJARAT STATE: A CASE STUDY OF VADODARA CITY

PARAG SUNIL SHUKLA*, PARIMAL HARIOM VYAS, and MADHUSUDAN N. PANDYA

Department of Commerce and Business Management, Faculty of Commerce, The Maharaja Sayajirao University of Baroda, Vadodara, Gujarat, India

Corresponding author. E-mail:parag.shukla.msu@gmail.com

ABSTRACT

In the omnichannel retail context, the mobile channel is emerging as the preferred medium for retailers to respond to its target customers. The proliferation of mobile devices in peoples lives have led to emergence of business potention of m-commerce. The challenge for e-retailers is to understand the shopping journey of the connected shopper and to understand their use of mobile devices for availing m-commerce services. India is gifted by factors such as young polpulation, rise of double-income families, uber culture, which has strong inclination for brands, and modern lifestyle. The young generation which use smartphones are fuelling consumption at a brisk pace. More than 235 million people in India access internet through mobile devices to avail various services, and it has created an opportunity for e-retailers to reach, stay in touch, and influence its target prospects continually that are located even in remote and rural areas.

In this research study, an attempt has been made to study the mobile commerce activities engaged into by shoppers who uses smartphones by measuring their actual experience, satisfaction, and its influence on their buying decisions. The researchers have collected primary data from 150 smartphone users, using Structured Nondisguised Questionnaire, who were conveniently drawn from the city of Vadodara located in the state of Gujarat. The results and findings of the research study are drawn based on use of statistical tools, namely, descriptive statistics, discriminant analysis, and hypothesis testing using chi-square test. The structural equation model was also developed by the researchers so as to offer crucial marketing and business implications to improve m-commerce services offered by e-retailers in near future.

12.1 PROLOGUE

M-commerce is e-commerce on mobile phones. *Mobile commerce* can be defined as any transaction in which mobile devices directly or indirectly enable the exchange of money for goods and services. The people's shift from e-commerce to m-commerce can be attributed to omnipresent feature of m-commerce and growth of m-commerce in India can be attributed to the expansion of low-cost smartphones availability and low mobile tariffs. People can do shopping and bidding without visiting shops as well as transfer the fund trough m-commerce and e-commerce practices with the help of laptops and desktop computers using internet, whereas m-commerce is conducted on mobile phones using internet. E-commerce can be called as anytime online transactions and m-commerce can be called as anytime anywhere online transactions. M-commerce is implemented through mobile applications and consequences can be observed in the form of increase in the number of mobile users and mobile internet subscribers in India. This is due to the ubiquity, reachability, mobility, and flexibility features of m-commerce. People preferred to make the use of mobile applications instead of web application for utility bill payment, ticket booking, fund transfer, e-mail, railway ticket reservations, movie ticket booking, and so on. Thus, m-commerce is gradually taking over the practice of e-commerce but it is not possible to replace the e-commerce fully considering the weaknesses of m-commerce such as tiny screen of device, weak processors, limited memory, poor resolutions, poor data entry, lack of WAP-enabled devices, problems in retrieval of data from website and search engines, and shortage of bandwidth. Mobile applications are also developed to give more security to the transactions. Increasingly people are using mobile applications instead of web applications. M-commerce is progressing and within some years, huge number of people will be using mobile applications (Naware, 2016). Smartphone usage in India is growing at an unprecedented rate and is still expected to grow further with new economic and banking reforms seen in the recent past. The entry of new service providers who promise to offer seamless data connectivity by offering various plans/schemes also propels this growth. This in turn will stimulate the growth of m-commerce in the near future as companies are striving to launch digital platforms and mobile-led applications for their consumers.

M-commerce industry is young in India. In total, 9% Indians are using smartphones for the purpose of rapidly consuming contents such as gaming, videos, songs, and entertainment on their smart devices and this leads to steady growth in mobile advertising and apps industry. Smartphone penetration has increased from 8% by end of the year 2014 to more than 21% in the year 2017. The growth in m-commerce has been fuelled by the availability of affordable smartphone and mobile data plans (Kaur and Singh, 2016; http://indianexpress.com).

This exponential increase in the smartphone penetration usage is likely to create more online shopper in India in the future. The e-retailers need to harness this opportunity by offering superior experience to shoppers for availing m-commerce services. Both, the global and local technology-based companies have been constantly evaluating the acceptance and the use of m-commerce. The need for understanding the reasons for mobile and electronic commerce adoption and factors affecting the successful adoption of m-commerce is important especially in developing countries. In this study, authors have attempted to capture and analyze the perception and adoption of m-commerce among respondents of Vadodara city of Gujarat state.

12.2 GROWTH TRAJECTORY OF M-COMMERCE IN INDIA

The number of mobile internet users in India is expected to rise to 314 million by the year 2017, nearly double the 159 million it had at the end of the year 2014, according to a report by industry body (The Internet and Mobile Association of India [IAMAI], 2014). With more than 300 million internet users, India has the second-largest internet user base in the world. But the internet penetration at 19% (approx.) is poor and limits the potential. The next wave of growth in penetration of internet will be driven by adoption of mobile internet. However, compared with key developed markets, which are almost nearing saturation, or other comparable developing markets such as Brazil and China, India has a long way to go with a sizeable portion of the population still devoid of internet access.

While the urban market in India has not reached its saturation point, the nonmetro and rural areas are the next wave of growth in the Indian internet. The advent of low-cost smartphone coupled with low mobile tariffs has empowered consumers to use data connectivity with the more usage of internet. Urban areas in India have wireless tele-density, or numbers of telephone connections for every hundred individuals living within an area, of 142.39%, while rural wireless tele-density was only 44.32%, as of September 2014.

The penetration of mobile devices in remote rural markets has to increase in order to be able to facilitate a rise in rural density; users accessing the internet through their mobile devices has become a key driver for increasing the overall internet subscriber base in India.

3G users' base constitutes about 36% of the overall mobile internet user base, while 4G is only beginning to set its foot in the market with a 2% share as of June 2014.

However, the 2G user base is expected to decline in the coming years, and growth will be led by 3G and 4G networks and more and more users will migrate to these higher speed networks (The Internet and Mobile Association of India [IAMAI], 2014).

12.3 FUTURE OF M-COMMERCE IN INDIA

The country has seen an overwhelming growth in day-to-day transactions carried out via mobile devices due to increasing access to smartphone and growing mobile internet penetration. According to report released in the year 2014 by market research firm Zinnov, it was estimated that India's market for mobile commerce would expand from $2 billion in the year 2014 to $19 billion by the year 2019.

A new wave of hyper-local commerce is observed in the market. At the hyperlocal level, many "mobile optional" businesses have taken the tried-and-tested model of e-commerce and opened it up to nearby products and services. If the person is hungry, ill, he or she want access to the closest restaurants, doctors, or plumbers, it is possible through the mobile option as quickly and effortlessly as possible, which is an emerging and the hottest new trend in Indian mobile commerce, that is, hyper-local delivery. In most cases, these businesses have developed desktop websites alongside mobile apps, as consumers get accustomed to navigating mobile discovery and payment platforms. Due to this, one can find several established e-commerce companies begin to acquire delivery start-ups or invest in their own hyper-local initiatives. To illustrate, Snapdeal has invested in grocery delivery start-up, namely, "PepperTap"; Paytm has invested in logistics company Loginext and acquired hyper-local service marketplace. Last year, Amazon India also began offering same-day delivery from local shops through Kirana and Flipkart rolled out its own grocery delivery service called Flipkart. Uber and Ola cabs have disrupted urban transportation by providing on-the-go cab booking. To grab this m-commerce opportunity, the retailers need to focus on mobile shopping applications to entice the connected Indian shoppers of today (Sharat, 2016).

At the city- and location-independent level, there are mostly web-first businesses that are naturally adapting to the mobile world. From e-commerce marketplaces (e.g., Flipkart) and travel aggregators (e.g., Make My Trip) to online classifieds (e.g., OLX) and media libraries (e.g., Saavn), many of these well-established digital businesses are continuing an inevitable transition to the rapidly growing channel of mobile (Sharat, 2016).

12.4 REVIEW OF LITERATURE

Attitudes to m-commerce services are formed by cultural and structural factors, and understanding of the cultural dimensions of a market can aid marketers immensely in developing appropriate m-commerce services, marketing these appropriately and in setting realistic adoption targets (Harris et al., 2005). Lee and Lee, (2007) studied the influence of mobile involvement and found that it positively predicts perceived benefits, but not perceived risks. Perceived benefits of m-commerce have a positive impact on the willingness to pay more money to use m-commerce and intention to use m-commerce, while perceived risks of m-commerce have a significant negative impact.

Mobile involvement is the significant mediating variable in the model to predict willingness to pay more money to use m-commerce and intention to use m-commerce (Lee and Lee, 2007).

Saneifard (2009) conducted empirical study on smartphone users and results showed that attitude toward use of m-commerce services has the strongest positive effect on determining behavioral intention to use and compatibility, perceived usefulness, perceived ease of use, innovation play the most significant role in mobile commerce adoption, only cost does not have a meaningful relation with attitude toward use (Saneifard, 2009). Kini

(2009) conducted research study in Chile and results show that the group studied uses electronic commerce extensively but is not comfortable using mobile commerce, and complain that mobile access speed, service quality, and price need improvement (Kini, 2009).

Huo et al. (2011) explored the social network factors, and analyzed the direct influences of m-commerce service on Chinese farmers' consumer behavior and the indirect effects of consumption network (Huo et al., 2011). Zheng et al. (2012) conducted research in China and result showed that consumers' attitude toward using mobile commerce is influenced significantly by perceived usefulness, perceived cost, perceived entertainment, and its own development of mobile commerce (Zheng et al., 2012). Hongjiang (2013) examined the validity of the research model and found that the selected demographic variables such as age, background, and economic level factors have impact on consumers' perceptions (Hongjiang, 2013).

Saleh and Mashhour (2014) conducted study and results showed that previous internet shopping experience associated with frequency and extent of mobile use have significant influence on trusting m-commerce and the intention to use it. Mobile transactions surpassed desktop as more retailers adopted consumer-friendly mobile sites and transaction-driving commerce apps (Saleh and Mashhour, 2014).

Increasingly people are using mobile applications instead of web applications. M-commerce is progressing and within some years, huge number of people will be using mobile applications (Naware, 2016).

Mobile commerce has gained a huge popularity although it is in its initial phase in india and plays an important role in affecting human life; however, mobile commerce needs some development in specific areas such as secure transaction, better shopping experience, and enhanced graphics. Deepika and Karpagam (2016) highlighted that user interface quality, information quality, and perceived privacy act as antecedents that drive to smartphone user satisfaction, trust, and user loyalty toward B2C mobile applications. The factor information quality does not have a significant influence toward the smartphone user satisfaction and trust (Deepika et al., 2016). The drivers of customer satisfaction in mobile commerce include a combination of merchandising, pricing, site navigation, etc. (Bustos, 2013).

12.5 RATIONALE OF THE RESEARCH STUDY

The use of smartphones and various handheld devices is growing at a rapid pace as apparent from the review of literature and the growth trajectory of mobile commerce in India as charted out in the previous section. Understanding the adoption of m-commerce services by the shoppers will play a pivotal role in realigning the marketing strategies by the marketers who are eying this potential. It is not only imperative for a firm to go "digital" but also to reconfigure the entire organization so as to meet the expectations of the shoppers by offering them a seamless hassle free experience. The recent government initiatives are also directed toward enabling citizens to use digital platforms over the conventional methods. The digital imperative is also seen in case of banking industry which will change the way consumers will transact as we usher in the new economy of digital banking and cashless economy. The motivation and justification of the study lies on these reasons.

The urban population will keep on evolving as the "young" and the "restless" prosumer owing to the unstoppable urge for smart devices and the power of the internet in their hands. Hence, it becomes crucial for marketers to study the facets of adoption vis-à-vis satisfaction of the consumers who are switching at a rapid pace to mobile applications instead of websites.

The authors in this chapter have made a humble attempt to study and highlight the various mobile commerce activities in which the shoppers' are engaged in the selected city of Vadodara in the state of Gujarat by measuring their actual experience, satisfaction, and its influence on their buying decisions.

12.6 RESEARCH METHODOLOGY

Research methodology mainly consists of following:

The researcher has used descriptive research design to study the adoption and satisfaction of selected m-commerce users of the Vadodara city who were conveniently drawn by applying nonprobability sampling design on the basis of convenience sampling method for the collection of the required primary data. By using survey research approach, data were collected from the representative sampling units from Vadodara city from total 150 respondents. In this research study, required primary data were collected using structured-non disguised questionnaire supported with personal interviewing of the selected customers. The rationale behind this is to explore and describe the underlying areas of use and adoption of mobile commerce of selected respondents in the city of Vadodara so as to draw meaningful inferences thereof.

The sampling frame to be adopted was not readily available so data was collected using convenient sampling method by approaching friends, relatives, colleagues, etc., as mostly all of these groups were using mobile applications to avail m-commerce services so as to lead to minimizations of biases in their response. Further, after screening all the questionnaires, only the duly filled qualifying questionnaires were considered for further data analysis so as to obtain the outcome that may be more meaningful to generalize the behavioral pattern of the target audience.

An attempt is made by the researchers to put forward the results and findings, based on statistical test applied to test the hypothesis, offered the implications in formulation and modifications of marketing strategies for improving m-commerce services to be provided by e-retailers in near future.

12.6.1 OBJECTIVES OF THE RESEARCH STUDY

In this chapter, an attempt has been made by the researchers to study the adoption and satisfaction of selected m-commerce users of Vadodara city.

The objectives of the research study were (1) to examine the activities performed by the selected respondents using smartphone, (2) to assess the actual experience of selected respondents using m-commerce based on selected criteria, and (3) to measure the overall satisfaction experienced by the selected respondents using m-commerce through use of smartphone.

12.6.2 DESIGNING OF STRUCTURED QUESTIONNAIRE

The structured questionnaire used neutrally worded questions. For question number 03 and 04 respondents were asked to rate on the 5-point Likert scale, namely, 01 as strongly disagree and 5 as strongly agree. In case of question number 05 the 5-point Likert scale was used 1 as highly dissatisfied and 5 as highly satisfied.

12.6.3 RELIABILITY OF THE STRUCTURED NONDISGUISED QUESTIONNAIRE

Reliability test was applied to determine how strongly the opinion of selected m-commerce users were related to each other, and also to compare its composite score. The overall Cronbach's alpha score of 0.880 and score ranging between 0.695 and 0.750 for group indicator items as shown in Table 12.1 showed internal reliability of the scale and reflected the degree of cohesiveness among the grouped indicator items (Malhotra, 2007; Nunnally, 1981).

TABLE 12.1 Summary of Indicators and Reliability Alpha Score.

S. no.	Grouped indicator items	Cronbach's alpha coefficient
1	Information	0.750
2	Usefulness	0.771
3	Comfort	0.763
4	Trust	0.695
5	Overall	0.880

Source: Compiled by authors.

12.7 DATA ANALYSIS OF THE RESEARCH STUDY

For the purpose of analyzing the data, the descriptive statistics and discriminant analysis are used. Chi-square test is also used to test the hypothesis and structural equation model (SEM) is developed for displaying the relationship between selected criteria and overall satisfaction of selected respondents. The findings of data analysis are as follows:

12.7.1 PROFILE OF THE RESPONDENTS

The demographic profile of selected respondents as given in Table 12.2 was prepared which revealed that the male and female respondents up to 30 years of age were with an average of 65% and 31 years and above were with an average of 35%. The analysis of the educational qualifications revealed that the below-graduation respondents up to 30 years of age were 87% and 31 years and above were 13%, whereas graduate respondents up to 30 years of age were 42% and 31 years and above were 58%.

Considering the occupation of the selected respondents, students up to 30 years of age were 80% and 31 years and above were 20%, whereas respondents engaged in occupation of service and business up to 30 years of age were with an average of 13% and 31 years and above were with an average 87%.

On analysis of the monthly income of the selected respondents, it was found that income below Rs. 20,000 up to 30 years of age were 77% and 31 years and above were 23%, whereas monthly income from 20,001 to 30,000 and above 30,001 up to 30 years of age were with an average of 58% and 31 years and above were with an average 42%.

TABLE 12.2 Profile of the Selected Respondents.

Sr. no.	Selected background variables of selected shoppers		Age group (in years)		
			Upto 30	31 and above	Total
			Number and percentages of respondents		
01	Gender	Males	38 (61.3)	24 (38.7)	62 (100)
		Females	61 (69.3)	27 (30.7)	88 (100)
03	Educational qualification	Below graduation	76 (87.4)	11 (12.6)	87 (100)
		Graduation	22 (41.5)	31 (58.5)	53 (100)
		Above graduation	1 (10.0)	9 (90.0)	10 (100)
04	Occupation	Student	89 (79.5)	23 (20.5)	112 (100)
		Service	3 (14.3)	18 (85.7)	21 (100)
		Business	1 (11.0)	8 (88.9)	9 (100)
		Others	6 (75.0)	2 (25.0)	8 (100)
05	Monthly family income	Below Rs. 20,000	50 (76.9)	15 (23.1)	65 (100)
		Rs. 20,001–30,000	33 (56.9)	25 (43.1)	58 (100)
		Above 30,001	16 (59.3)	11 (40.7)	27 (100)

Source: Compiled by authors.

12.7.2 BRANDS OF MOBILE PHONE USED BY RESPONDENTS

Considering the main five brands of mobile used by the selected respondents, it becomes clear from Table 12.3 that respondents up to 30 years of age preferred Samsung, Lenovo, Micromax, Nokia, and Lava, whereas respondents of 31 years and above age preferred Samsung, Nokia, Lenovo, Micromax, and Intex.

TABLE 12.3 Brand of Mobile Phone Used.

Brand of mobile phone	Age group (in years)		Total
	Upto 30	31 and above	
	Number and percentages of respondents		
Samsung	35 (35.4)	17 (33.3)	52 (34.7)
Nokia	6 (6.1)	7 (13.7)	13 (8.7)
HTC	1 (1.0)	0 (0.0)	1 (0.7)
Asus	0 (0.0)	2 (3.9)	2 (1.3)
Lenovo	8 (8.1)	5 (9.8)	13 (8.7)
Micromax	8 (8.1)	4 (7.8)	12 (8.0)
Vivo	0 (0.0)	1 (2.0)	1 (0.7)
Sony Ericsson	0 (0.0)	3 (5.9)	3 (2.0)
Spice	3 (3.0)	1 (2.0)	4 (2.7)
Intex	5 (5.1)	4 (7.8)	9 (6.0)
Microsoft Lumia	4 (4.0)	1 (2.0)	5 (3.3)
Karbon	1 (1.0)	1 (2.0)	2 (1.3)
Lava	6 (6.1)	0 (0.0)	6 (4.0)
Panasonic	4 (4.0)	0 (0.0)	4 (2.7)
Oppo	2 (2.0)	1 (2.0)	3 (2.0)
Gionee	4 (4.0)	2 (3.9)	6 (4.0)
Iball	1 (1.0)	0 (0.0)	1 (0.7)
LG	1 (1.0)	1 (2.0)	2 (1.3)
iPhone	1 (1.0)	0 (0.0)	1 (0.7)
Motrola	1 (1.0)	1 (2.0)	2 (1.3)
MI	4 (4.0)	0 (0.0)	4 (2.7)
Apple	4 (4.0)	0 (0.0)	4 (2.7)
Total	99 (100)	51 (100)	150 (100)

Source: Compiled by authors.

12.7.3 TIME SPENT BY RESPONDENTS ON MOBILE PHONE

As given in Table 12.4, the analysis of the time spent on mobile phone by respondents revealed that 82% of the respondents belonging to age group of 30 years preferred to spend about 4 h in a day on mobile, followed by 68% preffered to spend less than 2 h on mobile phone, and 57% who spent 2–4 h on mobile phone. The 43% of respondents belonging to age group of 31 years and above preferred to spend 2–4 h on mobile phone, followed by 33% preferred to spend less than 2 h, and 18% preferred to spend above 4 h on mobile phone.

TABLE 12.4 Time Spent on Mobile.

Time spent on mobile	Age group (in years)		
	Up to 30 years	31 years and above	Total
Less than 2 h	52 (67.5)	25 (32.5)	77 (100)
2–4 h	29 (56.9)	22 (43.1)	51 (100)
Above 4 h	18 (81.8)	4 (18.2)	22 (100)
Total	99 (66.0)	51 (34.0)	150 (100)

Source: Compiled by authors.

12.7.4 ACTIVITIES PERFORMED USING SMARTPHONE BY SELECTED AGE GROUP

TABLE 12.5 Mean Score of Activities Performed Using Smartphone by Selected Age Group.

Selected criteria	Upto 30 years ($N = 99$)		31 years and above ($N = 51$)	
	Mean	Standard deviation	Mean	Standard deviation
Communicate with friends and family members	4.16	1.095	3.84	1.405
Communication for placing orders	3.36	1.216	3.24	1.478
Information—browsing the internet in leisure time	3.56	1.136	3.31	1.364
Information for comparing various products	3.58	1.144	3.12	1.291
Information—downloading apps	4.01	1.025	3.61	1.457
Transactions—transferring money online	3.15	1.207	2.35	1.369
Transactions—paying utility bills	3.19	1.184	2.35	1.309
Transactions—mobile banking services	3.40	1.220	2.47	1.433
Transactions—booking movie tickets	3.33	1.116	2.43	1.389
Transactions—booking tickets for traveling	3.60	1.212	2.80	1.497

Source: Compiled by authors.

As given in Table 12.5, the key five activities performed by the respondents below 30 years of age using smartphone include communicate with friends and family members (4.16 mean); information—downloading apps (4.01 mean); transactions—booking tickets for traveling (3.60 mean); information for comparing various products (3.58 mean), and information—browsing the internet in leisure time (3.56 mean), whereas the key five activities performed by the respondents of 31 years and above age using smartphone include communicate with friends and family members (3.84 mean); information—downloading apps (3.61 mean); information—browsing the internet in leisure time (3.31 mean); communication for placing orders (3.24) and Information for comparing various products (3.12 mean).

In order to determine whether any significant difference exists between activities performed using smartphone by the respondents below 30 years of age and above 30 years of age, researcher has applied chi-square test and its values are given in Table 12.6.

TABLE 12.6 Frequency and Chi-square Score of Activities Performed Using Smartphone by Selected Respondents.

No.	Activities performed by respondents with mobile devices	Age group (in years)	SD	DA	NANDA	AG	SA	Chi-square "*p*" value
			Number and percentages of respondents					
01	Communicate with friends and family members	Up to 30	8 (05)	1 (0.7)	1 (0.7)	46 (31)	43 (29)	NS 0.390
		Above 30	8 (05)	2 (1.3)	0 (0.0)	21 (14)	20 (13.3)	
02	Communication for placing orders	Up to 30	10 (6.7)	18 (12)	10 (6.7)	48 (32)	13 (8.7)	NS 0.141
		Above 30	10 (6.7)	08 (5.3)	05 (3.3)	16 (10.7)	12 (08)	
03	Information—browsing the internet in leisure time	Up to 30	06 (04)	15 (10)	14 (9.3)	46 (30.7)	18 (12)	NS 0.337
		Above 30	08 (5.3)	08 (5.3)	04 (2.7)	22 (14.7)	09 (06)	
04	Information for comparing various products	Up to 30	07 (4.7)	13 (8.7)	13 (8.7)	48 (32)	18 (12)	NS 0.255

TABLE 12.6 *(Continued)*

		Above 30	07 (4.7)	12 (08)	06 (04)	20 (13.3)	06 (04)	
05	Information—downloading apps	Up to 30	06 (04)	03 (02)	06 (04)	53 (35.3)	31 (20.7)	NS 0.77
		Above 30	09 (06)	04 (04)	01 (0.7)	21 (14)	16 (10.7)	
06	Transactions—transferring money online	Up to 30	07 (4.7)	30 (20)	17 (11.3)	31 (20.7)	14 (9.3)	S 0.000
		Above 30	15 (10)	21 (14)	05 (3.3)	02 (1.3)	08 (5.3)	
07	Transactions—paying utility bills	Up to 30	07 (4.7)	27 (18)	18 (12)	34 (22.7)	13 (8.7)	S 0.000
		Above 30	15 (10)	20 (13.3)	04 (2.7)	07 (4.7)	05 (3.3)	
08	Transactions—mobile banking services	Up to 30	04 (2.7)	27 (18)	18 (10)	31 (20.7)	22 (17.7)	S 0.000
		Above 30	18 (12)	12 (08)	06 (04)	09 (06)	06 (04)	
09	Transactions—booking movie tickets	Up to 30	06 (04)	22 (14.7)	14 (9.3)	47 (31.3)	10 (6.7)	S 0.000
		Above 30	17 (11.3)	15 (10)	04 (2.7)	10 (6.7)	05 (3.3)	
10	Transactions—booking tickets for traveling	Up to 30	10 (6.7)	10 (6.7)	10 (6.7)	49 (32.7)	20 (13.3)	S 0.004
		Above 30	14 (9.3)	12 (08)	03 (02)	14 (9.3)	08 (5.3)	

AG, agree; DA, disagree; NANDA, neither agree nor disagree; NS, nonsignificant; S, significant; SA, strongly agree; SD, strongly disagree.

Source: Compiled by authors.

12.7.5 HYPOTHESIS

Ho: The activities performed on smartphone by selected respondents of below 30 years of age and above 30 years of age are similar.

As given in Table 12.6, the hypothesis related to the activities performed on smartphone by selected respondents of below 30 years of age and above

30 years of age, the nonsignificant results were found in case of criteria, namely, communicate with friends and family members, communication for placing orders, information gathering through browsing the internet in leisure time, Information for comparing various products and downloading apps. The significant result were observed in case of criteria related to doing transactions through mobile includes transferring money online, paying utility bills, mobile banking services, booking movie tickets, and booking tickets for traveling.

TABLE 12.7 Influence of Given Criteria on M-Commerce Adoption Through the Use of Smartphone.

Sr. no.	Criteria for measuring influence on buying decisions	Dissatisfied		Satisfied	
		Up to 30 years (N = 99)	31 years and above (N = 51)	Up to 30 (N = 99)	31 and above (N = 51)
		Number and percentages of respondents			
01	I am satisfied with the Information received	24 (16)	14 (09)	75 (50)	37 (25)
02	I feel doing mobile commerce transactions using smartphone useful in daily life	38 (25)	23 (15)	61 (41)	28 (19)
03	I feel ease while using smartphone technology for doing transactions	35 (23)	19 (13)	64 (43)	32 (21)
04	I trust on making m-commerce transactions using smartphone	46 (31)	24 (16)	53 (35)	27 (18)
05	I will continue to use m-commerce	37 (25)	22 (15)	62 (41)	29 (19)
06	I will recommend the use of m-commerce to others	40 (27)	22 (15)	59 (39)	29 (19)

Source: Compiled by authors.

As given in Table 12.7, considering the analysis of satisfaction of respondents with the use of m-commerce 50% of respondents who belonged to age group of 30 years expressed their satisfaction with the information received from mobile phone; 43% showed their satisfaction with the ease while using smartphone technology for doing transactions; 41% showed satisfaction with smartphone useful in daily life as well as showed their willingness to continue to use m-commerce. In total, 25% of the respondents who belonged to age group of above 30 years showed their satisfaction with the information received from mobile phone; 21% of respondents showed their satisfaction with the ease while using smartphone technology for doing transactions, and 19% of respondents showed their willingness to continue to use m-commerce as well as showed their intention to recommend the use of m-commerce to others.

12.7.6 OVERALL SATISFACTION OF RESPONDENTS FROM M-COMMERCE ADOPTION

Table 12.8 depicted the overall satisfaction of the selected respondents from the use m-commerce services. The results indicated that 45% of the respondents below 30 years of age and 22% of the respondents above 30 years of age expressed their satisfaction with use of m-commerce services, whereas 21% of the respondents below 30 years of age and 12% of the respondents above 30 years of age expressed their dissatisfaction related to m-commerce services.

TABLE 12.8 Overall Satisfaction for M-Commerce Adoption.

Overall satisfaction for m-commerce adoption	Age group (in years) (number and percentages of respondents)	
	Upto 30	**31 and above**
Highly dissatisfied	1 (0.7)	2 (1.3)
Dissatisfied	3 (02)	6 (04)
Neutral	27 (18)	10 (07)
Satisfied	59 (39)	28 (19)
Highly satisfied	9 (06)	5 (03)

Source: Compiled by authors.

12.7.6 DISCRIMINANT ANALYSIS

In order to explore the attributes/variables that are responsible for differentiating the groups, the discriminant analysis is applied based on the observed value as reported by the respondents. Through discriminant analysis, membership in mutually exclusive groups can be identified as well as linear combination of attributes which contribute most to group separation known as canonical discriminant functions (equations) are also identified. Discriminant analysis that involves the determination of a linear equation which is likely to be used in determining regression equation and the form of the equation or function is given as follows:

$$D = v_1X_1 + v_2X_2 + v_3X_3 = v_iX_i + a \quad (12.1)$$

where D = discriminate function; v = the discriminant coefficient or weight for that is variable; X = respondent's score for that variable; a = a constant; i = the number of predictor variables.

The results were obtained by the researcher after running discriminant analysis through the use of SPSS 15.0 software.

In order to predict a respondent's group membership, we first examined whether there are any significant differences between groups on each of the independent variables considering group means and analysis of variance (ANOVA) results.

It becomes clear that there exists mean differences between information, useful, and comfort described in Table 12.9 and these may be good discriminators. Table 12.10 provides statistical evidence of differences between means of cnformation and comfort producing value F's (information = 2.467 and comfort = 2.331). It is not worthwhile to proceed with further analysis if there are no group differences observed.

TABLE 12.9 Group Statistics of Respondents of Age Group up to 30 Years and 31 Years and Above for Selected Criteria.

Age group (in years)		Mean	Standard deviation	Valid *N* (listwise)	
		Unweighted	Weighted	Unweighted	Weighted
Upto 25	Information	22.2525	3.79942	99	99.000
	Useful	23.9697	4.23651	99	99.000
	Comfort	28.2929	4.60978	99	99.000
	Trust	24.5758	4.63591	99	99.000
25 and above	Information	21.0196	5.75323	51	51.000
	Useful	23.7451	6.02609	51	51.000
	Comfort	26.8431	6.94081	51	51.000
	Trust	23.8235	5.71211	51	51.000
Total	Information	21.8333	4.57659	150	150.000
	Useful	23.8933	4.89918	150	150.000
	Comfort	27.8000	5.53330	150	150.000
	Trust	24.3200	5.02119	150	150.000

Source: Compiled by authors.

TABLE 12.10 Tests of Equality of Group Means of Respondents of Age Group up to 30 Years and 31 Years and Above for Selected Criteria.

	Wilks' lambda	'*F*-Test' value	df1 Degree of freedom 1	df2 Degree of freedom 2	Significance
Information	0.984	2.467	1	148	0.118
Useful	1.000	0.070	1	148	0.791
Comfort	0.984	2.331	1	148	0.129
Trust	0.995	0.754	1	148	0.387

Source: Compiled by authors.

As given in Table 12.11, "The Pooled Within-Group Matrices" also supports use of these independent variables as intercorrelations between them found low. The ANOVA is based on the assumption that for each group the variances were equivalent but in discriminant analysis the basic assumption is that the variance–co-variance matrices are equivalent. The null hypothesis that the covariance matrices do not differ between groups is understood by Box's M tests. The researcher expects nonsignificant result from the Box's M test so that the null hypothesis of groups does not differ and can be retained. The log determinants should be equal if the assumption of equality to be held true. When this assumption is tested by Box's M, we are looking for a nonsignificant M to show similarity and lack of significant differences.

In our research, the log determinants does not appear similar and Box's M is 28.794 with F 2.782 which is significant at $p < 0.002$ (Tables 12.12 and 12.13). It indicates that the data differ significantly and null hypothesis is not accepted that the covariance matrices differ between groups.

TABLE 12.11 Pooled Within-groups Matrices Show Correlations Between Selected Criteria.

		Information	Useful	Comfort	Trust
Correlation	Information	1.000	0.498	0.611	0.348
	Useful	0.498	1.000	0.597	0.297
	Comfort	0.611	0.597	1.000	0.466
	Trust	0.348	0.297	0.466	1.000

Source: Compiled by authors.

TABLE 12.12 Log Determinants.

Age group (in years)	Rank	Log determinant
Upto 25	4	10.792
25 and above	4	12.763
Pooled within-groups	4	11.652

The ranks and natural logarithms of determinants printed are those of the group covariance matrices.
Source: Compiled by authors.

TABLE 12.13 Test Results of Box's M Test of Equality of Covariance Matrices.

Box's M		28.794
F	Approximately	2.782
	df1	10
	df2	48795.927
	Significance	0.002

Tests null hypothesis of equal population covariance matrices.
Source: Compiled by authors.

As given in Table 12.14, the eigen value indicates the proportion of variance which is sums of squares between-groups divided by sums of squares within-groups. If the outcome is in the form of large eigen value, it is considered as associated with a strong function but in our example, it is only 0.027.

The correlation between the discriminant scores and the levels of the dependent variable is known as canonical relation, and high correlation indicates a function that discriminates well.

The present correlation of 0.162 is not extremely high. The most perfect indicator of correlation is 1.00.

TABLE 12.14 Eigen Values—Proportion of Variance Explained.

Function	Eigen value	% of variance	Cumulative %	Canonical correlation
1	0.027[a]	100.0	100.0	0.162

[a]Canonical discriminant functions were used in the analysis.
Source: Compiled by authors.

Wilks' lambda indicates the significance of the discriminant function that is the ratio of within-groups sums of squares to the total sums of squares. It is the proportion of the total variance in the discriminant scores not explained by differences among groups.

Table 12.15 indicates a nonsignificant function ($p < 0.000$) and provides idea that the proportion of total variability is not explained, that is, it is 71.1% explained by differences among the groups.

TABLE 12.15 Wilks' Lambda Scores—of the Total Variance in the Discriminant Scores.

Test of function(s)	Wilks' lambda	Chi-square	df	Significance
1	0.974	3.879	4	0.423

Source: Compiled by authors.

Table 12.16 provides an index of the importance of each predictor and the sign indicates the direction of the relationship. Comfort score (0.725) was the strongest predictor, whereas information (0.639) was next in importance as a predictor. With large coefficients, these two variables, that is, comfort and information stand out as those that strongly predict allocation.

TABLE 12.16 Standardized Canonical Discriminant Function Coefficients.

	Function
	1
Information	0.639
Useful	−0.636
Comfort	0.725
Trust	0.064

Source: Compiled by authors.

The structure matrix Table 12.17 shows the correlations of each variable with each discriminate function and it is another way of indicating the relative importance of the predictors and it can be observed that the same pattern (Table 12.12) holds as 0.787 for information and 0.765 for comfort.

TABLE 12.17 Structure Matrix—Correlations of each Variable with each Discriminate Function.

	Function
	1
Information	0.787
Comfort	0.765
Trust	0.435
Useful	0.133

Pooled within-groups correlations between discriminating variables and standardized canonical discriminant functions.
Variables ordered by absolute size of correlation within function.
Source: Compiled by authors.

Table 12.18 indicates the unstandardized scores concerning the independent variables called as "canonical discriminant function coefficients." It is the list of coefficients of the unstandardized discriminant equation. It operates just like a regression equation, and in our example, we derive equation as follows.

$$D = (0.140 \times \text{respondent's score for information}) + (-0.130 \times \text{respondent's score for useful}) + (0.132 \times \text{respondent's score for comfort}) + (-0.013 \times \text{respondent's score for trust}) + 3.0934 \quad (12.2)$$

TABLE 12.18 Canonical Discriminant Function Coefficients.

Grouped indicator items	Function
	1
Information	0.140
Useful	−0.130
Comfort	0.132
Trust	0.013
(Constant)	−3.934

Unstandardized coefficients.
Source: Compiled by authors.

It becomes clear from the discriminant analysis that the information provided by the companies' application need to be consistent and precise, and then this will enlarge smartphone user satisfaction and trust which will guide the user to make the initial purchase.

In Figure 12.1, a simple regression model is presented where one observed variable, the overall satisfaction with m-commerce adoption is predicted as a linear combination of the other four observed variables, namely, information, useful, comfort, and trust with m-commerce adoption. There are some other variables (other than selected five variables) that also assumed to have an effect on overall satisfaction with m-commerce adoption for which the model assumes "1" as standardized regression weights. Each single-headed arrow represents a regression weight.

The value shown against two-sided arrows (0.47, 0.59, 0.50, 0.35, 0.62, and 0.30) is the correlation between four observed variables, information, useful, comfort, and trust with m-commerce adoption.

The values shown with single-sided arrow (0.34, 0.12, 0.15, and 0.12) are standardized regression weights. The value 0.35 is the squared multiple correlation of overall satisfaction with combined effect of four variables related with m-commerce adoption. It means the selected respondents' overall satisfaction considering four observed variables, namely, information, useful, comfort, and trust with m-commerce adoption is influenced by information (0.34) followed by comfort (0.15), and for other two factors useful and trust, the standardized regression weights were observed as 0.12 for both, respectively.

The study also suggests and supports the fact that the respondent's overall satisfaction is the result of the combined effect of four variables concerned with m-commerce adoption, namely, information, usefulness, comfort, and trust. The major predictors of overall satisfaction were found as information availability and comfort with adoption of m-commerce services. People preferred more information for mobile transaction and the comfort for making transactions.

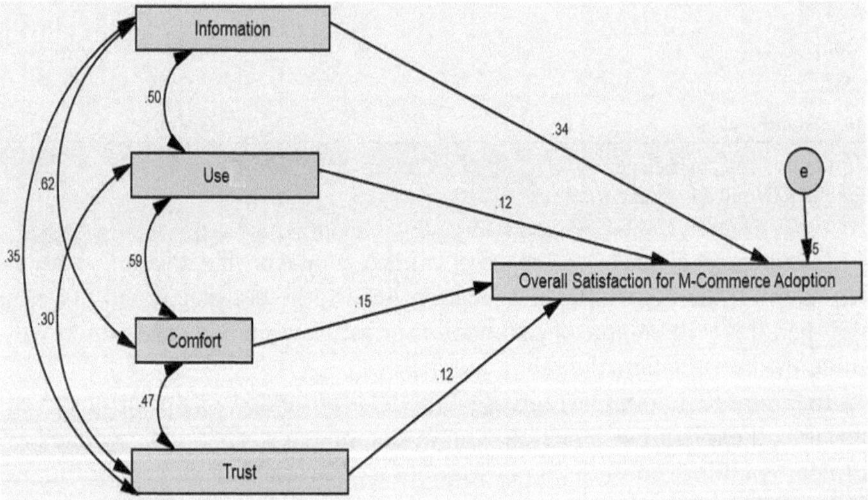

FIGURE 12.1 SEM showing relationship between selected attributes of transaction through smartphone and overall satisfaction.
Source: Compiled by authors.

12. 8 DISCUSSIONS AND MANAGERIAL IMPLICATIONS

The implications for the marketers can be regarded as counsel to those organizations who are exploring the mobile commerce idea targeted toward existing consumers or nonexistent users. Therefore, the implications given below are based on the empirical findings, analysis, and concussions conducted during the research.

Mobile commerce is no longer a question of if, or even when. It is in practice everywhere and all around. The question for marketer and IT personnel is, how to formulate right strategies for growth of m-commerce. Companies need to recognize the inhibitors and challenges involved in m-commerce which is reflected in consumer perceptions such as mobile commerce is unsafe, lack of security, confusion over business models related to money flow, insufficient connectivity and speed, and the immaturity of technology to integrate with existing e-commerce and IT back-end systems, and so on.

While collecting the data using structured nondisguised questionnaire and personal interview schedules with the respondents, it was revealed that the "secured transaction environment" is the most important aspect which mobile shoppers do consider while making a payment. With regards to this, the mobile shoppers also showed complacency for making payments involving a huge sum of money.

Some respondents also revealed that browsing an online catalogue with a mobile device is also a difficult and a complex task. Thus, marketers must develop a robust strategy to address this issue.

Considering the mean score of respondent's responses on activities performed using smartphone, it becomes clear that the use of smartphone by respondents of age group of below 30 years and above 30 years is similar and includes communication with friends and family members and collecting information by downloading apps, browsing internet in leisure time, and gathering information for comparing various products. So far as doing transactions through use of smartphone is concerned, the respondents below 30 years of age use smartphone for transferring money online; paying utility bills; using mobile banking services, and for booking movie and traveling tickets. The selected respondents above 30 years of age hardly use the smartphone for doing any monetary transactions.

Based on results of chi-square test, it can be inferred that the young people below 30 years of age and middle-aged people above 30 years of age make the use of smartphone for communication purpose and as a source of information. But the use of smartphone is more prominent amongst young respondents of below 30 years of age for doing transactions such as transferring money online, paying utility bills, mobile banking services, booking movie tickets, and booking tickets for traveling. The marketers need to focus on understanding the root cause of why the people above 30 years of age group are afraid of doing transactions through use of smartphone. What kind of changes or reform are required for attracting large number of people to transact through mobile apps? The answers to such questions lead to better formulation of the strategies to maintain and increase the transactions through mobile commerce.

The marketer needs to monitor the kind of information surfing about products and services done by the people through mobile transactions so that the desired information can be made available as well as up to date upgraded information can be maintained based on information requested and referred by the people. There is need to focus on providing the experience of ease while using smartphone technology for doing transactions through mobile apps.

Making the perfect mobile commerce app means not only designing in the right features to cover these bases but also taking advantage of the best enabling technology. The technologies close to equal in their importance for mobile commerce success include physical security built into mobile devices, methods to securely communicate payments, robust digital wallet security, and easier integration with payment gateways so that mobile devices can connect easily to terminals of mobile services providers. With such efforts, the marketer will definitely be able to strengthen the intention of users to make the continuous use of m-commerce as well as to spread positive word-of-mouth leading to further growth of m-commerce practices. Ultimately, mobile commerce is no different than other commerce. If the marketer wants people to buy their products, they need to make it convenient and fun, with a foundation of loyalty and trust. In order to provide satisfaction to m-commerce users, the companies need to think about the strategy from the perspective of the end user and consider what the experience of the customer will be like.

The companies should develop responsive design for their different devices on their sites as the users of the different devices are visiting the site not for same reason. User of tablet device may be visiting the site for a completely different purpose than that of web user through smartphone.

Responsive design to different devices does not solve the user experience or provide satisfaction but an attempt to develop responsive design will lead to a better experience. When a company invest in their sites, they need to consider what tasks or information the user of smartphone will need versus that of a web user and prioritizes which features are most prominent and easy to access, which dramatically increases the likelihood of creating a satisfying experience. Generally, apps provide a better experience because they are attuned to the frequent user who are aware of the capabilities of the app, more loyal, and committed to the brand because they have taken time to download the app to their device and such familiarity leads to higher satisfaction. However, the number of users who download the app is generally smaller than those who access the mobile site. Company needs its site to be as strong as the app if it wants to drive traffic to the app and secure long-term loyal customers.

12.9 CONCLUDING REMARKS

This research study pertains to the experiential value derived by the smartphone user satisfaction considering the selected attributes of information, usefulness, comfort, and trust toward mobile applications.

The findings suggested that quality of information and the ease and comfort with use of m-commerce are perceived as variables providing user satisfaction. The identified input factors perceived quality of information; comfort and ease of using smartphone and usefulness of mobile applications among the smartphone users were found to have a strong influence on user trust and satisfaction. The outcome of the study will guide companies that offer m-commerce-related products in the selection of digital products and in pursuance of future commercial opportunities.

The shopping experience in the virtual context plays a vital role for the mobile shopper. The marketers should make an effort to create and deliver an enriched online experience through web with a configuration of online functionality, information, and emotional appeals in the products or services offered. This will create a positive and engaging virtual interaction which will induce the shopper to use mobile shopping applications.

In recent years, the use of new information and communication technologies such as mobile telephony has experienced unprecedented growth, providing consumers with new ways of relating and more agile, personalized communication. The mobile's complementary nature and synergy with other direct sales systems, such as internet and television, have made it an excellent interactive marketing tool which helps companies to complement their communication and sales actions in other media.

Given the generalized use of mobiles and the scarcity of studies which analyze the profile of the users of this system, this piece of work has contributed to the research by providing an understanding of the factors influencing m-commerce adoption.

Companies are building the foundation for the Indian mobile commerce by enabling cashless, on-the-go financial transactions. From cab rides to movie tickets to utility bills, almost anything can be paid via simple mobile apps. However, mobile wallets in particular can add additional value by bridging payment with discovery and procurement to offer more seamless mobile shopping experiences—either through integration with existing ecommerce apps or by the creation of their own in-app features.

The innovations in mobile-first hyper-local business are emerging all over the world. India is especially powerful as it is a market of millions of newly connected mobile users, many of whom are leapfrogging web-based commerce altogether. Though the year 2015 has educated us about significant transition to mobile in India, many new innovations are tapping the economy for further growth. Answers to certain questions will determine the future of m-commerce such as which innovative mobile-first businesses are passing fads and which are here to stay? Will mobile experiences in India primarily be web-based, app-based, or some combination of the two? Can hyper-local commerce ever become hyper-profitable? The retailers in order to seek answers to such questions need to wait and see how the Indian mobile commerce ecosystem continues to evolve in future.

KEYWORDS

- **m-commerce**
- **m-shopping**
- **adoption**
- **satisfaction**
- **smartphone**

REFERENCES

Bustos, L. *The Secrets of Mobile Commerce Satisfaction*: An Interview with Larry Freed, 2013. http://www.getelastic.com/the-secrets-of-mobile-commerce-satisfaction-an-interview-with-larry-freed/ (accessed Jan 19, 2017).

Deepika, R.; Karpagam, V. Antecedents of Smartphone User Satisfaction, Trust and Loyalty Towards Mobile Applications. *Indian J. Sci. Technol.* **2016,** *9* (3), 1–8.

Harris, P.; Rettie, R.; Kwan, C. C. Adoption and Usage of M-Commerce: A Cross-cultural Comparison of Hong Kong and the United Kingdom. *J. Electron. Commer. Res.* **2005,** *6* (3), 210–224.

Hongjiang, X. M-commerce Development in Developing Country: Users' Perspective. *Scholarship and Professional Work—Business*, Paper 210, 2013. http://digitalcommons.butler.edu /cob_papers/210 (accessed Jan 19, 2017).

Huo, Y.; Zhang, B.; Ma, L.; Zhan, P. Research of Influences for M-commerce Service to Chinese Farmers' Consumption Network. *Int. J. Innov. Comput. Inf. Control* **2011,** *7* (4), 1995–2008.

Kini, R. B. Adoption and Evaluation of Mobile Commerce in Chile. *Electron. J. Inf. Syst. Eval.* **2009,** *12* (1), 75–88.

Lee, H.-H.; Lee, S.-E. Mobile Commerce: An Analysis of Key Success Factors. *Shopp. Cent. Res.* **2007,** *14* (2), 29–62.

Malhotra, N. K. *Marketing Research: An Applied Orientation*, 5th ed.; Pearson Prentice Hall: India, 2007; p 315.

Naware, A. M. M-commerce in India. *Int. J. Adv. Res. Comput. Commun. Eng.* **2016,** *5* (4), 913–915.

Nunnally, J. C. *Psychometric Theory*; Tata McGraw-Hill Publishing Ltd.: New Delhi, 1981.

Saleh, Z. I.; Mashhour, A. Consumer Attitude Towards M-commerce: The Perceived Level of Security and the Role of Trust. *J. Emerg. Trends Comput. Inf. Sci.* **2014,** *5* (2), 111–117.

Saneifard, R. Exploring Factors Affecting Mobile Commerce B2C Adoption in Iran. Masters Thesis, Lulea University of Technology, Continuation Courses Marketing and E-commerce, Department of Business Administration and Social Sciences, Division of Industrial marketing and E-commerce, 2009.

Sharat, P. How Indian Mobile Commerce will Change in 2016, 2016. https:// www.techinasia.com/talk/indian-mobile-commerce-change-2016 (accessed Jan 14, 2017).

The Internet and Mobile Association of India (IAMAI). 2014. http://www.iamai.in/media/details/3659#sthash.KZuhAEej.dpuf (accessed Jan 16, 2017).

Zheng, H.; Li, Y.; Jiang, D. Empirical Study and Model of User's Acceptance for Mobile Commerce in China. *Int. J. Comput. Sci. Issues* (*IJCSI*) **2012,** *9* (6), 278–283.

CHAPTER 13

M-COMMERCE: A BOON OR BANE FOR GROCERY SHOPPERS

ANUJA SHUKLA[1*] and SHIV KUMAR SHARMA[2]

[1]*Noida International University, Greater Noida, Gautam Budh Nagar, Uttar Pradesh, India*

[2]*Department of Management, Faculty of Social Sciences, Dayalbagh Educational Institute (Deemed University), Agra, Uttar Pradesh, India*

[*]*Corresponding author. E-mail: anuja.gshukla@gmail.com*

ABSTRACT

Mobile commerce has become an important topic, which has drawn attention in both industry and academia. India is soon to become the second-largest internet users' base in the world and has largest number of internet users in a free market democratic setup. The advancement of internet in mobile has opened new avenues for the marketers to reach the consumers. Many firms such as Grofers, PepperTap, and Amazon, etc., have launched their grocery shopping mobile applications (apps). The purpose of the study was to examine the role of mobile apps in shopping for grocery products and to evaluate the impact of demographical characteristics on mobile shopping. In-depth review of literature was done to analyze the diffusion of mobile technology in other fields. The objectives of the study were accomplished by using Technology Acceptance Model (TAM) proposed by Davis (1989)[1] to test how consumers perceive the use of mobile apps in shopping of grocery using mobile apps and performing demographical and content analysis. The data were collected from 94 consumers using standard questionnaire through online platform across India and were analyzed using partial least squares structural equation modeling. Firstly, confirmatory factor analysis (CFA) was conducted to test the outer validity of model which was followed by testing of hypotheses. The results are in agreement with the previous studies such as use of TAM in online grocery shopping, and understanding of students' intention to use e-learning. All of the constructs were proposed in TAM and their relationships were found to be significant in the study. Descriptive analysis was done to show that the mobile apps are considered useful for shopping yet is not very well adapted by the consumer in grocery sector.

[1]Perceived Usefulness, Perceived Ease of Use, and User Acceptance of Information Technology. *MIS Q.* **1989,** *13* (3), 319–340.

13.1 INTRODUCTION

India still practices the culture where the merchants deliver groceries and household staples to consumers' residence (Neilsen, April 2015). The similar concept of home delivery shopping is reemerging across the world but with a twist. In recent years, online shopping has seen an exponential growth by introduction of modern technology in various sectors of business. The home delivery model has shifted from the traditional *kirana* stores to the shopping through mobile devices. The application (app)-based shopping has changed the traditional shopping to the modern way of mobile shopping (m-shopping).

Mobile commerce is the use of mobile phone for making purchases. Wong and Sheng (2012) defined m-commerce as "any monetary transactions related to purchases of goods or services through internet enabled mobile phones or over the wireless telecommunication network." Mobile device seems to assist the consumer for doing his shopping (Yang, 2010). The shopping with a mobile device is an important topic and is given high attention in both industry and academia (Wang et al., 2015). The industry has marked a remarkable shift in consumer's intention of buying using mobile devices. Mobile apps are found to be the most significant way for Indians to access e-commerce (Livemint, 2016). Flipkart, an online shopping website gets 75% of its traffic through mobile apps. Amazon India emphasized the importance of app-based shopping by enabling their customers to shop anytime, anywhere, and anyway they want, using a mobile and a desktop site (Livemint, 2016). Therefore, mobile apps have been found to play a vital role in the consumers' buying process.

13.1.1 INDUSTRY PROFILE

According to AC Neilson Report (April, 2015), the number of Indians opting for shopping on mobile apps is on rise. The number of smartphone owners using mobile apps of e-commerce companies has gone up from 1680 (May 2014) to 4320 (May 2015), based on real-time mobile usage data across 8000 handsets in the country (Nielsen Informate Mobile Insights, July 2015). With India being the next big frontier for smartphone consumption, these numbers point to a larger shift in consumer behavior.

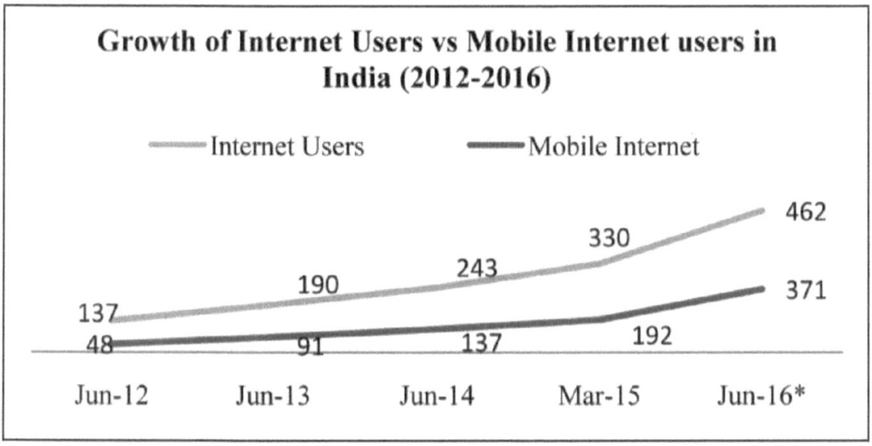

FIGURE 13.1 Growth of internet users versus mobile internet users in India (2012–2016). *Source:* Adapted from India@Digital Bharat, by IAMAI and BCG, Feb 2016 (Figures in million users, *estimated). https://www.academia.edu/34169219/India_at_Digital._bharat_CREATING_A_200_BILLION_INTERNET_ECONOMY)

Internet and Mobile Association of India (IAMAI, November 2015) reported that the number of users using internet on mobile is almost 58.5% of the entire internet accessed in India during March 2015, which is expected to cross 80% by June 2016 (Fig. 13.1). This increase in purchase and use of smartphone can be credited to the reducing average selling price of smartphones (Neeraj, 2016). In 2015, the consumers' share of using internet has increased from 54% to 64% as compared to previous year. The possible reasons could be the improvement in the infrastructure and improved internet speed. As shown in Figure 13.2, it can be concluded that the several dimensions which are related to m-commerce in grocery are showing increasing trend. Therefore, it is expected that m-commerce in grocery is to gain momentum soon and fast.

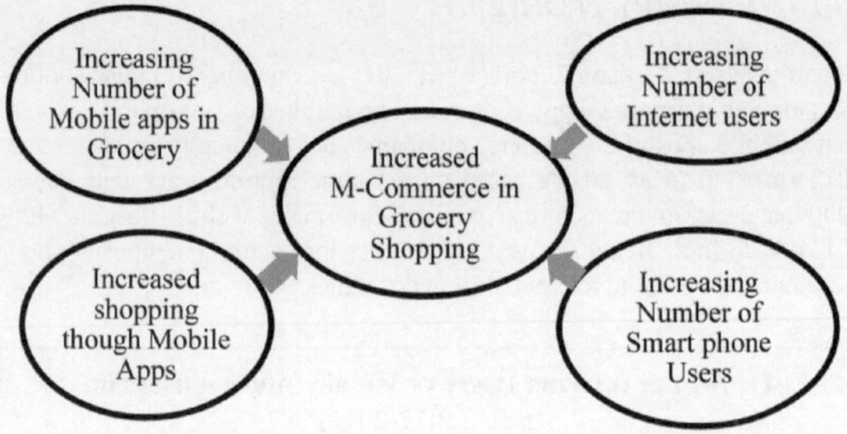

FIGURE 13.2 Holistic view of development of m-commerce in grocery.
Source: Compiled by authors.

13.1.2 ELECTRONIC GROCERY MARKET IN INDIA

Electronic grocery permits the consumers to buy grocery online through various payment mediums such as cash on delivery and online payments without any constraint of time and location. The e-grocery market covers large part of the online retail market. The rising disposable income and growing technical awareness among consumers is fueling the market growth in India. The increased internet penetration has been a fundamental driver behind the growth of the e-grocery market in India. Social media has impacted the growth of online consumers dynamically. For instance, BigBasket's Face-book page updates their customers about the extensive options and discounts available in different categories.

The top five e-grocery vendors in the Indian market are AaramShop, BigBasket, Jiffstore, LocalBanya, Nature's Basket, ZopNow (E-Grocery Market in India, 2015). Other vendors in the market include BazaarCart, Farm2Kitchen, Grofers, My Grahak, My247Market, PepperTap, Reliance Fresh Direct, and Veggie Bazaar, etc. These e-grocery vendors provide home delivery of various products including staples, edible oil, beverages, milk and dairy products, baby and child care, biscuits and cookies, laundry supplies, household and cleaning, cereals and breakfast, snacks, etc. Staples, edible oil, beverages, and milk and dairy products constitute nearly 46% of the market shares in the e-grocery market. These items are considered as essentials and are bought in high quantities (E-Grocery Market in India,

2015). One-quarter of global respondents are already ordering grocery products online for home delivery, and more than half (55%) are willing to use it in the future (Neilsen, April 2015). Mobile has become an important channel for shoppers as they research and purchase products (eMarketer, January 2016). The continued growth in smartphone subscriptions is expected to lead to an accelerated growth in data usage, whereas monthly mobile data consumption is expected to increase 18-fold by 2020 over current levels (Livemint, 14 Feb, 2016).

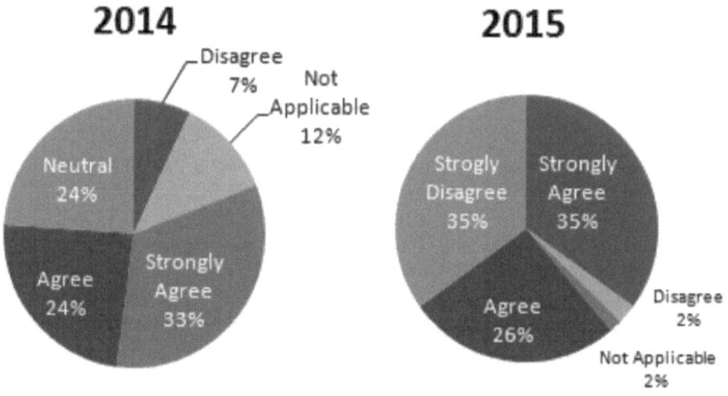

FIGURE 13.3 Retailers worldwide that have experienced major m-commerce growth.
Source: Adapted from "Key Business Drivers and Opportunities in Cross-Border Ecommerce: Entering an Omnichannel World 2015" by Payvision, Oct 30, 2015. https://hollandfintech.com/wp-content/uploads/2015/11/key-business-drivers-and-opportunities-2015.pdf).

The retailers worldwide strongly agree that they have experienced a major m-commerce growth both in 2014 and 2015 while retailers who agree also constitute a large share (Fig. 13.3). This data shows that there is recognition among retailers regarding growing m-commerce around the world.

13.1.3 NEED OF THE STUDY

India is soon to overtake the United States and become the world's second-largest market for smartphones exceeding 200 million smartphone users (eMarketer, January, 2016). The consumers are using the smartphone as a tool for doing shopping sitting in their comfort zone. Also, India's current government is pushing digitization through an ambitious project called Digital India. The various sectors have received an overwhelming response in fashion industry, consumer durable, electronic items, etc., but m-shopping is not so popular and is in nascent stage in the grocery sector. This study aims to fulfill this gap by testing usage of mobile technology in the field of grocery shopping and users attitude toward it.

13.2 LITERATURE REVIEW

With increasing number of grocery apps, the way grocery shopping is done in India has changed. It is an emerging market generating high revenues (eMarketer, January 2016). So, it becomes important to study what motivates the consumer for turning to mobile apps for grocery shopping. Shopping through mobile apps is gaining high importance in the field of marketing and retailing (Wang et al., 2015). This is a comparative new area of study (Fig. 13.4) but still it poses an immense growth potential in future and literature is found limited in this field. (Holmes et al., 2013; Groß, 2015). It was found during literature review that most studies are conducted in the United States, Taiwan, China, etc., and are found very less in other parts of the world (Groß, 2015). This in-depth review of literature has identified that the studies related to the adoption mobile phone toward grocery shopping has not been conducted for Indian consumers. This was identified as a gap that motivated the researcher to conduct this study. Lu and Su (2009) investigated the impact of purchase intention on m-shopping websites and found that the influence of mobile as a tool is limited in grocery sector as compared to other sectors. Since there is a growth in mobile technology in India and also due to increased penetration of smartphones, it is expected that it will certainly affect grocery sector in India. This study explains the role of technology (mobile apps) in grocery shopping among Indian consumers.

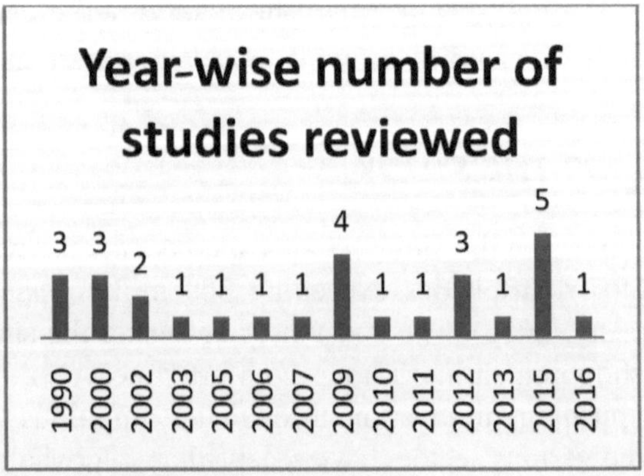

FIGURE 13.4 Year-wise number of studies reviewed.
Source: Compiled by authors.

Technology Acceptance Model, popularly known as TAM was designed to predict the likelihood that a new technology will be adopted with in a group of individuals or organizations (Davis et al., 1989). TAM was designed on the basis of Theory of Reasoned Action (TRA) given by Fishbein and Ajzen (1975). The aim of TAM was to provide a model to the researchers that can help them analyze the cause of rejection or acceptance of technology. TAM establishes a relationship between intention, attitude, and the actual behavior derived from TRA and two new variables: perceived ease of use and perceived usefulness (Table 13.1). TAM was meant to explain the behavior of computer usage and was found to be well suited for modeling computer acceptance (Davis, 1989). TRA is a well-established model in explaining human behavior developed in 1975 by Fishbein and Ajzen. According to TRA, an individual's behavior is result of behavioral intention which is the result of his attitude toward it. This relation is influenced by subjective norm and attitude.

TABLE 13.1 Definition of Constructs.

Model	Construct	Definitions in context to study	Author
TAM	Perceived usefulness (U)	The extent to which potential users expect that mobile app will benefit their grocery shopping	Davis (1989)
	EOU	The degree to which a consumer believes that using a mobile app would be free of effort	
TRA	BI	How likely one is to buy grocery using mobile application	Fishbein and Ajzen (1975)
	Attitude (A)	It measures feeling of an individual (positive or negative) about buying grocery through mobile application	

BI, behavioral intention; EOU, perceives ease of use; TAM, technology acceptance model; TRA, theory of reasoned action.
Source: Compiled by authors.

Mobile commerce is a relatively new area of research. With the advancement of internet on mobile, the consumers are shifting toward mobile commerce or the use of mobile phones for shopping. Customers use mobile devices because the technology provides convenient access, which leads them to incorporate m-shopping into their habitual routines (Wang et al., 2015). Shopping through mobile is examined in various fields (Kim et al.,

2016). Similar study was conducted on Hong Kong consumers to test the use of mobile commerce which showed that attitude of users toward mobile commerce services was found to be impactful (Fong and Wong, 2015).

The literature on particular use of mobile for grocery shopping was very limited. Fröhlke and Pettersson (2015) used TAM to analyze the intention of Swedish consumers for using mobile as a tool for purchasing grocery. The researcher identified the factors that influenced the intention of consumers. TAM has successfully explained adoption of numerous technologies that range from software package to different online services (Al-Rahimi and Musa, 2013). Moon and Domina (2015) conducted an exploratory study focused on American and South Korean consumers' intention to use fashion mobile apps (FMA) to purchase fashion products and how that intention is affected by cultural differences. Attitude toward using FMA was the most significant predictor of intention to purchase fashion products through FMAs for both South Korean and American participants. Hu et al. (1999) used TAM to explain physicians' decisions to accept telemedicine technology in the healthcare context. McKechnie et al. (2006) studied TAM to understand consumers' behavior toward use of internet in area of financial services. Ko et al. (2009) examined the potential of a consumer technology adoption model in fashion products. Park (2009) tested adoption of e-learning by students based on TAM.

Conceptual studies were conducted by Davis et al. (1989), Pavlou (2003), Zhou et al. (2007), and Fishbein and Ajzen (1975). Ajzen (2002) clarified conceptual and methodological ambiguities surrounding the concept of perceived behavioral control. Perceived behavioral control, perceived self-efficacy, and perceived controllability were discussed in detail using concept of theory of planned behavior (TPB). It is shown that perceived control over performance of a behavior, though composed of separable components that reflect beliefs about self-efficacy and controllability, can nevertheless be considered a unitary latent variable in a hierarchical factor model. It was further argued that there is no necessary correspondence between self-efficacy and internal control factors, or between controllability and external control factors. Pavlou (2003) predicted consumer acceptance of e-commerce by proposing a set of key drivers for engaging consumers in on-line transactions. The results suggested that ease of use may not be an important factor that directly influences consumer perceptions to accept e-commerce. Liu et al. (2016) examined the relationship between experience and repurchase intention. Amaro and Duarte (2015) proposed a model to analyze the reasons that affect intentions of consumer to purchase travel online using TAM, Innovation Diffusion theory (IDT), and TPB constructs.

Venkatesh and Davis (2000) developed and tested a theoretical extension of the Technology Acceptance Model (TAM2) that explains perceived usefulness and usage intentions in terms of social influence and cognitive instrumental processes. The study was conducted on accepting of technology in doing work by floor supervisor, people in financial services, accounting services, and investment banking. The study incorporated additional theoretical constructs spanning social influence processes (subjective norm, voluntariness, and image) and cognitive instrumental processes (job relevance, output quality, result demonstrability, and perceived ease of use). Questionnaires were administered to potential users at three points in time: after initial training (T1), 1 month after implementation (T2), and 3 months after implementation (T3). Self-reported usage behavior was measured at T2 and T3, and also 5 months after implementation (T4). Multiple regression analysis showed that TAM2 was strongly supported across four organizations and three points of measurement (preimplementation, 1 month post implementation, and 3 months post implementation). Li and Huang (2009) extended TAM model by adding perceived risk in existing TAM. The results suggest the need for consideration of perceived risk as an antecedent in the technology acceptance. Tseng and Hsu (2010) applied e-WOM communications and TAM toward consumers' intention toward Eee PC (innovative technology). Kim (2012) tried to understand the purchase mechanism in online shopping. Their study established an integrated model of initial trust and TAM. The results reinforced the theory that a relationship between belief and intention was better explained when it was mediated by attitude. Wong et al. (2012) analyzed the various reasons that motivated consumers to adopt m-shopping. Al-Rahimi et al. (2013) used TAM to analyze the students' perception toward use of social media for collaborative learning by testing TAM for students in Malaysia. Wu and Wang (2005) presented an extended technology acceptance model (TAM) that integrates innovation diffusion theory, perceived risk, and cost into the TAM to investigate what determines user mobile commerce acceptance. Extensive literature reviews were conducted by Surendran (2012) and Chuttur (2009). An extensive survey of extant related studies was synthesized into OSAM (Online Shopping Acceptance Model) to explain consumer acceptance of online shopping by Zhou et al. (2007). Shankar and Balsubramanian (2009) found in their literature review that marketers would have to work diligently to enhance customer receptiveness to mobile marketing in the broader population, and, in the short run, need to focus on customers who are receptive to these initiatives.

13.3 RESEARCH METHODOLOGY

For the purpose of current study, descriptive research design with survey technique was undertaken. The main objectives of the study were:

1. To analyze the role of technology in using mobile commerce for grocery shopping.
2. To analyze the consumer buying behavior based on consumers' demographic characteristics.

13.3.1 *PROPOSED RESEARCH MODEL AND HYPOTHESIS*

The objectives and hypotheses of the study are based on the conceptual framework of TAM by Davis (1989) (Fig. 13.5). The researcher aims to test TAM in adoption of mobile apps for grocery shopping among Indian consumers.

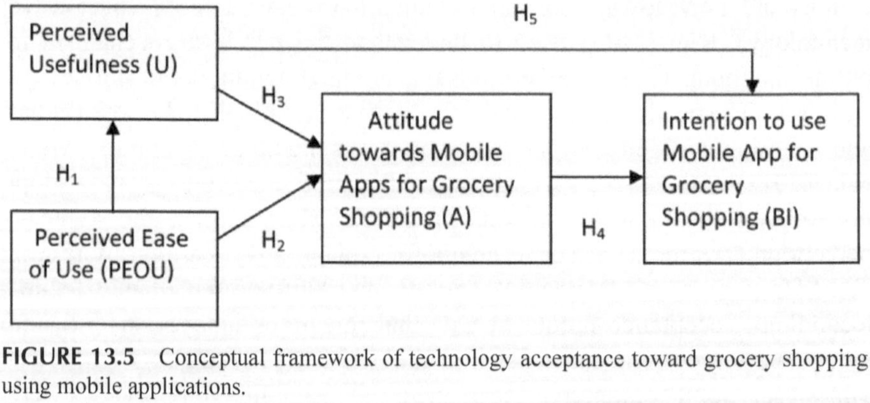

FIGURE 13.5 Conceptual framework of technology acceptance toward grocery shopping using mobile applications.
Source: Adapted from Davis (1989).

13.3.1.1 *PERCEIVED EASE OF USE AND PERCEIVED USEFULNESS*

TAM suggests that perceived ease of use have strong linkage with perceived usefulness. (Davis et al., 1989; Venkatesh and Davis, 2000; Venkatesh and Morris, 2000). Thus, it is hypothesized that

H1: Perceived ease of use is positively related to perceived usefulness.

13.3.1.2 PERCEIVED EASE OF USE AND ATTITUDE TOWARD MOBILE APP FOR GROCERY SHOPPING

Perceived usefulness has been found to impact attitudes toward all technologies while previous research have shown that impact of ease of use on attitudes toward using technology depends on the specific technology or situation (Wu and Wang, 2005). Since m-shopping is a fairly new technology, ease of use might contribute toward the attitude toward using mobile app. Thus, it is hypothesized that

H2: Perceived ease of use is positively related to attitude toward mobile app for grocery shopping.

13.3.1.3 PERCEIVED USEFULNESS AND ATTITUDE TOWARD MOBILE APP FOR GROCERY SHOPPING

In previous researches, it has been found that perceived usefulness has a strong relationship to user acceptance, attitude, and behavior (Ko et al., 2009). Also, it was found that perceived usefulness was positively related to attitude toward using new technology, such as the mobile internet (Kim et al., 2007). Thus it is hypothesized that

H3: Perceived usefulness is positively related to attitude toward mobile app for grocery shopping.

13.3.1.4 ATTITUDE TOWARD MOBILE APP FOR GROCERY SHOPPING AND INTENTION TO USE MOBILE APP FOR GROCERY SHOPPING

Davis (1989) found that attitude has minor impact on behavior intention while Fong and Wong (2015) found attitude to be the most important factor in determining behavior of consumer toward use of mobile apps. Also, attitude toward using fashion merchandise was found to be the most significant predictor of intention to purchase fashion products for both South Korean and American participants (Moon and Domina, 2015). Thus, it is hypothesized that

H4: Attitude toward mobile app for grocery shopping is positively related to intention to use mobile app for grocery shopping.

13.3.1.5 PERCEIVED USEFULNESS AND INTENTION TO USE MOBILE APP FOR GROCERY SHOPPING

Prior studies on consumer behavior found that a consumer's perceived ease of use and perceived usefulness of a system affect behavior intention of the consumer toward the system (Gao and Bai, 2014; Li, 2009). Thus it is hypothesized that

H5: Perceived Usefulness is positively related to intention to use mobile app for grocery shopping.

13.3.2 SAMPLE DESIGN

The target population for study comprised consumers who use internet on smartphone and are above 18 years as supported by Wong et al. (2012). Purposive sampling was used to draw appropriate representative sample from the population. The reason for selection of purposive sampling was to make sure that the consumers that were sampled should know about m-shopping and should be able to evaluate the same as adapted by McKechnie et al. (2006).

13.3.3 MEASUREMENT INSTRUMENTS

The current study focuses on how the consumers use technology for buying products. Here, we are analyzing the role of mobile app (technology) in facilitation of the grocery buying among Indian consumers. The questionnaire items are adapted from the standard scales in context of technology usage. Perceived usefulness is measured by four items, perceived ease of use is measured by four items, attitude toward mobile app for grocery shopping is measured by three items, and intention to use mobile app for grocery shopping is measured by four items (Table 13.2). The respondents marked their responses on a 5-point Likert scale ranging from strongly agree (5) to strongly disagree (1).

TABLE 13.2 Overview of Questionnaire Items.

Constructs	Items	Notation	Developer
Perceived usefulness	I believe that the use of a mobile device would make my grocery shopping process more effective	PU1	Fröhlke and Pettersson (2012)
	I believe that the use of a mobile device would make my grocery shopping process more convenient	PU2	
	I think that I would save time by using a mobile device while shopping for groceries	PU3	
	I believe that, in general, using a mobile device in my grocery shopping process would have been useful	PU4	
Perceived ease of use	It is easy to purchase grocery through mobile shopping	PEOU1	Fröhlke and Pettersson (2012)
	It is easy to learn how to purchase through mobile shopping	PEOU2	
	It is easy to use mobile shopping	PEOU3	
	I think it would be easy to use a mobile device in the grocery shopping process	PEOU4	
Attitude toward mobile app for grocery shopping	I believe it would be advantageous to use my mobile device in my grocery shopping process	A1	Fishbein and Ajzen (1975)
	I think it would be a good idea to use a mobile device when shopping for groceries	A2	
	I think it would be positive to be able to use my mobile device when shopping for groceries	A3	
Intention to use mobile app for grocery shopping	I intend to keep using of mobile app for buying groceries in the future	BI1	Al-Rahimi et al. (2013)
	I would not mind to switch over to another technology if it has better functionalities	BI2	
	I intend to increase the use of my mobile app in the future	BI3	
	I intend to recommend my friends to using of mobile app for grocery shopping in the future	BI4	

Source: Compiled by authors.

13.3.4 DATA COLLECTION

Pretest: The questionnaire was pretested on a sample size of 30 respondents. The overall Cronbach's alpha was 0.952 which was acceptable (>0.7). The responses were taken by interview method to check if respondents faced any problem in answering the questions. As the questionnaire was shared online, need was felt to identify the location of the respondents. So, one question was added in the questionnaire inquiring for residence of respondents. Also, need was felt to identify possible reasons of purchase/nonpurchase by customer, therefore one more question was added for the same. The revised questionnaire was used for data collection.

Primary data was collected using self-administered questionnaire using online platform in two phases. This method is considered quite appropriate as the time frame for study is fixed and is a fast and cheap method for data collection from infinite population (Malhotra, 2010). In first phase, the questionnaire link was shared on Facebook account of researcher, Facebook pages of grocery apps such as PepperTap, Askme grocery, LocalBanya, etc. In the second phase, the questionnaire link was sent as e-mail to contacts, colleagues, and other fellow researchers. The data was automatically recorded on spreadsheet linked to Google forms. The data collection period ranged from 6 May to 7 July 2016.

13.4 DATA ANALYSIS

13.4.1 DATA CLEANING

Total 101 responses were collected. The completely blank four forms were deleted reducing the data to 97 usable questionnaires. Then the forms of people who do not possess smartphone were removed leading to 94 questionnaires. Further it was seen that, all the 94 forms included respondents who were using internet on their smartphones. Table 13.3 shows that variables had certain percentage of missing values, except gender and age. Total number of missing values was 22; howsoever, the percent of values missing for each variable was maximum 2.1% which is less than 10% and does not significantly affect the data (Hair, 2011). It was important to see if the missing values show any pattern. Little MCAR (missing completely at random) test was used to test pattern of missing values. Little MCAR is like testing of hypothesis to see if the values missing are at random or there is any pattern. The results of Little MCAR test measured 0.768 (Chi-square

= 31.372, degree of freedom (df) = 38, and significance = 0.768) which is greater than 0.05, and hence is nonsignificant (Table 13.4). It means that observed pattern of missing values is completely random and there is no such specific cause behind it.

TABLE 13.3 Missing Values in Data.

Items	Missing values		Items	Missing values	
	Count	Percent		Count	Percent
PU1	2	2.1	A1	1	1.0
PU2	1	1.0	A2	1	1.0
PU3	1	1.0	A3	2	2.1
PU4	2	2.1	BI1	2	2.1
PEOU1	1	1.0	BI2	2	2.1
PEOU2	1	1.0	BI3	2	2.1
PEOU3	1	1.0	BI4	2	2.1
PEOU4	1	1.0	Gender	0	0.0
Age	0	0.0			

Source: Compiled by authors.

TABLE 13.4 EM Means.

PU1	PU2	PU3	PU4	PEOU1	PEOU2	PEOU3	PEOU4	A1	A2	A3	BI1	BI2	BI3	BI4
2.63	2.47	2.24	2.48	2.46	2.07	2.02	2.34	2.43	2.58	2.54	2.75	2.14	2.16	2.6

Source: Compiled by authors.

13.4.2 IMPUTATION OF MISSING VALUES

As discussed, the missing data was classified to be MCAR (missing completely at random). There are two approaches which can be used for treatment of missing data. Complete case approach was not found appropriate here because sample size is small (<250). So the treatment of values was done statistically by using the expectation maximization method (EM) using statistical package for social sciences. EM is the most suitable method of assigning values to missing items as it leads to least bias under the conditions of random missing data. In this method, the expectation step and maximization step are repeated several times until maximum likelihood estimates are obtained.

13.4.3 DESCRIPTIVE ANALYSIS

The demographic profile of respondents is subcategorized as respondents' gender, age, marital status, occupation, and location. A brief description of respondents' demographic characteristics is presented in Table 13.5. There are 52.1% females as compared to 47.9% male respondents. In terms of age, most significant respondents lie in the age group of 18–29 constituting about 70.2% of the respondents. The data was skewed as 56.4% of respondents were married. The respondents belong to various occupations showing diversity but the respondents majorly were salaried. In terms of state of residence, the large share of responses was from National Capital Region (NCR) constituting about 30.9% of responses.

TABLE 13.5 Descriptive Statistics.

Variable	Category	Frequency *n* (94)	Percentage (100%)
Gender	Female	49	52.1
	Male	45	47.9
Age	18–29	66	70.2
	30–39	23	24.5
	40–49	5	5.3
Marital status	Bachelor	53	56.4
	Married	41	43.6
Occupation	Private job	45	47.9
	Government job	2	2.1
	Student	31	33
	Homemaker	9	9.6
	Self-employed	7	7.4
	Other	2	2.1
Tier wise	Tier I	23	24.5
	Tier II	69	73.4

Source: Compiled by authors.

The questionnaire was shared by online platform; therefore, the responses received were across the country from 28 cities as shown in cartogram (Fig. 13.6).

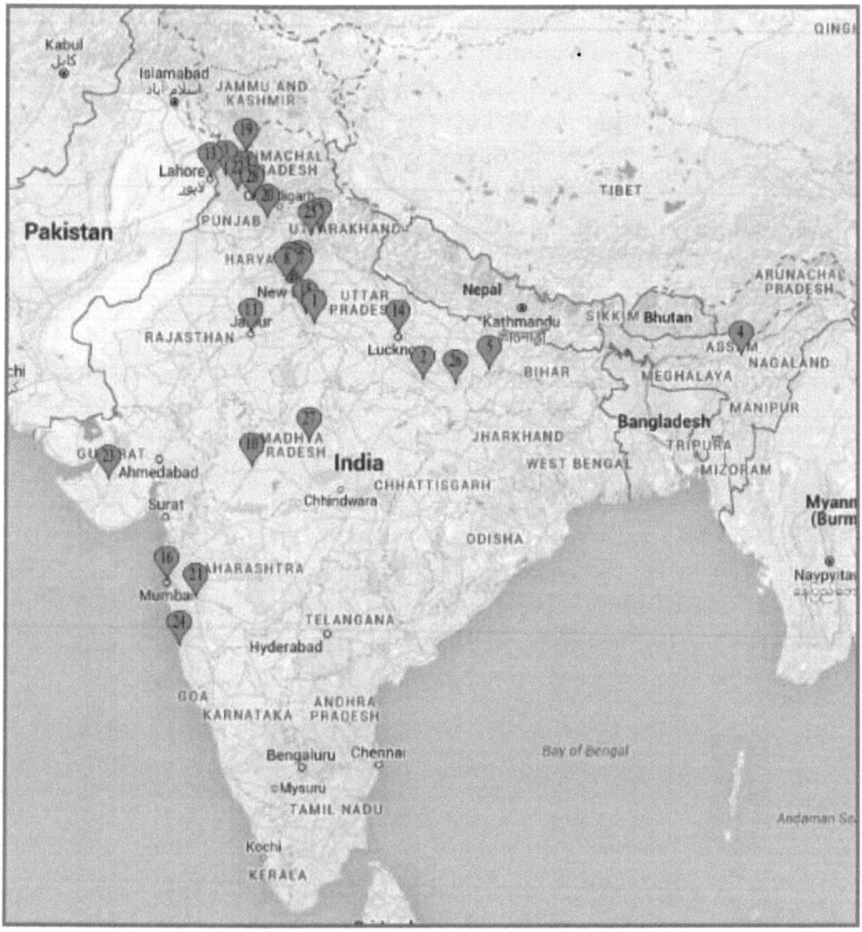

FIGURE 13.6 (See color insert.) Cartogram representing location of respondents.
Source: Compiled by authors (created on https://www.mapcustomizer.com).

13.4.4 INFERENTIAL ANALYSIS

The data collected through online survey was analyzed to get results in alignment of the objectives of study. The data was analyzed using partial least squares structural equation modeling (PLS-SEM). PLS-SEM was used for testing of hypothesis as it has few assumptions about data distribution (Hair et al., 2012; Hair et al., 2012; Hair et al., 2011). PLS does not hold assumptions of normality for data as it uses method of bootstrapping for estimating parameters (Gefen et al., 2011; Henseler et al., 2012). Normality was therefore not checked. It is also considered better for testing constructs with fewer variables. Since all the constructs have been measured by using 3 items or 4 items, this characteristic seems to be relevant. It is more advanced technique as it means testing regression equation multiple times. The research under taken is based on a strong theoretical support and the aim is to test and validate the theory. The constructs were reflective in nature and the items were predefined. PLS-SEM is considered to be a soft modeling approach as it has delivered good results with fewer assumptions.

13.4.4.1 JUSTIFICATION OF SAMPLE SIZE

The sample size that can be used for testing a model in PLS-SEM is obtained by either finding out the highest number of formative constructs or by identifying highest numbers of antecedents leading to that construct (Barclay et al., 1995). The greater one is used to calculate minimum sample size by multiplying it by 10. There are no formative constructs in model. The highest number of arrows point toward behavioral intention which is 2. Thus, minimum size necessary is 10 cases per predictor, which means 20. So, the sample size 94 is a good number for analyzing in PLS-SEM.

13.4.4.2 MEASUREMENT MODEL

The research model was evaluated using the PLS as hypothesized in the theory as shown in Figure 13.5. At the first step, PLS algorithm was calculated. All factor loadings seemed significant (>0.6) except BI2 (0.47) which made it a case for deletion. The PLS algorithm was run and significant factor loadings were obtained. Further bootstrapping was done to check that the test on sample is the true representation of population. Bootstrapping was run on 5000 subsamples (Hair, 2014) at two-tail test. The values of t were found to be >1.96 at 0.5 level of significance. Thus, the values were found significant.

13.4.4.3 ASSESSMENT OF MEASUREMENT

The variables of study were identified as reflective constructs. The reflective constructs are measured by determining indicator reliability, convergent validity, internal consistency, and discriminant validity (Hair et. 2011; Henseler et al., 2009).

Cronbach's alpha: The reliability of measurements was measured by Cronbach's alpha. All constructs seem to satisfy reliability conditions (refer Table 13.7). The value of Cronbach's alpha for perceived usefulness is equal to 0.861, perceived ease of use is equal to 0.823, attitude is equal to 0.918, and behavioral intention is equal to 0.719 which seem to be robust as these values are higher than 0.7.

Composite reliability: Many researchers such as Hair et al. (2011) and Henseler et al. (2009) consider composite reliabilities as more suitable for PLS-SEM than Cronbach's alpha. The composite reliability of constructs ranges from 0.948 to 0.834 (PU = 0.905, PEOU = 0.878, A = 0.948, and BI = 0.834). The values between 0.70 and 0.95 are considered "satisfactory to good" (Hair et al., 2014).

Convergent validity: To evaluate convergent validity average variance extracted of each variable was calculated, as suggested by Fornell and Larcker (1981) (Table 13.6). The AVE (average variance extracted) for all items show convergent validity as the value exceeds 0.50 (PU = 0.705, PEOU = 0.643, A = 0.859, and BI = 0.633).

Indicator reliability: The factor loadings were found to be significant >0.5 as shown in Table 13.7. The values were also found significant when the *t* values were checked using Bootstrapping.

Discriminant validity: Discriminant validity is used to check "if the construct is more strongly related to its own measures than with any other constructs by examining the overlap in variance by comparing the AVE of each construct with the squared correlations among construct" (Fornell and Larcker, 1981) (Table 13.6). The diagonal elements represent AVE of construct and are shown in bold. The other values represent the correlation between constructs. The square root of each construct's AVE was found to be larger than its correlations with any other construct. Therefore, it can be concluded that the constructs share high variances among themselves as compared to other constructs; thus, supporting the adequate discriminant validity of the scales (Henseler et al., 2009).

TABLE 13.6 Discriminant Validity of the Constructs (Fornell–Larcker Criteria).

	A	BI	PEOU	PU
A	**0.926**			
BI	0.644	**0.796**		
PEOU	0.733	0.589	**0.802**	
PU	0.788	0.681	0.683	**0.839**

Source: Compiled by authors.

TABLE 13.7 Measurement Statistics of Constructs.

Construct	Mean	SD	Indicator loading	*t* value	Composite reliability	Cronbach's alpha	AVE
Perceived usefulness					0.905	0.861	0.705
PU1	3.340	1.027	0.830	17.802			
PU2	3.521	1.079	0.864	21.001			
PU3	3.755	1.235	0.831	15.844			
PU4	3.543	1.059	0.833	27.145			
Perceived ease of use					0.878	0.823	0.643
PEOU1	3.553	1.208	0.820	19.495			
PEOU2	3.947	1.152	0.767	11.370			
PEOU3	3.979	1.167	0.749	10.727			
PEOU4	3.660	1.077	0.866	29.110			
Attitude					0.948	0.917	0.858
A1	3.585	1.162	0.908	38.346			
A2	3.404	1.114	0.934	60.754			
A3	3.468	1.127	0.937	59.069			
Behavioral intention							
BI1	3.213	1.071	0.884	34.039	0.834	0.719	0.633
BI3	3.851	1.120	0.583	4.970			
BI4	3.383	1.177	0.882	34.343			

AVE, average variance extracted; SD, standard deviation.

Note: *t* values are obtained by conducting bootstrapping on 5000 samples and at 0.05 % level of significance.

Source: Compiled by authors.

Test of collinearity: Collinearity is the group of points present on a single line. It is a phenomenon in which two or more predictor variables in a multiple regression model are highly correlated, meaning that one can be linearly predicted from the others with a substantial degree of accuracy. The VIF (variance inflation factor) values are greater than 5 which is indicator of collinearity among the indicators (Table 13.8). In the given model, all the VIF values were less than 5 which shown absence of collinearity.

TABLE 13.8 Test of Collinearity.

	A	BI	PEOU	PU
A		2.637		
BI				
PEOU	1.876			1.00
PU	1.876	2.637		

Note: In the given model, all the VIF values were less than 5 which shown absence of collinearity.
Source: Compiled by authors.

13.4.4.3 ASSESSMENT OF STRUCTURE

The outer model provided acceptable results for reliability and validity. The inner model was examined to assess the relationship between variables by testing of hypotheses. The analysis on internal structure supported all five hypotheses as shown in Table 13.9 as the *t* values are found to be significant ($t > 1.96$).

TABLE 13.9 Results of PLS-SEM.

Notation	Relation	Hypotheses	*t* value	*p* value	Beta	Result	Rank
H1	PEOU → PU	Perceived ease of use is positively related to perceived usefulness	12.336	0.000	0.683	Supported	1
H2	PEOU → A	Perceived ease of use is positively related to attitude toward mobile app for grocery shopping	4.344	0.000	0.365	Supported	4

TABLE 13.9 *(Continued)*

Notation	Relation	Hypotheses	*t* value	*p* value	Beta	Result	Rank
H3	PU → A	Perceived usefulness is positively related to attitude toward mobile app for grocery shopping	6.455	0.000	0.539	Supported	2
H4	A → BI	Attitude toward mobile app for grocery shopping is positively related to intention to use Mobile App for grocery shopping	1.965	0.050	0.284	Supported	5
H5	PU → BI	Perceived usefulness is positively related to Intention to use mobile app for grocery shopping	3.813	0.001	0.457	Supported	3

Source: Compiled by authors.

The strongest relationship was found between perceived ease of use and perceived usefulness (β = 0.683, p = 0.000) (Fig. 13.7). This means that if the consumer finds the use of mobile phone easy for purchasing grocery, he is more likely to find it useful as well. The second strongest relationship was found between perceived usefulness and attitude (β = 0.539, p = 0.000). It indicates that if the consumer finds the mobile technology to be useful for his grocery shopping, he will develop a favorable attitude toward it. The hypothesis regarding perceived usefulness and behavioral intention was also supported (β = 0.683, p = 0.000). This means that if a user of mobile app perceives mobile app as useful, he will have positive intention toward the purchase of grocery using it. The relationship of perceived ease of use and attitude was found significant (β = 0.365, p = 0.000). This indicates that if the consumer finds the technology to be easy for usage, he develops a positive attitude toward it. The last hypothesis examined the relationship between attitude and behavioral intention which was also supported (β = 0.284, p = 0.050). This indicates that if a consumer develops a positive attitude for cause of a technology, he may intend to use it in near future.

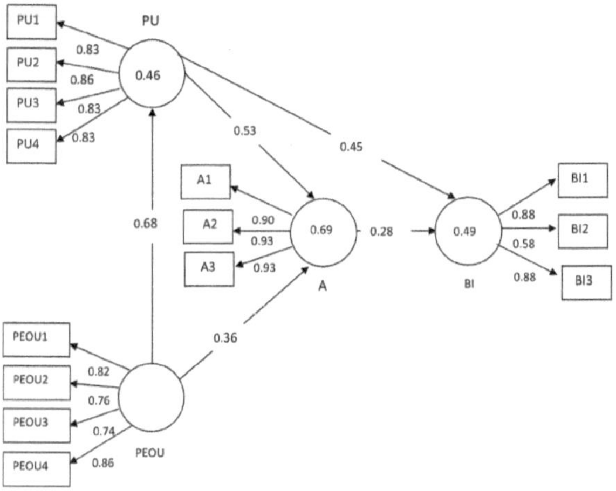

FIGURE 13.7 Regression values of constructs.
Source: Compiled by authors.

13.4.5 DEMOGRAPHICAL ANALYSIS

13.4.5.1 GENDER-WISE ANALYSIS

Out of 94 respondents, only 85 respondents used mobile for shopping through apps which indicates 90.4% of population (Table 13.10). Out of 45 males, only 40 males use mobile for shopping through apps which means only 88.8% of the total. Out of 49 females, only 45 females use mobile for shopping through apps comprising about 91.8%. Out of 94 respondents, only 36 respondents use mobile for shopping through apps which accounts for 38.2%. Out of 45 males, only 16 males use mobile for shopping through apps (35.5%). Out of 49 females, only 20 reported to use mobile apps for shopping (40.8%) (Fig. 13.8). Thus it can be interpreted that

1. Since percentage of females opting for m-shopping is higher than males, it means that females prefer to use mobile apps over males for doing shopping.
2. Since number of females opting for grocery shopping though mobile is less than males, it means for grocery shopping, males prefer to use mobile apps over females.
3. Number of females opting shopping on mobile declines when considered for grocery. The same trend is seen for men too. So it can be concluded that grocery shopping through mobile apps is less preferred as compared to general shopping through mobile apps.

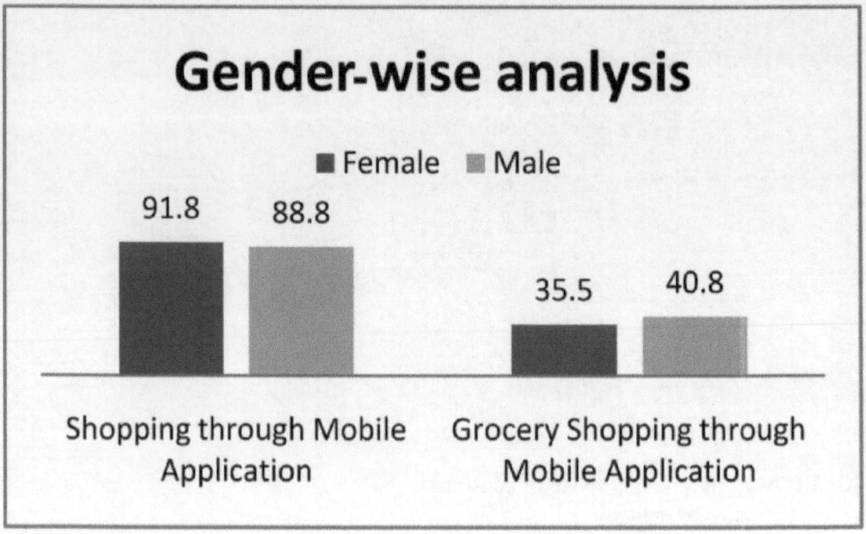

FIGURE 13.8 Gender-wise analysis of consumers' shopping behavior.
Source: Compiled by authors.

13.4.5.2 AGE-WISE ANALYSIS

The younger population is showing high interest in purchasing goods through mobile apps by contributing 92.4% of entire sample, followed by senior age group but when it comes to grocery shopping, the senior age group reported no interest. Therefore, it can be concluded that, age group 18–29 years prefer shopping through mobile for general and grocery items. (Table 13.10 and Fig. 13.9)

TABLE 13.10 Demographical Analysis.

Factor		Total	Shopping through mobile application			Grocery shopping through mobile application		
			No	Yes	% (Yes)	No	Yes	% (Yes)
Gender	Female	49	4	45	91.8	29	20	35.5
	Male	45	5	40	88.8	28	16	40.8
Age group	18–29	66	5	61	92.4	36	30	45
	30–39	23	3	20	86.9	16	6	26
	40–49	5	1	4	90	5	0	0

TABLE 13.10 *(Continued)*

Factor		Total	Shopping through mobile application			Grocery shopping through mobile application		
			No	Yes	% (Yes)	No	Yes	% (Yes)
City	Tier I	55	7	48	87.3	35	20	36.4
	Tier II	27	1	26	96.3	16	11	40.7
	Tier III	10	02	8	80.0	8	2	20.0
Marital status	Married	41	4	37	90.2	24	18	43.9
	Unmarried	53	5	48	90.5	34	18	33.9
Occupation	Govt. employee	2	0	2	100	1	1	50.0
	Home-maker	9	2	7	78	6	3	33.3
	Private job	45	4	41	91	24	2	4.4
	Self-employed	7	0	7	100	6	1	14.3
	Student	31	3	28	90	20	11	35.5
	Total	94	9	85	90.4	57	36	38.2

Source: Compiled by authors.

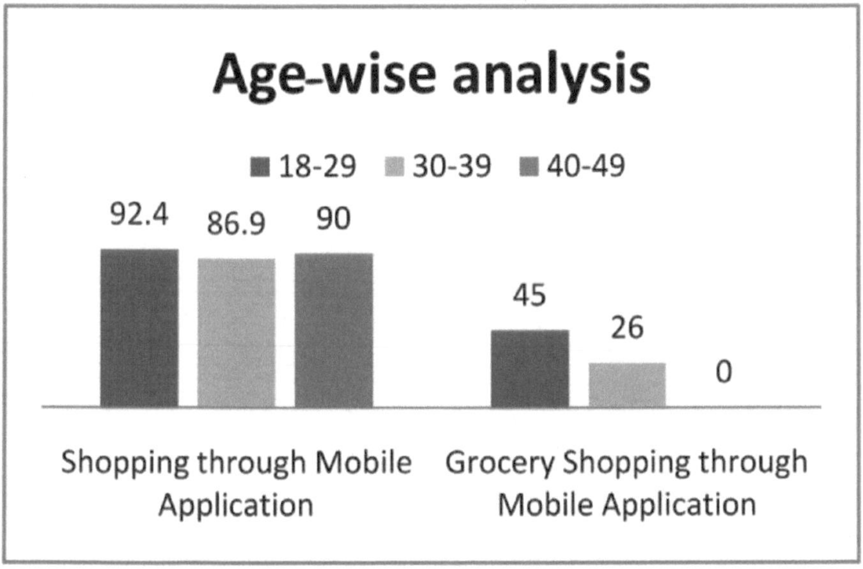

FIGURE 13.9 Age-wise analysis of consumers' shopping behavior.
Source: Compiled by authors.

13.4.5.3 LOCATION-WISE ANALYSIS

Tier II c0ity consumers prefer to shop using mobile apps in comparison to Tier I consumers. Same trend was reflected in purchase of grocery through mobile app, yet a sharp decline was noticed in shopping of grocery in comparison to general shopping from 90.4% to 38.2%. (Fig. 13.10)

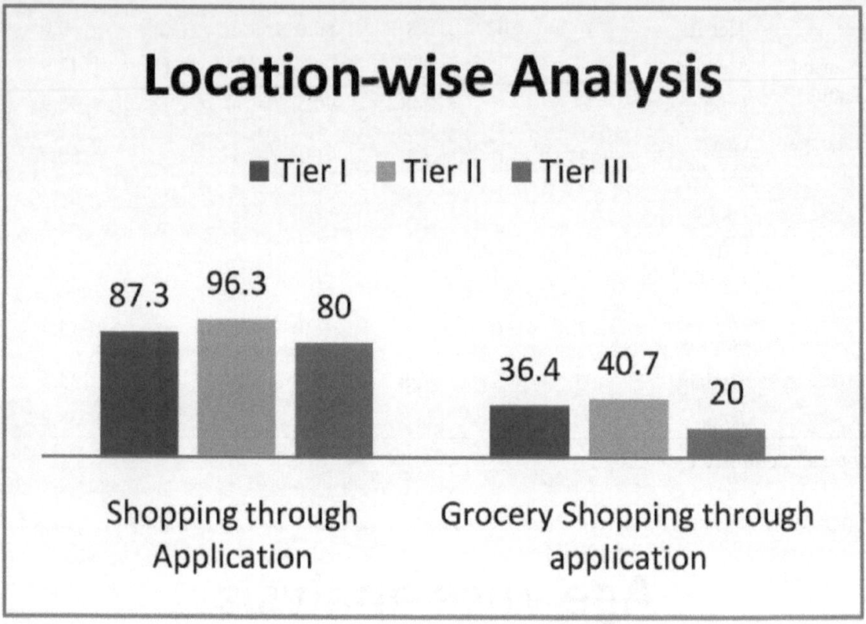

FIGURE 13.10 Location-wise analysis of consumers' shopping behavior.
Source: Compiled by authors.

13.4.5.4 MARITAL STATUS-WISE ANALYSIS

There was no significant difference in purchasing habit of married and unmarried consumer for shopping through mobile apps in general. But in m-shopping for grocery items, married people agreed to use it more as compare to unmarried people (Fig. 13.11).

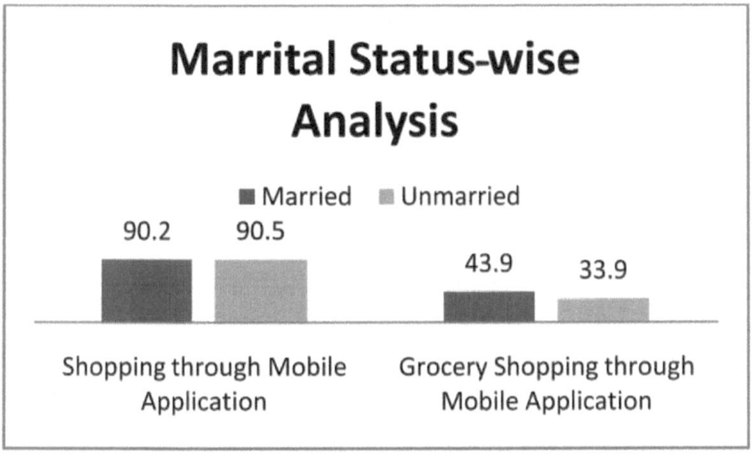

FIGURE 13.11 Marital status-wise analysis of consumers' shopping behavior.
Source: Compiled by authors.

13.4.5.5 OCCUPATION-WISE ANALYSIS

The respondents engaged in private job constituted the largest segment of people shopping using mobile, followed by students. However, in case of grocery shopping, the largest portion was constituted by students followed by housewife (Fig. 13.12).

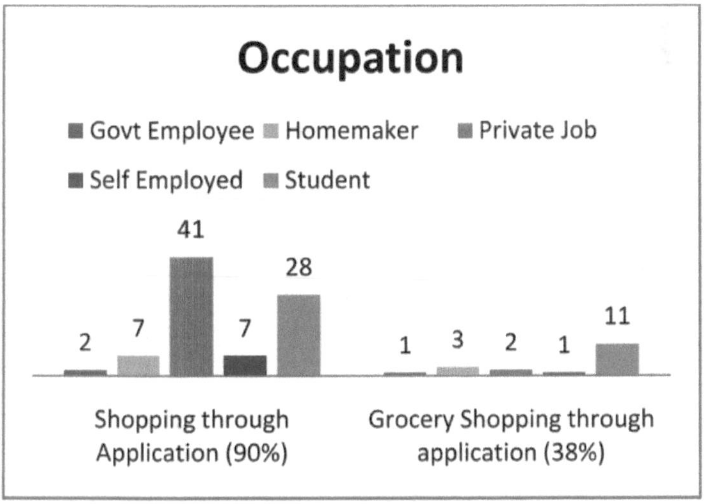

FIGURE 13.12 Occupation-wise analysis of consumers' shopping behavior.
Source: Compiled by authors.

13.4.5.6 GENDER AND MARITAL STATUS

Cross tabulation was done by taking two variables—gender and marital status together to see how the consumers attitude differ in their shopping behavior for normal goods and grocery shopping. (Table 13.11 and Fig. 13.13).

Shopping through mobile App

Unmarried female > Married male >Married female > Unmarried male

Grocery shopping through mobile Apps

Married female > Unmarried male > Married male > Unmarried female

TABLE 13.11 Gender- and Marital Status-wise Analysis of Consumers.

		Shopping through application				**Grocery shopping through application**			
Gender	Marital status	Yes	No	Total	Percent	Yes	No	Total	Percent
Female (49)	Married	23	3	26	88.5	13	13	26	50.0
	Unmarried	22	1	23	95.7	7	16	23	30.4
Male (45)	Married	14	1	15	93.3	5	10	15	33.3
	Unmarried	26	4	30	86.7	11	19	30	36.7

Source: Compiled by authors.

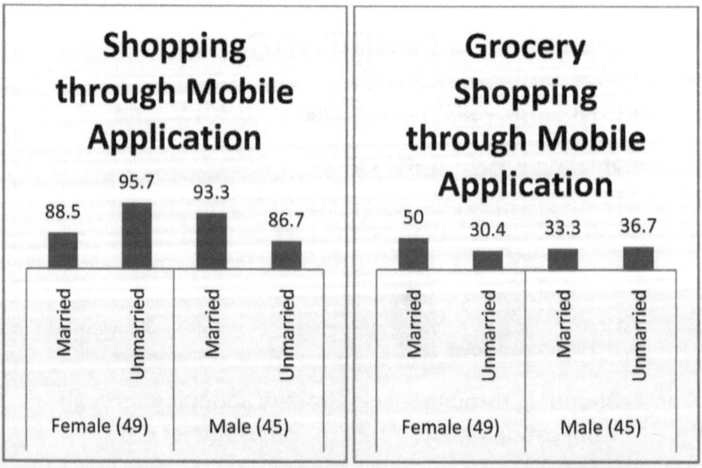

FIGURE 13.13 Gender and marital status-wise analysis of consumers' shopping behavior.
Source: Compiled by authors.

13.4.6 QUALITATIVE ANALYSIS

Qualitative analysis was performed to understand the reasons behind the purchase/nonpurchase of grocery using mobile apps by customer. The respondents provided reasons for their preference or nonpreference of grocery shopping through mobile apps. Content analysis was performed and following results were obtained (Table 13.12).

TABLE 13.12 Suggestions by Consumers.

Favorable reasons for doing grocery shopping through mobile	Favorable reasons for doing grocery shopping through mobile	Expectation
Convenience	Proximity to market	Offers and discounts
Ease of use	Quality concerns	
Time saving	Inconvenient	
	Better service by kirana shop	
	App not available	
	Problem of choice	

Source: Compiled by authors.

13.5 CONCLUSION

The aim of the study was to look into the usage of mobile apps for grocery shopping among Indian consumers. For the purpose of accomplishing the objectives, three-level analyses were conducted. At first level, TAM was tested to evaluate the acceptance of m-shopping for grocery items in India. The results are in agreement with the previous studies such as use of TAM in online grocery shopping (Kurnia and Chien, 2003) and understanding of students' intention to use e-learning (Park, 2009). All of the constructs proposed in TAM and their relationships were found to be relevant in this study. This study will enrich the literature in the field of electronic commerce-enabled adoption of technology using TAM which is currently still limited (Gefen and Straub, 2000; Chien et al., 2003). TAM model was successful in explaining adoption of technology in the various fields, yet this model was not tested in grocery sector (Park, 2009). The relationship between behavioral intention and attitude was exactly on the border line which is consistent with the theory that role of attitude in explaining behavioral intention is very limited, so it can be dropped from TAM as presented in TAM2 (Venkatesh and Davis, 1996; Venkatesh and Davis, 2000; Venkatesh et al., 2003).

The successful validation of TAM indicates that the consumers are very well adapted toward the use of mobile as a technology for shopping. The second-level testing was done to analyze the difference of usage behavior across demographical variables. The descriptive results show that 90% of the consumers use mobile for shopping but still only 38% consumers agreed to shop for grocery items. This difference was significant across genders, tier, and marital status (Table 13.10).

Further, qualitative study was conducted to identify the underlying reasons as to why the consumer does not use mobile for grocery shopping. Content analysis summarized the various problems faced by consumers (Table 13.12) and their expectations (Table 13.13).

TABLE 13.13 Suggestions to Marketers.

Product	**Price**
Improve quality of products	Reduce prices in comparison to kirana stores
Increase variety	
Improve services offered	
Place	**Promotion**
Increase market presence and availability across country	Increase offers and discounts

Source: Compiled by authors.

The contribution of research can be summarized as a validation of the TAM given by Davis in 1989, in a completely new empirical context in terms of country (India), population (sample is a mix of student, salaried, and homemakers), area (grocery shopping), and technology used (mobile). The study was able to test the adoption of mobile apps (technology) for purchasing grocery items in the context of Indian grocery retailing industry as all the relationships pertaining to TAM were found significant. This can be linked with the empirical findings that consumer find mobile technology as useful and easy to use, but do not have positive attitude of using it, therefore, the relation of attitude and behavioral intention was found weak. These results show that the m-shopping from grocery is not very well adopted by the people although the technology is not a problem. The e-retailers need to analyze the hidden reasons that stop consumers from buying grocery using apps. The results are consistent with the market trends that such grocery shopping apps are not in profits, even PepperTap (one newly launched mobile grocery shopping apps) was shut down and many others are in loss.

13.5.1 SUGGESTIONS

Based on the qualitative analysis of the data collected from the respondents, following suggestions are extended:

1. In technological context, as tested by TAM, few relationships were stronger than others. Perceived ease of use affects perceived usefulness and attitude toward using mobile app. Therefore, the marketers should try to make mobile apps more easy to use. This will increase the users' perception about easy use of app and will in turn, increase the attitude and behavioral intention to use mobile app for grocery shopping.
2. The consumers find the mobile app usage as easy, convenient, and time saving but few consumers stated that mobile apps were not available in there city. Therefore, the mobile app m-retailers should increase their reach across the country.
3. Consumers have an expectation of better offers and discount on mobile apps as compared to *kirana* shops. The mobile retailers can improve their sales by offering attractive bundled pricing and BOGO (buy one get one) offers.
4. Consumers were concerned about the quality of the products received. In terms of quality, the branded grocery (tea, masala, dal, rice, flour, etc.), could not be a major issue, but green vegetables and fruits loose freshness during time of delivery. So the m-retailers should work upon improving their brand image and improve shopping experience with customers.
5. Consumers were satisfied by *kirana* shopping as they found better service from them. The m-retailers should try to improve their services and deliver a better shopping experience to the customer.
6. Consumers were also concerned that there is a problem of choice on mobile apps. Although the m-retailers cover the large varieties, yet few products needed by consumers were not available on the app. Therefore, the m-retailers can focus upon increasing the variety.

In a nutshell, it can be said that the consumers concerns were very basic in nature considering 4P's of marketing—product, place, price, and promotion.

KEYWORDS

- **m-commerce**
- **PLS-SEM**
- **grocery shopping**
- **consumer behavior**
- **adoption of technology**

REFERENCES

Ajzen, I. Perceived Behavioral Control, Self-efficacy, Locus of Control, and the Theory of Planned Behavior. *J. Appl. Soc. Psychol.* **2002,** *32* (4), 665–683.

Al-Rahmi, W.; Othman, M. Using TAM Model to Measure the Use of Social Media for Collaborative Learning. *Int. J. Eng. Trends Technol.* **2013,** *5* (2), 90–95.

Amaro, S.; Duarte, P. An Integrative Model of Consumers' Intentions to Purchase Travel Online. *Tour. Manag.* **2015,** *46*, 64–79.

Barclay, D.; Higgings, C.; Thompson, R. The Partial Least Squares (PLS) Approach to Casual Modeling: Personal Computer Adoption and Use as an Illustration. *Technol. Stud.* **1995,** *2* (2), 285–309.

Chien, A. W.; Kurnia, S.; von Westarp, F. In *The Acceptance of Online Grocery Shopping*, BLED 2003 Proceedings, 2003, p 52.

Chuttur, M. Y. Overview of the Technology Acceptance Model: Origins, Developments and Future Directions. *Work. Pap. Inf. Syst.* **2009,** *9* (37), 9–37.

Davis, F. D. Perceived Usefulness, Perceived Ease of Use, and User Acceptance of Information Technology. *MIS Q.* **1989,** *13* (3), 319–340.

Davis, F. D.; Bagozzi, R. P.; Warshaw, P. R. User Acceptance of Computer Technology: A Comparison of Two Theoretical Models. *Manag. Sci.* **1989,** *35* (8), 982–1003.

E-Grocery Market in India—Market Research 2015–2019. 2015. https://www.technavio.com/report/e-grocery-market-in-india-market-research-2015–2019 (accessed Aug 21, 2016)

eMarketer. Mobile Commerce Round Up. 2016. https://www.emarketer.com/public_media/docs/eMarketer_Mobile_Commerce_Roundup.pdf (accessed Aug 14, 2016)

Fishbein, M.; Ajzen, I. *Belief, Attitude, Intention, and Behavior: An Introduction to Theory and Research*. 1975.

Fong, K. K. K.; Wong, S. K. S. Factors Influencing the Behavior Intention of Mobile Commerce Service Users: An Exploratory Study in Hong Kong. *Int. J. Bus. Manag.* **2015,** *10* (7), 39.

Fornell, C.; Larcker, D. F. Evaluating Structural Equation Models with Unobservable Variables and Measurement Error. *J. Market. Res.* **1981,** *18* (1), 39–50.

Fröhlke, M.; Pettersson, L. What Factors Influence a Consumer's Intention to Use a Mobile Device in the Grocery Shopping Process? 2015.

Gao, L.; Bai, X. A Unified Perspective on the Factors Influencing Consumer Acceptance of Internet of Things Technology. *Asia Pac. J. Market. Logist.* **2014,** *26* (2), 211–231.

Gefen, D.; Straub, D. W.; Rigdon, E. E. An Update and Extension to SEM Guidelines for Administrative and Social Science Research. *Manag. Inf. Syst. Q.* **2011,** *35* (2), iii–xiv.

Groß, M. Mobile Shopping: A Classification Framework and Literature Review. *Int. J. Retail Distribut. Manag.* **2015,** *43* (3), 221–241.

Hair, J. F.; Ringle, C. M.; Sarstedt, M. PLS-SEM: Indeed a Silver Bullet. *J. Market. Theory Pract.* **2011,** *19* (2), 139–152.

Hair, J. F.; Sarstedt, M.; Pieper, T. M.; Ringle, C. M. The Use of Partial Least Squares Structural Equation Modeling in Strategic Management Research: A Review of Past Practices and Recommendations for Future Applications. *Long Range Plan.* **2012,** *45* (5), 320–340.

Hair Jr, J. F.; Hult, G. T. M.; Ringle, C.; Sarstedt, M. *A Primer on Partial Least Squares Structural Equation Modeling (PLS-SEM)*; Sage Publications: California, 2016.

Henseler, J.; Ringle, C. M.; Sinkovics, R. R. The Use of Partial Least Squares Path Modeling in International Marketing. In *New Challenges to International Marketing*; Emerald Group Publishing Limited: Bradford, UK, 2009; pp 277–319.

Henseler, J.; Ringle, C. M.; Sarstedt, M. Using Partial Least Squares Path Modeling in International Advertising Research: Basic Concepts and Recent Issues. In *Handbook of Research in International Advertising;* Okazaki, S. Ed.; Edward Elgar Publishing: Cheltenham, UK, 2012; pp 252–276.

Holmes, A.; Byrne, A.; Rowley, J. Mobile Shopping Behaviour: Insights into Attitudes, Shopping Process Involvement and Location. *Int. J. Retail Distribut. Manag.* **2013,** *42* (1), 25–39.

Hu, P. J.; Chau, P. Y.; Sheng, O. R. L.; Tam, K. Y. Examining the Technology Acceptance Model Using Physician Acceptance of Telemedicine Technology. *J. Manag. Inf. Syst.* **1999,** *16* (2), 91–112.

IAMAI. Report. 2015. http://www.iamai.in/media/details/4486,%20New%20Delhi,%20November%2017,%202015 (accessed Jan 26, 2016).

Kim, J. B. An Empirical Study on Consumer First Purchase Intention in Online Shopping: Integrating Initial Trust and TAM. *Electron. Commer. Res.* **2012,** *12* (2), 125–150.

Kim, H. W.; Chan, H. C.; Gupta, S. Value-based Adoption of Mobile Internet: An Empirical Investigation. *Decis. Support Syst.* **2007,** *43* (1), 111–126.

Kim, M. J.; Chung, N.; Lee, C. K.; Preis, M. W. Dual-route of Persuasive Communications in Mobile Tourism Shopping. *Telemat. Informat.* **2016,** *33* (2), 293–308.

Ko, E.; Kim, E. Y.; Lee, E. K. Modeling Consumer Adoption of Mobile Shopping for Fashion Products in Korea. *Psychol. Market.* **2009,** *26* (7), 669–687.

Li, Y. H.; Huang, J. W. Applying Theory of Perceived Risk and Technology Acceptance Model in the Online Shopping Channel. *World Acad. Sci. Eng. Technol.* **2009,** *53* (1), 919–925.

Liu, Y.; Pu, B.; Guan, Z.; Yang, Q. Online Customer Experience and Its Relationship to Repurchase Intention: An Empirical Case of Online Travel Agencies in China. *Asia Pac. J. Tour. Res.* **2016,** *21* (10), 1085–1099.

Livemint. Mobile Internet Users in India to Double by 2017. Livemint, 2016. http://www.livemint.com/ (accessed Feb 30, 2016)

Lu, H. P.; Yu-Jen Su, P. Factors Affecting Purchase Intention on Mobile Shopping Web Sites. *Internet Res.* **2009,** *19* (4), 442–458.

Malhotra, N. K. *Marketing Research an Applied Orientation*; Pearson Education: India, 2007.

McKechnie, S.; Winklhofer, H.; Ennew, C. Applying the Technology Acceptance Model to the Online Retailing of Financial Services. *Int. J. Retail Distribut. Manag.* **2006,** *34* (4/5), 388–410.

Moon, E.; Domina, T. Willingness to Use Fashion Mobile Applications to Purchase Fashion Products: A Comparison Between the United States and South Korea. *J. Text. Appar. Technol. Manag.* **2015,** *9* (3), 1–15.

Neeraj. Mobile Internet Users in India 2016: 371 Mn by June, 76% Growth in 2015. 2016. www.dazeinfo.com: http://dazeinfo.com/2016/02/08/mobile-internet-users-in-india-2016-smartphone-adoption-2015/ (assessed Feb 10, 2016).

Neilsen, A. C. Mobile Shoppers Turn App- Happy. 2015 http://www.nielsen.com/in/en/insights/reports/2015/mobile-shoppers-turn-app-happy.html (accessed Feb 10, 2016).

Neilsen. The Future of Grocery E Commerce, Digital Technology and changing shopping preferences around the world. A C Neilsen. 2015. https://www.nielsen.com/content/dam/nielsenglobal/vn/docs/Reports/2015/Nielsen%20Global%20Ecommerce%20and%20The%20New%20Retail%20Report%20APRIL%202015%20(Digital).pdf (accessed July 15, 2016).

Park, S. Y. An Analysis of the Technology Acceptance Model in Understanding University Students' Behavioral Intention to Use E-learning. *J. Edu. Technol. Soc.* **2009,** *12* (3), 150.

Pavlou, P. A. Consumer Acceptance of Electronic Commerce: Integrating Trust and Risk with the Technology Acceptance Model. *Int. J. Electron. Commer.* **2003,** *7* (3), 101–134.

Shankar, V.; Balasubramanian, S. Mobile Marketing: A Synthesis and Prognosis. *J. Interact. Market.* **2009,** *23* (2), 118–129.

Surendran, P. Technology Acceptance Model: A Survey of Literature. *Int. J. Bus. Soc. Res.* **2012,** *2* (4), 175–178.

Tseng, F. M.; Hsu, F. Y. In *The Influence of eWOM Within the Online Community on Consumers' Purchasing Intentions: The Case of the Eee PC*, Proceedings of the 2010 International Conference on Innovation and Management, Malaysia, July, 2010.

Venkatesh, V.; Davis, F. D. A Model of the Antecedents of Perceived Ease of Use: Development and Test. *Decis. Sci.* **1996,** *27* (3), 451–481.

Venkatesh, V.; Davis, F. D. A Theoretical Extension of the Technology Acceptance Model: Four Longitudinal Field Studies. *Manag. Sci.* **2000,** *46* (2), 186–204.

Venkatesh, V.; Morris, M. G. Why Don't Men Ever Stop to Ask for Directions? Gender, Social Influence, and Their Role in Technology Acceptance and Usage Behavior. *MIS Q.* **2000,** *24* (1), 115–139.

Venkatesh, V.; Morris, M. G.; Davis, G. B.; Davis, F. D. User Acceptance of Information Technology: Toward a Unified View. *MIS Q.* **2003,** 425–478

Wang, R. J. H.; Malthouse, E. C.; Krishnamurthi, L. On the Go: How Mobile Shopping Affects Customer Purchase Behavior. *J. Retail.* **2015,** *91* (2), 217–234.

Wong, R.; Sheng, S. Y. A Business Application of the System Dynamics Approach: Word-of-mouth and Its Effect in an Online Environment. *Technol. Innov. Manag. Rev.* **2012,** *2* (6), 42.

Wu, J. H.; Wang, S. C. What Drives Mobile Commerce? An Empirical Evaluation of the Revised Technology Acceptance Model. *Inf. Manag.* **2005,** *42* (5), 719–729.

Yang, K. Determinants of US Consumer Mobile Shopping Services Adoption: Implications for Designing Mobile Shopping Services. *J. Consum. Market.* **2010,** *27* (3), 262–270.

Zhou, L.; Dai, L.; Zhang, D. Online Shopping Acceptance Model: A Critical Survey of Consumer Factors in Online Shopping. *J. Electron. Commer. Res.* **2007,** *8* (1), 41.

PART IV

M-COMMERCE: MISCELLANY

CHAPTER 14

MEASURING MOBILE COMMERCE SERVICE QUALITY: A REVIEW OF LITERATURE

AMIT SHANKAR* and BIPLAB DATTA

Vinod Gupta School of Management, Indian Institute of Technology, Kharagpur 721302, India

**Corresponding author. E-mail: amitshankar@iitkgp.ac.in*

ABSTRACT

Service quality has captured the attention of many service marketing researchers over past three decades. Globalization and information technology development have witnessed the metamorphosis from traditional to digital business, and service quality has been replaced by mobile service quality. Mobile commerce (m-commerce) refers to any transaction trough mobile devises. Siau et al. (2001) defined m-commerce as innovative and new version of electronic commerce, in which transactions are conducted through wireless telecommunication network. This chapter undertakes a systematic review of mobile commerce service quality (m-CSQ)-related articles to critically analyze scale development methodology and to know major determinants of measurement scales in different contexts of mobile commerce. An extensive qualitative literature review of 68 m-CSQ measurement scale-related articles have been conducted to fulfill above objectives. Articles were selected from EBSCOhost, ABI/INFORM, Google Scholar, ProQuest Direct, Wiley, Scopus, ScienceDirect, and Emerald Insight databases by manual search. Initially, 122 articles were identified by this process. After further refining, a total of 68 articles were included in this study. Data were analyzed in two steps; in first step m-CSQ scale development process adopted by different authors has been explored and in second step different determinants of measurement scale have been identified in four major mobile commerce contexts. Furthermore, countries of origin of the studies were also explored for better understanding of growth and development of different context of mobile commerce and overall conceptual framework of m-CSQ has been developed. M-commerce was divided into different contexts such as m-retailing, m-heath, m-internet, m-entertainment, m-telecommunication, m-service, m-banking, and m-payment. Privacy and security, content, responsiveness, efficiency, reliability, ease of use, and usefulness were identified as consistent determinants of m-CSQ measurement scales in most of the contexts. This chapter provides framework of m-CSQ scale to guide managers and scholars to identify crucial determinants of m-CSQ in different contexts.

14.1 INTRODUCTION

Advancement in information technology has witnessed the emergence of new platform for service delivery, named mobile commerce. This new channel is extension of electronic commerce with some distinctive characteristic (Balasubramanian et al., 2002). M-commerce is new format of business in which all the services are delivered irrespective of time and location. Diffusion of mobile phone and availability of affordable internet service motivate service provider as well as consumers to choose m-commerce. This platform is cost effective for both the parties, service providers and consumers. With the help of m-commerce producers increase their reach among users and consumers can get customize service. Mobility is the crucial factor which metamorphosed the structure of the traditional business. Nowadays, mobile devices are being used as strategic tool for delivering services in every service context. So, understanding of consumer perception about mobile commerce service quality (m-CSQ) is required for betterment of services. Measurement of m-CSQ got maximum attention by many m-commerce researches and many authors have developed determinants of m-CSQ measurement scale in different contexts and cultures.

Mobile retailing, mobile health, mobile internet, mobile entertainment, mobile telecommunication, mobile banking, mobile payment, and mobile service are different subfields of m-commerce in which scales were developed to measure service quality.

This chapter aims to explore m-CSQ measurement scale in above contexts. A detailed literature review has been performed to identify crucial determinants of m-CSQ measurement scale in every context.

14.2 METHOD

A systematic search was conducted to identify the articles related to m-CSQ measurement scale. Articles were sourced from EBSCOhost, Google Scholar, Scopus Science Direct, Sage, ProQuest Direct, Wiley, and Emerald Insight databases. M-CSQ, mobile service quality, mobile retailing quality, and quality word with specific m-commerce context has been searched for the identification of the relevant articles. For example, to find mobile retailing service quality-related papers, mobile retailing + quality or m-retailing + quality were used as keywords to search articles. Initially, total 122 articles were found with this search process. Finally, 68 articles were included in the study after excluding case study, book reviews, and book chapters. Some of the articles which were not accessed were also excluded from the study. Data were analyzed in two steps; in first step m-CSQ scale development process adopted by different authors has been explored and in second step different determinants of measurement scales were identified in major m-commerce contexts. Furthermore, countries of origin of the studies were also explored for better understanding of growth and development of different contexts of m-commerce and finally, overall conceptual framework of m-CSQ has been developed.

14.3 M-CSQ MEASUREMENT SCALE DEVELOPMENT PROCEDURE

In most of the previous studies, both qualitative and quantitative methods were used for the purpose of scale development in the m-CSQ contexts. Nunnally and Bernstein (1994), Arnold and Reynolds (2003), and Hosseini et al. (2013) adopted the procedure suggested by Churchill (1979).Most of the previous related studies have been following the steps described below.

14.3.1 ITEM GENERATION

Items were identified either through extensive literature review or unstructured interview and focus group studies. Some studies have used modified SERVQUAL and E-S-QUAL scale to measure m-CSQ in different contexts.

14.3.2 MEASUREMENT SCALE

In majority of the studies, five-point Likert scale was used as measurement scale. However, some studies used seven-point Likert scale to collect the responses (Chang, 2011; Rafiq et al., 2012). In recent studies, authors argued that in comparison to five-point Likert scale, seven-point Likert scale has better variability and reliability (Boshoff, 2007; Dawes, 2008). Some of the studies used coding scales (Kim et al., 2006) and C-OAR-SE (Rossiter, 2002) scales to collect the data for scale development. Most of the studies used five- and seven-point Likert scales for measurement.

14.3.3 SAMPLING METHOD AND DATA COLLECTION

In most of the m-CSQ measurement scale-related studies, data were collected for more than 200 samples. However, some of the studies have collected data through less than 200 respondents (Marimon et al., 2010; Ho and Lin, 2010; Chang, 2011). In most of the previous studies, data were collected through either online questionnaire or offline survey. Cross-sectional data have been collected in most of the studies.

14.3.4 MEASURES PURIFICATION

In most of the previous studies, Exploratory Factor analysis method was implemented to reduce factors (Parasuraman et al., 2005; Connolly et al., 2010; Liang and Pei-Ching, 2015). However, Confirmatory Factor Analysis was implemented in most of the studies to determine confirmations of the factors (Sun et al., 2009; Rafiq et al., 2012).

14.3.5 RELIABILITY

Cronbach's alpha was used in several studies to test the reliability of the constructs. Findings suggested that Cronbach's alpha coefficient value more than 0.7 indicates significant reliability (Nunnally and Bernstein, 1994). However, in some of the studies value of Cronbach's alpha coefficient ranging between 0.5 and 0.7 was also considered reliable (Meng and Mummalaneni, 2010; Sukasame, 2004).

14.3.6 VALIDITY

In most of the studies, content, convergent, discriminant, and criterion validity were analyzed. Convergent and discriminant validity were frequently used in most of the studies. Study indicates that different studies have different views on criteria for validity.

Some of the studies (Gounaris et al., 2010; Al-Hawari, 2015) adopted the method suggested by Fornell and Larcker (1981) whereby factor loading of each item should be more than 0.5 and value of all construct reliability coefficients should be more than 0.7. Some of the studies adopted the method suggested by Hair et al. (2016), by examining high and significant factor loading of respective items (Wu and Ding, 2007; Rafiq et al., 2012). To establish the relationship with outcome variables (satisfaction, loyalty, perceived value, and trust) of service quality, most of the studies used criterion validity analysis (Boshoff, 2007; Chiou et al., 2009).

14.4 DIMENSIONS OF M-CSQ MEASUREMENT SCALE IN DIFFERENT CONTEXTS

M-commerce has been divided into mobile retailing, mobile health, mobile internet, mobile entertainment, mobile telecommunication, mobile banking, mobile payment, and mobile service for better understanding of the measurement scales. All the contexts are inter-related with significant difference in the environmental characteristics. Crucial determinants of service quality measurement scale in each context have been explored. Importance of the determinate was ranked as per their appearance frequencies in the literature.

14.4.1 MOBILE RETAILING (M-RETAILING)

M-retailing is not only a channel for buying and selling of the goods and services but also it is a comprehensive medium for providing information and receiving feedback. In the past, many studies have identified crucial measures of m-retailing service quality (Wu and Wang, 2006; Wang and Liao, 2007; Huang et al., 2015). Responsiveness (Lin, 2012; Huang et al., 2015) and perceived ease of use (Wang and Liao, 2007; Lin, 2011; Özer et al., 2013) were found as crucial determents of m-retailing service quality scale followed by content of the applications (Wang and Liao, 2007; Huang et al., 2015). Some of the studies also found tangibility (Lin, 2012) as crucial determinants of m-retailing service quality as physical presence is not available in case of virtual retailing (Fig. 14.1).

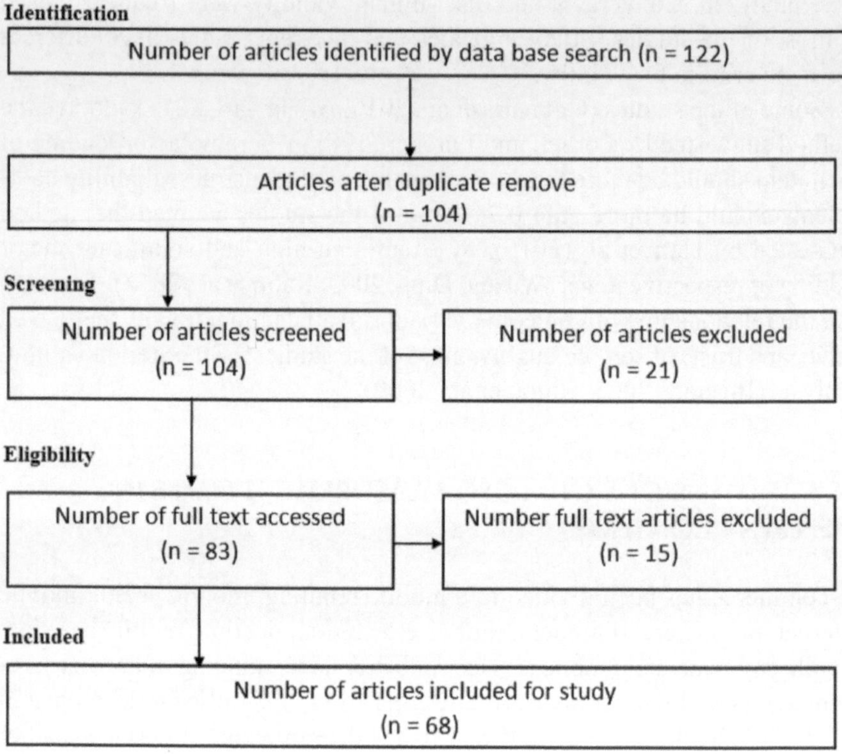

FIGURE 14.1 Flow diagram of selection process of articles for systematic review.
Source: Compiled by authors.

14.4.2 MOBILE HEALTH (M-HEALTH)

M-health is an innovative channel in the healthcare sector in which healthcare services are provided to the consumers through mobile devices. According to Mobile Healthcare Alliance, m-health is system that provides healthcare services at the point-of-care. Nowadays, every healthcare organization has its mobile application. All medical facilities including checkup, consulting, and medicine prescription are provided through m-health. In the context of m-health, privacy and security is an important determinant of measurement scale (Akter et al., 2010; Meigounpoory et al., 2014) as users do not wish to reveal their personal information. Social influence has crucial impact on using m-health application, so social influence is also found as a crucial determinant (Zhang et al., 2014; Dwivedi et al., 2016) of measurement scale. Reliability (Akter et al., 2010; Meigounpoory et al., 2014) and system efficiency (Akter et al., 2010; Meigounpoory et al., 2014) were also found crucial measures of m-health service quality.

14.4.3 MOBILE INTERNET (M-INTERNET)

When wireless internet is accessed through a portable modem, mobile phone, Universal Serial Bus (USB) wireless modem, tablet, or other mobile devices, it is named as m-Internet (Chae et al., 2002). Increasing wireless teledensity and economical internet plans motivate users to adopt m-Internet. It is need of time to identify determinants of m-internet, because of high competition in the industry. Previous studies found perceived usefulness as an important determinant of m-Internet service quality (Lu et al., 2009; Kim et al., 2007; Tan and Chou, 2008; Vlachos et al., 2011). Most of the studies also found content quality (Chae et al., 2002; Cheong and Park, 2005; Kuo et al., 2009) and ease of use (Cheong and Park, 2005; Tan and Chou, 2008) as crucial measures of m-Internet service quality. Price (Cheong and Park, 2005; Kim et al., 2007) of the internet connection was also found as an important construct of m-Internet service quality followed by personalization (Tan and Chou, 2008; Vlachos et al. 2011) and enjoyment (Cheong and Park, 2005; Kim et al., 2007).

14.4.4 MOBILE TELECOMMUNICATION (M-TELECOMMUNICATION)

Mobile telecommunication is a type of communication from a distance through mobile device using electrical signals. Nowadays, there is metamorphic shift in m-telecommunication paradigm from voice-centered communication and short messaging services (SMS) to a combination of high-speed data communication and multimedia (Kim et al., 2004). Introduction of smart phones has restructured the m-telecommunication market. In the context of m-CSQ, m-telecommunication was most explored context. Network quality was found as the most important determinant of m-telecommunication measurement scale (Seth et al., 2008; Nimako et al., 2012; Dubey and Malik, 2013; Hosseini et al., 2013). Reliable image of telecommunication service providers in the mind of users was crucial determinant of m-telecommunication service quality (Seth et al., 2008; Blery et al., 2009; Negi, 2009; Gautam and Tiwari, 2012). Previous studies also found user convenience (Kim et al., 2004; Seth et al., 2008; Gautam and Tiwari, 2012) and responsiveness (Seth et al., 2008; Blery et al., 2009; Kothari et al., 2011; Al-Aali et al., 2011) of the service providers as crucial measures. Assurance (Seth et al., 2008; Blery et al., 2009; Al-Aali et al., 2011), customer services (Santouridis and Trivellas, 2010; Hosseini et al., 2013), empathy (Negi, 2009; Al-Aali et al., 2011), price (Kim et al., 2004; Santouridis and Trivellas, 2010; Paulrajan and Rajkumar, 2011), and tangibility (Seth et al., 2008; Nimako et al., 2012) were other important determinants of m-telecommunication.

14.4.5 MOBILE SERVICES (M-SERVICES)

Nowadays, most of the services including value added, public utilities, portal services, postal, and entertainment services are being provided through mobile devices. Quality of information available on the application was found as the most important determinant of m-service quality scale (Stiakakis and Georgiadis, 2011; Basoglu et al., 2014; Sarrab et al., 2016). Usefulness of the application is also crucial determinant of application as users will continue to use the application only if they found it useful in fulfilling their daily needs (Basoglu et al., 2014; Kim and Lee, 2013; Sarrab et al., 2016). Reliability of mobile service-providing applications was also identified as a crucial measure of m-services service quality (Stiakakis and Georgiadis, 2011; Malhotra and Malhotra, 2013; Sarrab et al., 2016). Customization (Stiakakis and Georgiadis, 2011; Basoglu et al., 2014), ease of use (Basoglu et al., 2014; Kim and Lee, 2013), and flexible services (Malhotra and Malhotra, 2013; Sarrab et al., 2016) were also found as crucial measure of m-services service quality scale in previous studies.

14.4.6 MOBILE BANKING (M-BANKING)

When consumers interact with bank through mobile devices and avail banking facilities, it is called m-banking (Barnes and Corbitt, 2003). SMS, Interactive Voice Response (IVR), Mobile Application, and Wireless Application Protocol (WAP) are the major channel by which banks are providing their services. Ease of use (Gu et al., 2009; Yu and Fang, 2009; Lin, 2011; Hanafizadeh et al., 2014; Mohammadi, 2015) and usefulness (Kumar and Ravindran, 2012; Talukder et al., 2014; Mohammadi, 2015; Shankar and

Kumari, 2016; Tran and Corner, 2016) of m-banking application were found as crucial determinants of m-banking service quality scale. Consumers will continue to use m-banking only if they found it less complex and more useful than tradition banking. Most of the studies found consumer trust (Gu et al., 2009; Kim et al., 2009; Lin, 2013; Shankar and Kumari, 2016) as crucial measure of m-banking service quality scale. Quality of the system design (Lee and Chung, 2009; Yu and Fang, 2009; Lin, 2013) and its compatibility (Lin, 2011; Dzogbenuku, 2013; Hanafizadeh et al., 2014) with consumers were also found as important determinants of the scale. Other than above factors, credibility (Kumar and Ravindran, 2012; Talukder et al., 2014; Tran and Corner, 2016), relative advantage (Yu and Fang, 2009; Lin, 2011), perceived security (Yu and Fang, 2009; Lin, 2013), information quality (Chung and Kwon, 2009; Lee and Chung, 2009; Zhou, 2012), and cost (Lin, 2013; Tran and Corner, 2016) were also found as crucial measures of m-banking service quality.

14.4.7 MOBILE PAYMENT (M-PAYMENT)

Initiation, authorization, and completion of payment transaction through mobile application are called m-payment (Mallat, 2007). In terms of financial transactions, m-banking and m-payment are similar financial transaction platforms but different in terms of parties involved in it. In m-payment system, third party is additional member along with bank and user. Previous related studies found that perceived usefulness (Schierz et al., 2010; Keramati et al., 2012; Koenig-Lewis et al., 2015) and ease of use (Kim et al., 2010; Koenig-Lewis et al., 2015) were crucial major of m-payment service quality measurement scale. These constructs were also found crucial in m-banking measurement. Social influence (Slade et al., 2014) was also found as a crucial measure of m-payment service quality measurement scale. Moreover, perceived risk (Slade et al., 2014) and compatibility (Kim et al., 2010; Schierz et al., 2010) were also found important constructs of measurement scale.

14.5 CONCEPTUAL FRAMEWORK

Exploration of outcome variables of service quality is necessary for better understanding of the concept. The findings indicate that consumer satisfaction, perceived value, loyalty, purchase intention, trust, and adoption intention were found as major outcomes of m-CSQ. Most of the studies found crucial impact of m-CSQ on satisfaction (Chae et al., 2002; Chung and Kwon, 2009; Akter et al., 2010; Nimako et al., 2012; Kim and Lee, 2013; Sarrab et al., 2016) in every context. If consumers have positive perception about m-commerce services, they will feel satisfied. Loyalty was the other factor which was identified as outcome variable of m-CSQ in previous studies (Kim et al., 2004; Kumar and Lim, 2008; Kumar and Ravindran, 2012; Lin, 2012). However, some studies found that m-CSQ has significant impact on perceived value and satisfaction and these factors lead to customer loyalty. Some of the studies found significant impact of m-CSQ on trust (Lee and Chung, 2009; Lin et al., 2011; Zhou, 2012; Lee and Park, 2013). Some of the studies also found mediating impact of trust between m-CSQ and consumer satisfaction (Kim et al., 2009). Consumer perceived value was also found criterion variable of m-CSQ in some of the previous studies (Kumar and Lim, 2008; Kuo et al., 2009; Chen and Cheng, 2012). The results also indicate that perceived value mediates relationship between m-CSQ and its outcome variables (Lim et al., 2006). Usefulness, ease of use, responsiveness, security, content, network quality, tangibility, application design, and system availability were identified as consistent measures of m-CSQ scale in most of the contexts (Fig.14.2). Consumer satisfaction, loyalty, perceived value, purchase intention, and trust were major outcome variables in previous studies.

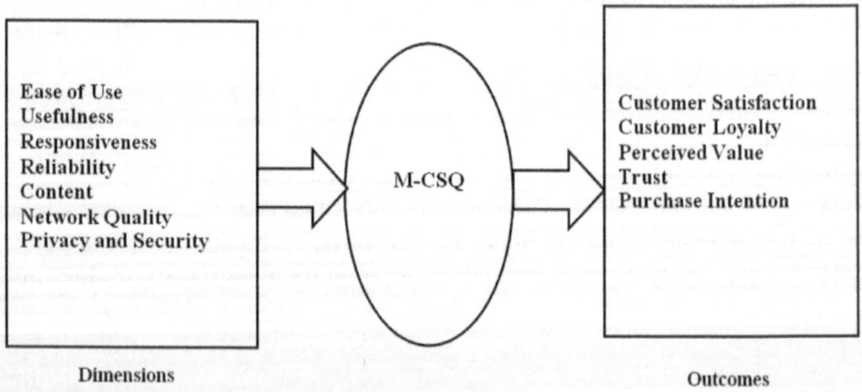

FIGURE 14.2 Conceptual framework.
Source: Compiled by authors.

14.6 COUNTRY-WISE ORIGINATION OF M-CSQ MEASUREMENT STUDIES

Most of the m-CSQ studies were undertaken in Korea (Chae et al., 2002; Kim et al., 2004; Cheong and Park, 2005; Lee and Chung, 2009; Kim and Lee, 2013; Lee and Park, 2013) and Taiwan (Wu and Wang, 2006; Kuo et al., 2009; Yu and Fang, 2009; Huang et al., 2015) followed by India, Iran, United States, and China (Fig.14.3). Seth et al. (2008), Kothari et al. (2011), Kumar and Ravindran (2012), and Dubey and Malik (2013) have explored measures of m-CSQ scale in Indian context. Some of the studies identified determinants of m-CSQ measurement scale in China (Lu et al., 2009; Zhou, 2012) and Iran (Keramati et al., 2012; Hosseini et al., 2013; Meigounpoory et al., 2104). Some USA-based studies have also explored crucial measures of m-CSQ scale in different contexts (Lu et al., 2003; Lim et al., 2006; Malhotra and Malhotra, 2013). Some authors have determined the crucial measures of m-CSQ measurement scale through cross-cultural studies (Vlachos et al., 2011; Dwivedi et al., 2016).

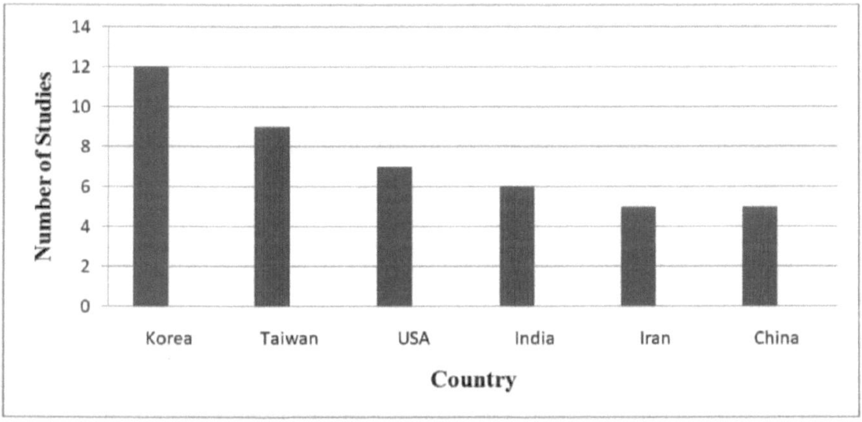

FIGURE 14.3 Country-wise origination of M-CSQ measurement studies.
Source: Compiled by authors.

14.7 CONCLUSION

In recent years, many researchers explored m-CSQ measurement scale. M-banking, m-Internet, m-health, m-retailing, m-service, m-telecommunication, and m-payment are major contexts in which service quality measurement scale were developed.

Usefulness, ease of use, responsiveness, security, content, network quality, tangibility, application design, and system availability have been identified as consistent determinants of m-CSQ measurement scale in most of the contexts. Content quality and ease of use were found as crucial determinants of m-Internet service quality measurement scale. In measuring m-telecommunication service quality, network quality and reliability were found as consistent determinants. Quality of information available on mobile application and usefulness of m-service applications have been found crucial constructs of m-service service measurement scale. In context of m-banking and m-payment, usefulness and ease of use were determined as most important determinants of service quality measurement scale.

Consumer satisfaction, loyalty, perceived value, trust, and purchase intention were found as major outcome variables of m-CSQ in different contexts. In most of the studies, these variables were used as dependent variables. In some studies, trust and perceived value were studied as mediating variable between m-CSQ and its outcome variables. Most of m-CSQ-related studies were conducted in Korea and Taiwan. Some of the studies were based on China, India, Iran, and United States.

This study explores determinants of m-CSQ scale in different contexts, which is uniqueness of the study. Findings of this study could provide a direction to the researches who are working in the area of m-CSQ scale development.

14.8 RECENT TRENDS AND FUTURE DIRECTION

In recent years, many studies developed m-CSQ measurement scale. In those scales, m-banking, m-Internet, m-health, m-retailing, m-service, m-telecommunication, and m-payment are major contexts in which scale were developed.

However, in most of the mobile-based service-related studies, main focus is on adoption and uses of the behavior, some of the studies try to measure mobile service quality in different context and culture. In developed countries, most the studies are based on e-service quality, but, in developing market such as India, China, Iran, and Taiwan studies are focused on m-service quality.

Future study may be in the direction to measure behavioral outcome of the M-CSQ. Movement in the direction related to mobile services study is indicating that in future, mobile service satisfaction, mobile service loyalty, mobile services perceived value, and mobile service co-creation measurement will be the new arena of research. A comprehensive, globally accepted mobile m-CSQ measurement scale can also be developed in future.

KEYWORDS

- **SERVQUAL**
- **mobile commerce**
- **mobile commerce service quality**
- **measurement scale**
- **m-telecommunication**

REFERENCES

Akter, S.; D'Ambra, J.; Ray, P. Service Quality of M-health Platforms: Development and Validation of a Hierarchical Model Using PLS. *Electron. Mark.* **2010,** *20* (3–4), 209–227.

Al-Aali, A.; Khurshid, M. A.; Nasir, N. M.; Al-Aali, H. Measuring the Service Quality of Mobile Phone Companies in Saudi Arabia. *King Saud Univ. J. Adm. Sci.* **2011,** *22* (2), 43–55.

Al-hawari, M. A. How the Personality of Retail Bank Customers Interferes with the Relationship Between Service Quality and Loyalty. *Int. J. Bank Mark.* **2015,** *33* (1), 41–57.

Arnold, M. J.; Reynolds, K. E. Hedonic Shopping Motivations. *J. Retail.* **2003,** *79* (2), 77–95.

Balasubramanian, S.; Peterson, R. A.; Jarvenpaa, S. L. Exploring the Implications of M-commerce for Markets and Marketing. *J. Acad. Mark. Sci.* **2002,** *30* (4), 348–361.

Barnes, S. J. and Corbitt, B. Mobile Banking: Concept and Potential. *Int. J. Mobile Commun.***2003,** *1* (3), 273–288.

Basoglu, N.; Daim, T.; Polat, E. Exploring Adaptivity in Service Development: The Case of Mobile Platforms. *J. Prod. Innov. Manag.* **2014,** *31* (3), 501–515

Blery, E.; Batistatos, N.; Papastratou, E.; Perifanos, I.; Remoundaki, G.; Retsina, M. Service Quality and Customer Retention in Mobile Telephony. *J. Target. Measure. Anal. Mark.* **2009,** *17* (1), 27–37.

Boshoff, C.; A Psychometric Assessment of ES-QUAL: A Scale to Measure Electronic Service Quality. *J. Electron. Commer. Res.* **2007,** *8* (1), 101–114.

Chae, M.; Kim, J.; Kim, H.; Ryu, H. Information Quality for Mobile Internet Services: A Theoretical Model with Empirical Validation. *Electron. Mark.* **2002,** *12* (1), 38–46.

Chang, W. L.; A Mixed-initiative Model for Quality-based E-services Pricing. *Total Qual. Manag. Bus. Excell.* **2011,** *22* (9), 975–991.

Chen, C. F.; Cheng, L. T.. A Study on Mobile Phone Service Loyalty in Taiwan. *Total Quality Management & Business Excellence*, 2012; Vol. 23 (7–8), pp 807–819.

Chiou, J. S.; Wu, L. Y.; Sung, Y. P.; Buyer Satisfaction and Loyalty Intention in Online Auctions: Online Auction Web Site Versus Online Auction Seller. *J. Serv. Manag.* **2009,** *20* (5), 521–543.

Chung, N.; Kwon, S. J. Effect of Trust Level on Mobile Banking Satisfaction: A Multi-group Analysis of Information System Success Instruments. *Behav. Inf. Technol.* **2009,** *28* (6), 549–562.

Churchill Jr, G. A. A Paradigm for Developing Better Measures of Marketing Constructs. *J. Mark. Res.* **1979,** 64–73.

Connolly, R.; Bannister, F.; Kearney, A.; Government Website Service Quality: A Study of the Irish Revenue Online Service. *Eur. J. Inf. Syst.* **2010,** *19* (6), 649–667.

Dawes, S. S. The Evolution and Continuing Challenges of E-governance. *Public Adm. Rev.* **2008,** *68* (1), 86–102.

Dubey, S. K.; Malik, R. K. Comparative Analysis of Rural Consumers' Preferences for Mobile Service Providers. *Amity Manag. Rev.* **2013,** *3* (1), 30–39.

Dwivedi, Y. K.; Shareef, M. A.; Simintiras, A. C.; Lal, B.; Weerakkody, V. A. Generalised Adoption Model for Services: A Cross-country Comparison of Mobile Health (m-health). *Gov. Inf. Q.* **2016,** *33* (1), 174–187.

Dzogbenuku, R. K. Banking Innovation in Ghana: Insight of Students' Adoption and Diffusion. *J. Internet Bank. Commer.* **2013,** *18* (3), 1.

Fornell, C.; Larcker, D. F. Structural Equation Models with Unobservable Variables and Measurement Error: Algebra and Statistics. *J. Mark. Res.* **1981,** *18* (3), 382–388.

Gautam, V.; Tiwari, A. An Empirical Study to Identify Service Quality Dimensions in the Indian Mobile Telecommunications Services Sector. *Ethics Crit. Think. J.* **2012,** *2*, 107–142.

Gounaris, S.; Dimitriadis, S.; Stathakopoulos, V. An Examination of the Effects of Service Quality and Satisfaction on Customers' Behavioral Intentions in E-shopping. *J. Serv. Mark.* **2010,** *24* (2), 142–156.

Gu, J. C.; Lee, S. C.; Suh, Y. H. Determinants of Behavioral Intention to Mobile Banking. *Exp. Syst. Appl.* **2009,** *36* (9), 11605–11616.

Hair, Jr, J. F.; Hult, G. T. M.; Ringle, C.; Sarstedt, M. *A Primer on Partial Least Squares Structural Equation Modeling (PLS-SEM)*; Sage Publications, 2016.

Hanafizadeh, P.; Behboudi, M.; Koshksaray, A. A.; Tabar, M. J. S. Mobile-banking Adoption by Iranian Bank Clients. *Telemat. Inform.* **2014,** *31* (1), 62–78.

Cheong, J.; Park, M. C. Mobile Internet Acceptance in Korea. *Internet Res.* **2005,** *15* (2), 125–140.

Ho, C. T. B.; Lin, W. C. Measuring the Service Quality of Internet Banking: Scale Development and Validation. *Eur. Bus. Rev.* **2010,** *22* (1), 5–24.

Hosseini, S. Y.; BahreiniZadeh, M.; ZiaeiBideh, A. Providing a Multidimensional Measurement Model for Assessing Mobile Telecommunication Service Quality (MS-Qual). *Iran. J. Manag. Stud.* **2013,** *6* (6), 7–29.

Huang, E. Y.; Lin, S. W.; Fan, Y. C. MS-QUAL: Mobile Service Quality Measurement. *Electron. Commer. Res. Appl.* **2015,** *14* (2), 126–142.

Keramati, A.; Taeb, R.; Larijani, A. M.; Mojir, N. A Combinative Model of Behavioural and Technical Factors Affecting 'Mobile'-payment Services Adoption: An Empirical Study. *Serv. Indus. J.* **2012,** *32* (9), 1489–1504.

Kim, J. Y.; Lee, H. S. Key Factors Influencing Customer Satisfaction in Korea's Mobile Service Sector. *J. Int. Bank. Commer.* **2013,** *18* (3).

Kim, M. K.; Park, M. C.; Jeong, D. H. The Effects of Customer Satisfaction and Switching Barrier on Customer Loyalty in Korean Mobile Telecommunication Services. *Telecommun. Policy* **2004,** *28* (2), 145–159.

Kim, M.; Kim, J. H.; Lennon, S.J.; Online Service Attributes Available on Apparel Retail Web Sites: An ESQUAL Approach. *Manag. Serv. Qual.***2006,** *16* (1), 51–77.

Kim, H. W.; Chan, H. C.; Gupta, S. Value-based Adoption of Mobile Internet: An Empirical Investigation. *Decis. Support Syst.* **2007,** *43* (1), 111–126.

Kim, J.; Jin, B.; Swinney, J. L. The Role of Etail Quality, E-satisfaction and E-trust in Online Loyalty Development Process. *J. Retail. Consum. Serv.* **2009,** *16* (4), 239–247.

Kim, C.; Mirusmonov, M.; Lee, I. An Empirical Examination of Factors Influencing the Intention to Use Mobile Payment. *Comput. Human Behav.* **2010,** *26* (3), 310–322.

Koenig-Lewis, N.; Marquet, M.; Palmer, A.; Zhao, A. L. Enjoyment and Social Influence: Predicting Mobile Payment Adoption. *Serv. Indus. J.* **2015,** *35* (10), 537–554.

Kothari, R.; Sharma, A.; Rathore, J. Service Quality in Cellular Mobile Services: An Empirical Study of Cellular Mobile Users. *Vidwat* **2011,** *4* (1), 11.

Kumar, A.; Lim, H. Age Differences in Mobile Service Perceptions: Comparison of Generation Y and Baby Boomers. *J. Serv. Mark.* **2008,** *22* (7), 568–577.

Kumar, G. R. Ravindran, D. S. An Empirical Study on Service Quality Perceptions and Continuance Intention in Mobile Banking Context in India. *J. Int. Bank. Commer.* **2012,** *17* (1), 1–22.

Kuo, Y. F.; Wu, C. M.; Deng, W. J. The Relationships Among Service Quality, Perceived Value, Customer Satisfaction, and Post-purchase Intention in Mobile Value-added Services. *Comput. Human Behav.***2009,** *25* (4), 887–896.

Lee, K. C.; Chung, N. Understanding Factors Affecting Trust in and Satisfaction with Mobile Banking in Korea: A Modified DeLone and McLean's Model Perspective. *Interact. Comput.* **2009,** *21* (5), 385–392.

Lee, J.; Park, M. C. Factors Affecting the Smartphone Users to Use the Mobile Portal Services: Focusing on Korean Mobile Portals. *Inf. Syst. e-Bus. Manag.* **2013,** *11* (2), 235–252.

Liang, C. C.; Pei-Ching, W. Internet-banking Customer Analysis Based on Perceptions of Service Quality in Taiwan. *Total Qual. Manag. Bus. Excell.* **2015,** *26* (5–6), 550–568.

Lim, H.; Widdows, R.; Park, J. M-loyalty: Winning Strategies for Mobile Carriers. *J. Consum. Mark.* **2006,** *23* (4), 208–218.

Lin, H. F. An Empirical Investigation of Mobile Banking Adoption: The Effect of Innovation Attributes and Knowledge-based Trust. *Int. J. Inf. Manag.* **2011,** *31* (3), 252–260.

Lin, H. H. The Effect of Multi-channel Service Quality on Mobile Customer Loyalty in an Online-and-mobile Retail Context. *Serv. Indus. J.* **2012,** *32* (11), 1865–1882.

Lin, H. F. Determining the Relative Importance of Mobile Banking Quality Factors. *Comput. Stand. Interfaces* **2013,** *35* (2), 195–204.

Lu, Y.; Zhang, L.; Wang, B. A Multidimensional and Hierarchical Model of Mobile Service Quality. *Electron. Commer. Res. Appl.* **2009,** *8* (5), 228–240.

Malhotra, A.; Kubowicz Malhotra, C. Exploring Switching Behavior of US Mobile Service Customers. *J. Serv. Mark.* **2013,** *27* (1), 13–24.

Mallat, N. Exploring Consumer Adoption of Mobile Payments–A Qualitative Study. *J. Strateg. Inf. Syst.* **2007,** *16* (4), 413–432.

Marimon, F.; Vidgen, R.; Barnes, S.; Cristóbal, E. Purchasing Behaviour in an Online Supermarket. *Int. J. Mark. Res.* **2010,** *52* (1), 111.

Meigounpoory, M. R.; Sajadi, S. M.; Danehzan, I. Conceptualization of the Factors Affecting the Quality of Mobile Health Services of Active SMEs in Healthcare System. *Economics* **2014,** *1* (4), 311–321.

Meng, J.; Mummalaneni, V. Measurement Equivalency of Web Service Quality Instruments: A Test on Chinese and African American Consumers. *J. Int. Consum. Mark.***2010,** *22* (3), 259–269.

Mohammadi, H. A Study of Mobile Banking Loyalty in Iran. *Comput. Hum. Behav.* **2015,** *44*, 35–47.

Negi, R. Determining Customer Satisfaction Through Perceived Service Quality: A Study Of Ethiopian Mobile Users. *Int. J. Mobile Mark.* **2009,** *4* (1), 31–38.

Nimako, S. G.; Azumah, F. K.; Donkor, F.; Adu-Brobbey, V. Confirmatory Factor Analysis of Service Quality Dimensions Within Mobile Telephony Industry in Ghana. *Electron. J. Inf. Syst. Evaluat. Vol.* **2012,** *15* (2), 197–215.

Nunnally, J. C.; Bernstein, I. H.; The Assessment of Reliability. *Psychom. Theory* **1994,** *3*, 248–292.

Özer, A.; Argan, M. T.; Argan, M. The Effect of Mobile Service Quality Dimensions on Customer Satisfaction. *Procedia-Soc. Behav. Sci.* **2013,** *99*, 428–438.

Parasuraman, A.; Zeithaml, V.; Malhotra, A.; E-S-QUAL: A Multiple-item Scale for Assessing Electronic Service Quality. *J. Serv. Res*. **2005,** *7* (3), 213–234.

Paulrajan, R.; Rajkumar, H. Service Quality and Customers Preference of Cellular Mobile Service Providers. *J. Technol. Manag. Innov.* **2011,** *6* (1), 38–45.

Rafiq, M.; Lu, X.; Fulford, H.; Measuring Internet Retail Service Quality Using ES-QUAL. *J. Mark. Manag*. **2012,** *28* (9–10), 1159–1173.

Rossiter, J. R. The C-OAR-SE Procedure for Scale Development in Marketing. *Int. J. Res. Mark.* **2002,** *19* (4), 305–335.

Santouridis, I.; Trivellas, P. Investigating the Impact of Service Quality and Customer Satisfaction on Customer Loyalty in Mobile Telephony in Greece. *TQM J.* **2010,** *22* (3), 330–343.

Sarrab, M.; Elbasir, M.; Alnaeli, S. Towards a Quality Model of Technical Aspects for Mobile Learning Services: An Empirical Investigation. *Comput. Hum. Behav.* **2016,** *55*, 100–112.

Schierz, P. G.; Schilke, O.; Wirtz, B. W. Understanding Consumer Acceptance of Mobile Payment Services: An Empirical Analysis. *Electron. Commer. Res. Appl.* **2010,** *9* (3), 209–216.

Seth, A.; Momaya, K.; Gupta, H. M. Managing the Customer Perceived Service Puality for Cellular Mobile Telephony: An Empirical Investigation. *Vikalpa* **2008,** *33* (1), 19 (Research).

Shankar, A.; Kumari, P. Factors Affecting Mobile Banking Adoption Behavior in India. *J. Internet Bank. Commer.* **2016,** *21* (1), 1–24.

Siau, K.; Ee-Peng, L.; Shen, Z. Mobile Commerce: Promises, Challenges, and Research Agenda. *J. Database Manag.* **2001,** *12* (3), 4–13.

Slade, E. L.; Williams, M. D.; Dwivedi, Y. K. Devising a Research Model to Examine Adoption of Mobile Payments: An Extension of UTAUT2. *Mark. Rev.* **2014,** *14* (3), 310–335.

Stiakakis, E.; Georgiadis, C. K. A. In *Model to Identify the Dimensions of Mobile Service Quality*, Mobile Business (ICMB), 2011 Tenth International Conference on IEEE, 2011; pp 195–204.

Sukasame, N.; The Development of E-service in Thai Government. *BU Acad. Rev*. **2004,** *3* (1), 17–24.

Sun, Q.; Wang, C.; Cao, H. In *Applying ES-QUAL Scale to Analysis the Factors Affecting Consumers to Use Internet Banking Services*, IITA International Conference on Services Science, Management and Engineering, 2009 (SSME'09); IEEE, 2009; pp 242–245.

Talukder, M.; Quazi, A.; Sathye, M. Mobile Phone Banking Usage Behaviour: An Australian Perspective. *Australas. Account. Bus. Finance J.* **2014,** *8* (4), 83–104.

Tan, F. B.; Chou, J. P. The Relationship Between Mobile Service Quality, Perceived Technology Compatibility, and Users' Perceived Playfulness in the Context of Mobile Information and Entertainment Services. *Int. J. Hum. Comput. Interact.* **2008,** *24* (7), 649–671.

Tran, H. T. T.; Corner, J. The Impact of Communication Channels on Mobile Banking Adoption. *Int. J. Bank Mark.* **2016,** *34* (1), 78–109.

Vlachos, P. A.; Giaglis, G.; Lee, I.; Vrechopoulos, A. P. Perceived Electronic Service Quality: Some Preliminary Results from a Cross-national Study in Mobile Internet Services. *Int. J. Hum. Comput. Interact.* **2011,** *27* (3), 217–244.

Wang, Y. S.; Liao, Y. W. The Conceptualization and Measurement of M-commerce User Satisfaction. *Comput. Hum. Behav.* **2007,** *23* (1), 381–398.

Wu, J. H.; Wang, Y. M. Development of a Tool for Selecting Mobile Shopping Site: A Customer Perspective. *Electron. Commer. Res. Appl.* **2006,** *5* (3), 192–200.

Wu, K. W.; Ding, M. C. Validating the American Customer Satisfaction Index Model in the Online Context: An Empirical Study of US Consumer Electronics E-tailers. *Int. J. Bus. Inf.* **2007,** *2* (2), 199–220.

Yu, T. K.; Fang, K. Measuring the Post-adoption Customer Perception of Mobile Banking Services. *Cyberpsychol. Behav.* **2009,** *12* (1), 33–35.

Zhang, X.; Guo, X.; Lai, K. H.; Guo, F.; Li, C. Understanding Gender Differences in M-health Adoption: A Modified Theory of Reasoned Action Model. *Telemed. e-Health* **2014,** *20* (1), 39–46.

Zhou, T. Understanding Users' Initial Trust in Mobile Banking: An Elaboration Likelihood Perspective. *Comput. Hum. Behav.* **2012,** *28* (4), 1518–1525.

CHAPTER 15

IS INDIA POISED FOR M-COMMERCE IN THE CASHLESS MILIEU?

ARINDAM CHAKRABARTY*

Department of Management, Rajiv Gandhi University (Central University), Rono Hills, Doimukh 791112, Itanagar , Arunachal Pradesh, India

**E-mail: arindam.management@gmail.com*

ABSTRACT

Mobile commerce (m-commerce) may be orchestrated as the subset and advanced form of e-commerce. It has been emerging as the popular means of business throughout the globe. India is witnessing rapid changes in the demographic and lifestyle variables with the revolution of technological development. It essentially reinforces higher penetration of mobiles including smart phones and growing affinity toward the use of internet through mobile gadgets. M-commerce has been evolving as the new generation business trajectory to play a potentially leading role in Indian economy. The State policy has been consistent to develop public infrastructure including information and communication technology in the country by the successive governments. In recent times, the State has been emphasizing on strengthening digital movements and cashless ecosystem. This chapter will attempt to explore India's preparedness to grab the opportunity for excelling m-commerce.

15.1 INTRODUCTION

The policy-makers of India have been committed to accord top priority to the development of public infrastructure for fuelling the engine of economic growth in the country. It is predominantly reflected in the recent Economic

Survey Report 2015–2016. Emphasizing on the development of rural infrastructure, Government of India had taken a time-bound action plan called "Bharat Nirman (2005–2009)." Development of communication infrastructure is an essential prerequisite to extend the fruits of liberalizations to every citizen of India and particularly to the rural population. The central point of the state policy encircles the agenda that every person should have equitable access to the telecommunication platform. The thrust on increasing tele-density in the country has been consistent irrespective of changes in the political ideologies among the successive governments.

Now, telecom network has become a basic infrastructure such as electricity, road, water, etc. It has become the integral part of the ecosystem for accelerating economic development of the country. However, the urban India has witnessed a phenomenal telecom revolution, whereas the rural "Bharat" is far behind from its counterpart. Rural Telephony Program was earmarked as one of the important strategic interventions under the Bharat Nirman Plan to ensure the influx of information for agriculture, welfare schemes, emergency services, etc., for improving health, education, employment, and business opportunities in the rural areas. This is evident in Andhra Pradesh, where job cardholders under Mahatma Gandhi National Rural Employment Guaranty Act were provided biometric smartcards to disburse their wages directly. Earlier in the manual system, there were complaints and possibilities of siphoning of funds in fictitious names. Muralidharan et al. (2016) reported that such digital ecosystem efficiently diminished these leakages by 35% and optimized in reduction of delays in wage payments by 19%. This electronic platform facilitated to enhance their wages up to 24%. The return on investment of this smartcard infrastructure was found significant with an estimation of seven times of the implementation cost. Around 90% workers favored this biometric smartcard system (Muralidharan et al., 2016). The Economic Survey (2015–2016) Report mentioned that the smartcards program was a tremendous success. As a part of global trend, India witnessed a phenomenal growth in the field of computer, software, information and communication technology (ICT), etc., in past few decades. As a corollary to this, the concept of electronic commerce (e-commerce) introduced in Indian market. The concept of e-commerce has been developed to facilitate immobile users connected with wired internet network through phone lines or local area network.

With the rapid increase of mobile penetration, internet usage through mobile gadgets and usage of smart phones in India, the new-generation business model is emerging which will be value added by a new e-commerce applications platform. It is known as wireless e-commerce or mobile commerce, popularly known as m-commerce. Mobile commerce is a browsing platform for buying and/or selling of goods and services online using the mobile phones or tablets. It is a complete and convenient shopping or trading experience using the handset. Mobile commerce being the subset and advanced form of e-commerce has been emerging as popular means of business throughout the globe. As far as internet access is concerned, mobiles are not the alternative or substitute for computer rather it is indeed more powerful instrument to effectively communicate with all the clients either for interactions, sharing database, or performing online transactions (Parveen et al., 2012). Paavilainen (2002) constructed a distinct differentiation between m-commerce and the traditional "e-commerce" reinforcing special characters or features of mobile-led commerce like the concepts of ubiquity, interactive mechanism, that is, intimacy, time sensitivity, and location access.

TABLE 15.1 Definitions and Classifications of M-Commerce.

Classification	Definition	References
1. Subset/revolutionized form of e-commerce	M-commerce is a subset or a new form of e-commerce and all the aspects involved can be extended and applied to m-commerce	Kwon and Sadeh, 2004; Coursaris and Hassanein, 2002; Scharl et al.; 2005
2. Transactions	A transaction having a definite value or utility, administered through any mobile terminal equipment on the mobile telecommunication network, can be considered a part of mobile commerce	Yan, 2005; Barnes, 2002; Tsalgatidou and Pitoura, 2001
3. Business ecosystem	M-commerce is an interactive ecology system of people and organizations based on social and technological effects	Mylonopoulos and Doukidis, 2003
4. Combination of classifications 1, 2, and 3	M-commerce is an extension of e-commerce in which products or goods are managed through wireless mobile equipment without time or place constraints in order to increase the profitability/efficiency of business processes	Kuo and Yu, 2006; Ming, 2008

Source: Adapted from Mehmood (2015).

The inherent constraints such as time and space limitations are conquered by m-commerce as the users can access and participate in the economic activities any time anywhere through internet without a plug-in terminal (Golden and Regi, 2013). In recent times, the State has been emphasizing on digital movements and cashless ecosystem. This essentially brings tremendous opportunities for growth of mobile commerce or mobile electronic commerce (MEC) in India. There are variety of mobile commerce services which include online banking, bills payment, delivery and tracking information, shopping, selling, and so on.

15.2 REVIEW OF LITERATURE

Mobile commerce was first introduced in 1997 with the use of SMS and Wireless Application Protocol (WAP)-based payments as an alternative to cash or card transactions. Coca-Cola vending machine in Helsinki, Finland began accepting payments by SMS text message. In 1998, Finnish mobile operator Radiolinja started selling ringtones, the first digital content, to be downloadable via mobile phone. Ericsson and Telenor Mobil introduced a service to buy movie tickets by mobile phone in 1998. In the same year, iMode in Japan and Smart Money in Philippines started the first major m-commerce platform (Golden and Regi, 2013).

M-commerce-enabled services began to increase from 2000. In 2000, KLM Royal Dutch Airlines allowed air travelers to use mobile phones to arrange travel and to confirm loyalty points. International Retail Giants such as HMV, Amazon, etc., started m-commerce application in the internet-enabled cell phones in 2000. In the same year, the concept of cell phones as mobile wallet emerged in Africa. In 2001, Motorola tested Pizzacast, a mobile commerce system, for ordering Domino's Pizza (Brown, 2014). Norway introduced mobile payment system for car parking services. Austria began train ticketing service via mobile. Japan resumed mobile-enabled air tickets purchase facility. In July 2001, the first conference on mobile commerce was held in London. In 2002, the first book on mobile commerce was published by Tomi Ahonen titled "*M-profits*." The University of Oxford in 2003 started a short-term course to discuss mobile commerce for the first time in 2003 (Golden and Regi, 2013).

According to Ericsson Mobility Report, June 2014 and KPMG in India Analysis 2015, the access to mobile phones throughout the globe was recorded 93% in the first quarter of 2014 while in India, the figure was measured somehow lower with a penetration of 73% during the same time period. However, the access to mobile phone in India is gradually growing. (KPMG-IAMAI, 2015)

In 2016, India's mobile phone penetration has crossed 1 billion users mark, according to the recent data released by the country's telecom regulator. China crossed the 1 billion mobile users landmark in 2012. The growth trajectory for the increase in mobile subscription may be contributed by introduction of 3G network, new-generation and reasonably priced devices, and continuous decrease in mobile tariffs.

According to European Mobile Industry Observatory Report of 2011, the mobile penetration results in economic growth of countries (European Mobile Industry Observatory, 2011). It is estimated, a 10% increase in penetration could increase the GDP growth rate by up to 1.2% to 2%. The economic growth factor is higher for developing countries and their adoption rates are also higher (Banerjee and Ros, 2002). In India the encouraging fact is that the large number of Indians have mobile in comparison to the number of people having desktops or personal computers. A substantial portion (around 39%) of Indian mobile users will switch over to smart phone by 2019 as projected by e-Marketer Statista 2016. So the logical corollary is that internet services will be accessed by the people of India primarily through mobile gadgets in near future. This is evidenced in the FICCI-KPMG Report 2014 on "Indian Media and Entertainment Sector." This report showcased that India has become the third-largest internet user block across the globe. Most surprisingly, above 50% of such internet users exclusively rely on mobile reinforced access. The penetration of internet access in India is just 19% which is relatively low if compared to other developed and even few developing economies (FICCI-KPMG, 2014).

TABLE 15.2 Internet Penetration 2014.

Name of the country	Internet penetration 2014
Australia	89.6%
The United States	86.8%
Japan	86.0%
Brazil	53.4%
China	46.0%
India	19.2%

Source: Adapted from Internet Live Stats; e-Marketer; KPMG in India Analysis, 2015.

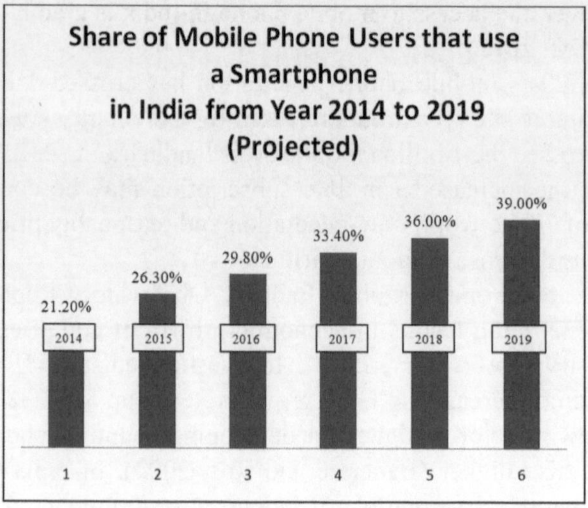

FIGURE 15.1 Share of mobile phone users that use a smartphone in India from 2014 to 2019.
Source: Adapted from Statista (2016). Retrieved on Mar 28, 2017 from https://www.statista.com/statistics/257048/smartphone-user-penetration-in-india/.

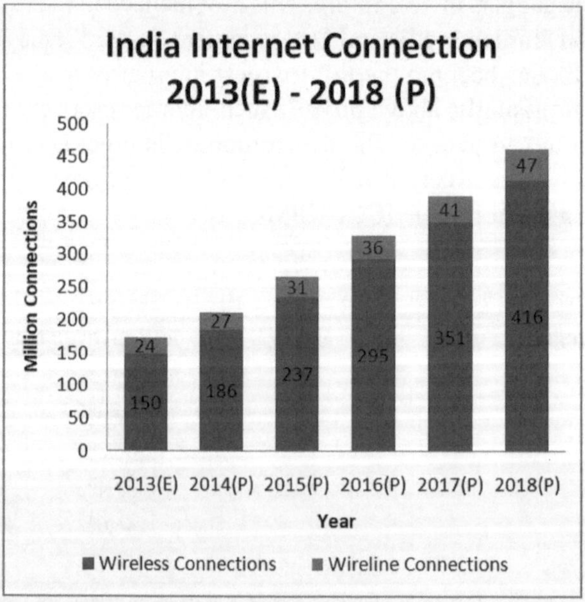

FIGURE 15.2 India internet connections 2013(E)–2018 (P).
Source: Adapted from KPMG in India Analysis Reported in FICCI-KPMG (2014). *The Stage is Set: Indian Media and Entertainment Industry Report*, 2014, p 96. Retrieved on Mar 28, 2017 from http://aibmda.in/sites/default/files/FICCI-KPMG%20Report%202014.pdf.

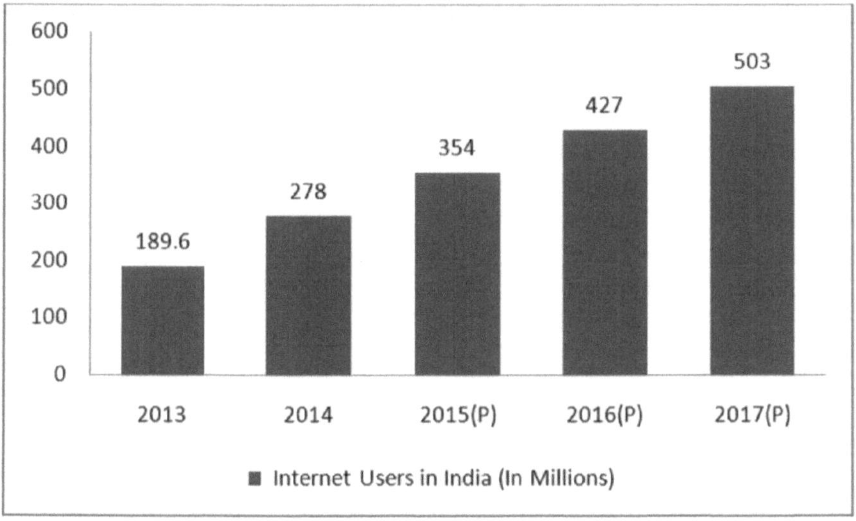

FIGURE 15.3 Internet users in India (in millions).
Source: Adapted from IAMAI-IMRB Mobile Internet in India 2014 Report and KPMG-FICCI M&E Industry Report 2015 reported in KPMG-IAMAI. *India on the Go: Mobile Internet Vision 2017*; July, 2015, p. 14.

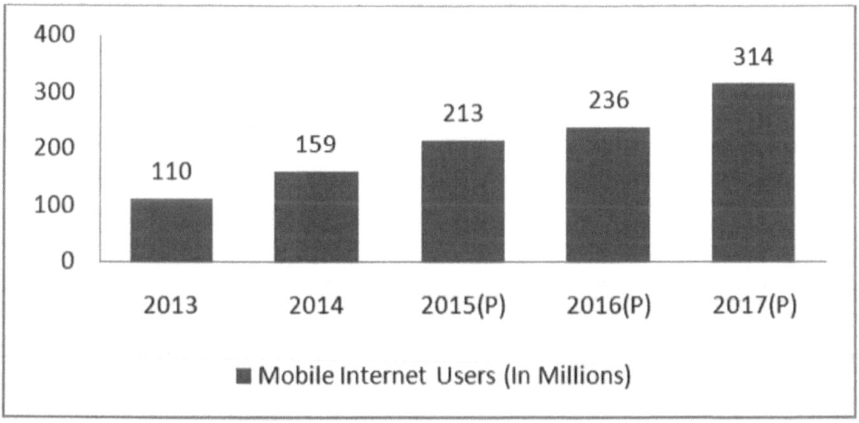

FIGURE 15.4 Mobile internet users in India 2013–2017 [E].
Source: Adapted from *IAMAI-IMRB Mobile Internet in India Report*, 2014; KPMG-IAMAI, 2015.

According to the report (2014) titled "India on the Go: Mobile Internet Vision 2017" published by Internet and Mobile Association of India (IAMAI) and KPMG, the figure of mobile internet users in the country is going to reach 236 million and 314 million by the end of 2016 and 2017, respectively.

The report indicates that over 500 million Indians would access internet by 2017 which was just above 350 million users as on June 2015. The commitments of the government toward rapid and deepening the penetration of internet services across the nation have been consistent throughout. In 2014–2015, the Government of India allocated 5 billion INR for building adequate infrastructure to foster broadband network even in the villages of India to reinforce "Digital India" initiative under the aegis of National Rural Internet and Technology Mission. This gigantic program was to promote indigenous manufacturing of hardware and development of software (Indian Express, 2014). It has been further augmented by an additional sanction of 1 billion INR budget for improving e-governance setup in India. The National Optical Fiber Network (NOFN) initiative by Digital India is committed to provide broadband connectivity to cover 250,000 Gram Panchayats to target rural population. This may motivate to increase the adoption of data-enabled devices by the people in these areas (BCG and Google, 2016). In 2015–2016 Union budget declaration, the Government rationalized import duty of mobile handsets and tablets to stimulate the indigenous manufacturers. As a part of the "Make in India" program, Department of Electronics and Information Technology has undertaken a time-bound initiative by establishing a joint task force drawing representatives both from industry as well as Government in order to rejuvenate and strengthen the manufacturing ecosystem in the country to achieve a targeted production of 500 million mobile gadgets and their components by 2019 (EY-FICCI Report, 2015).

The Deloitte Global Mobile Consumer Survey, 2015 analyzes the mobile usage patterns of over 2000 Indian consumers as part of a worldwide survey conducted on 49,000 respondents across 30 countries. As of June, 2015, 86% of those surveyed owned smart phones, which is 12% higher than in 2013 globally. According to this research, phone owners in India are highly interested in in-store mobile payments and mobile money transfers. Given an option of in-store mobile payments, 71% of those surveyed said they would use the facility. 54% of them said they are interested in mobile money transfers. However, barriers for in-store payments still exist. The most common reason given for not using phone to make a payment was fear about security cited by 39% of respondents (The Deloitte Global Mobile Consumer Survey, 2015).

The literature including recent reports and research outcomes indicate that there is indeed a growing trend of mobile-led internet usage and affinity toward m-commerce throughout the globe. Similar pattern is also observed in India.

15.3 OBJECTIVES OF THE STUDY

The objectives of the study are:

1. To study whether population size of a country has any impact on the level of implementation of cashless model.
2. To explore India's preparedness to grab the opportunity for excelling m-commerce as the new-generation instruments for economic development.
3. To explore opportunities and inherent challenges of m-commerce in the cashless milieu in India.
4. To suggest proactive roles of the State to promote m-commerce in India.

15.4 METHODOLOGY

The study was performed using data from secondary sources particularly published reports from reputed national as well as international agencies. For this study, descriptive statistics, parametric and nonparametric tests have been used with the help of suitable statistical packages.

15.5 OBSERVATION AND OVERVIEW

To make an indicative understanding on the relationship between population of a country and percentage of cashless transactions, top 15 countries in terms of percentage of cashless transactions are ranked from highest to lowest. The population of those countries is tabulated. Since the population census in each country has not been conducted in same base year, the ranking of the study countries in terms of population are performed.

TABLE 15.3 Top 15 Cashless Countries with Corresponding Population Profiling.

Sl. no.	Name of the country	Population[a]	% of cashless transactions[b]	Rank in cashless transactions	Corresponding population rank
1.	Singapore	5,603,740	61%	1	15
2.	Netherland	16,877,352	60%	2	12
3.	France	66,259,012	59%	3	6
4.	Sweden	9,723,809	59%	4	14
5.	Canada	34,834,840	57%	5	10
6.	Belgium	10,449,361	56%	6	13
7.	United Kingdom	63,742,976	52%	7	7
8.	The United States	318,892,096	45%	8	2
9.	Australia	22,507,616	35%	9	11
10.	Germany	80,996,688	33%	10	5
11.	South Korea	49,039,984	29%	11	8
12.	Spain	47,737,940	16%	12	9
13.	Brazil	202,656,784	15%	13	3
14.	Japan	127,103,392	14%	14	4
15.	China	1,355,692,544	10%	15	1

[a]*Source*: Adapted from CIA World Factbook. Population–Country Comparison–Top 100 as of January 1, 2014.
Retrieved on Mar 22, 2017 from http://www.indexmundi.com/g/r.aspx?t=100 and http://www.indexmundi.com/g/r.aspx?t=100andv=21andl=en.
[b]*Source*: MasterCard Advisors Measuring Progress Toward a Cashless Society, *BT Online* December 29, 2016.

Using Spearman's rank correlation method, it is attempted to understand whether there is any relationship between size of population of a country and the corresponding percentage in cashless transaction.

TABLE 15.4 Association Between Cashless Transaction and Corresponding Population Size of the Nations.

Correlations

		Rank in cashless transactions	Corresponding population rank
Pearson correlation	Rank in cashless transactions	1.000	−0.711
	Corresponding population rank	−0.711	1.000
Significance (1-tailed)	Rank in cashless transactions	0.0	0.001
	Corresponding population rank	0.001	0.0
N	Rank in cashless transactions	15	15
	Corresponding population rank	15	15

N, top 15 cashless countries.

Model summary

Model	R	R square	Adjusted R square	Standard error of the estimate	Change statistics				
					R square change	F change	df1	df2	Significance F change
1	−0.711[a]	0.505	0.467	3.265	0.505	13.269	1	13	0.003

df, degree of freedom; *F*, *F* statistic; *R*, correlation coefficient.

[a]Predictors: (constant), corresponding population rank.

ANOVA[a]

Model		Sum of squares	Df	Mean square	F	Significance
1	Regression	141.432	1	141.432	13.269	0.003[b]
	Residual	138.568	13	10.659		
	Total	280.000	14			

Df, degree of freedom; *F*, *F* statistic.
[a]Dependent variable: rank in cashless transactions.

[b]Predictors: (constant), corresponding population rank.

Model B		Unstandardized coefficients		Standardized coefficients	*t*	Signifi-cance	95.0% confidence interval for B	
		Standard error	Beta			Lower bound	Upper bound	
	(constant)	13.686	1.774		7.715	0.000	9.853	17.518
1	Corresponding population rank	−0.711	0.195	−0.711	−3.643	0.003	−1.132	−0.289

B, regression coefficient (unstandardized); *t*, *t* statistic.

Coefficient correlations[a]			
Model			**Corresponding population rank**
1	Correlations	Corresponding population rank	1.000
	Covariances	Corresponding population rank	0.038

[a]Dependent variable: rank in cashless transactions.

Colinearity diagnostics[a]					
Model	**Dimension**	**Eigen value**	**Condition index**	**Variance proportions**	
				(constant)	**Corresponding population rank**
1	1	1.880	1.000	0.06	0.06
	2	0.120	3.956	0.94	0.94

[a]Dependent variable: rank in cashless transactions.

Chi-square tests			
	Value	**Df**	**Asymp. significance (two-sided)**
Pearson Chi-square	210.000[a]	196	0.234
Likelihood ratio	81.242	196	1.000
Linear-by-linear association	7.072	1	0.008
N of valid cases	15		

[a]225 cells (100.0%) have expected count < 5. The minimum expected count is 0.07.

Symmetric measures					
		Value	**Asymp. standard error**[a]	**Approx. T**[b]	**Approx. significance**
Interval by interval	Pearson's *R*	−0.711	0.128	−3.643	0.003[c]
Ordinal by ordinal	Spearman correlation	−0.711	0.150	−3.643	0.003[c]
N of valid cases		15			

[a]Not assuming the null hypothesis.

[b]Using the asymptotic standard error assuming the null hypothesis.

[c]Based on normal approximation.

Source: Compiled by author.

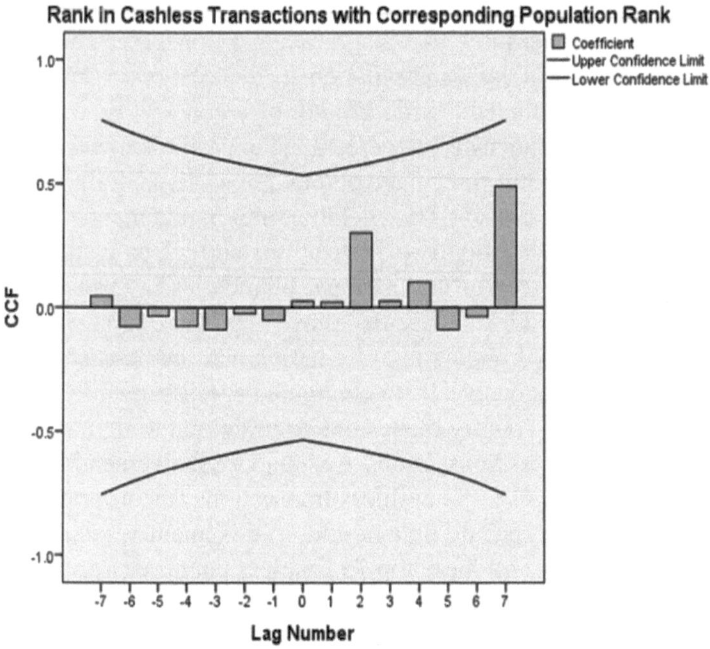

FIGURE 15.5 Rank in cashless transactions with corresponding population rank.
Source: Compiled by author.

The results indicate that larger the size of population of a nation, lesser is the percentage of cashless transaction. This could be supplemented by the following understanding that the higher population, countries are unlikely to provide adequate infrastructure to be used by the countrymen, that is, accessibility of services, ease of using services, knowledge of its usefulness, etc. (Fishbein and Ajzen, 1975; Davis, 1986). Higher populated nations are facing more difficulty to distribute equitable share of wealth among the citizens and fail to provide uniform growth to all the people in the key social priority derivatives such as health, education, employment, etc.

In India, it is estimated that 2% transactions are cashless (BT Online, 2016) while in China, being the highest populated country of world, cashless transactions just touch to 10%. At this juncture, it is evident that 100% cashless economy is indeed daydreaming for any of the nations of the world. At the same time, it is also visible that the developed and most developing nations are moving toward less cash model. If India is committed to move forward as economic superpower and should become more competitive, there is no doubt that India needs economic reforms such as greater financial inclusion, banking for every citizen, increasing roles of banking, and ensuring welfare and social security measures like insurance, pensions, etc. The increasing and meaningful role of larger population in the economic activities shall essentially accelerate the economic growth of a country. As a corollary to this view, India should enhance its percentage of cashless transactions. Now the question remains what percentage of total transactions, India may project to be cashless from its present 2% level in a time-bound manner.

In the era of liberalization, privatization, and globalization, we live in a global village devoid of any national boundaries in terms of free and fair trade practices. Indicative results show that, there is an inverse relation between the size of population of a nation and the percentage of cashless transaction in the country. If it is imagined and assumed hypothetically that every person of a country makes one transaction during a specific time period then 61% cashless transactions in Singapore shall essentially generate 3,418,281.4, that is, 3,418,282 cashless transactions (incorporating indivisibility factor) during a specific time period. In this manner, total numbers of cashless transactions of all these top 15 cashless countries during a specific time period will be generated. This may be divided by the total cumulative transactions of all the study countries during a specific time period with the same assumption that every person of each of these countries makes one transaction during a specific time period.

TABLE 15.5 Top 15 Cashless Countries Involvement of Population with Computations.

Sl. no.	Name of the country	Population[a]	% of cashless transactions[£]	Total no. of transactions assuming that every person of a country makes one transaction during a specific time period (incorporating indivisibility factor)	Total no. of cashless transactions assuming that every person makes one transaction during a specific time period (incorporating indivisibility factor)	Weighted average of percentage of cashless transactions considering all the study (top 15 cashless countries)
1	Singapore	5,603,740	61%	5,603,740	3,418,282	20.768% ≈ 21%
2	Netherland	16,877,352	60%	16,877,352	10,126,412	
3	France	66,259,012	59%	66,259,012	39,092,818	
4	Sweden	9,723,809	59%	9,723,809	5,737,048	

TABLE 15.5 *(Continued)*

5	Canada	34,834,840	57%	34,834,840	19,855,859
6	Belgium	10,449,361	56%	10,449,361	5,851,643
7	United Kingdom	63,742,976	52%	63,742,976	33,146,348
8	The United States	318,892,096	45%	318,892,096	143501444
9	Australia	22,507,616	35%	22,507,616	7,877,666
10	Germany	80,996,688	33%	80,996,688	26,728,908
11	South Korea	49,039,984	29%	49,039,984	14,221,596
12	Spain	47,737,940	16%	47,737,940	7,638,071
13	Brazil	202,656,784	15%	202,656,784	30,398,518
14	Japan	127,103,392	14%	127,103,392	17,794,475
15	China	1,355,692,544	10%	1,355,692,544	135,569,255
Σ		2,412,118,134		2,412,118,134	500,958,349

[a]*Source:* Adapted from the CIA World Factbook. Population–Country Comparison–Top 100 as of January 1, 2014. Retrieved on Mar 22, 2017 from http://www.indexmundi.com/g/r.aspx?t=100 and http://www.indexmundi.com/g/r.aspx?t=100andv=21andl=en.
[b]*Source:* MasterCard Advisors Measuring Progress Toward a Cashless Society. *BT Online,* Dec 29, 2016.

Assuming that every person participates in the financial transaction equitably out of which a certain percent transactions are cashless. If every person makes one transaction during a specific time period, then it is depicted that

Weighted average of percentage of cashless transactions with respect to population size considering all the study countries

$$= \frac{\sum \text{Number of cashless transactions across the study countries}}{\sum \text{Total number of transactions in the study countries}} \times 100$$

$$= \frac{500{,}958{,}349}{2{,}412{,}118{,}134} \times 100\,\%$$

$$= 20.7684\%$$

$$\approx 21\%.$$

It illustrates that among the top 15 cashless countries, the weighted average of percentage of cashless transactions with respect to population size is 21%. India may fix 21% as the targeted percentage for cashless transactions within a specific timeframe in order to remain competitive, particularly within the top 15 cashless economies of the world.

There is no doubt that increasing level of cashless transactions would necessarily enhance efficiency of a state and it could be more effective for the elephantine economy of India as the cashless model leads to reinforce and consolidate the economy with manifold benefits highlighted as follows:

- Greater transparency of financial transactions
- Widening the taxpayers' bracket
- Cost minimization for high denomination currency preparation and its maintenance
- Higher efficiency in economic activities
- Reduction of impacts of parallel economy
- Decrease in leakage of resources in social and developmental projects

On the other hand, India should take the advantage of its deep mobile penetration and agent networks by making greater use of mobile payments technology. Mobiles not only help to transfer money quickly and securely but also improve the quality and convenience of service delivery (Economic Survey 2015–2016).

TABLE 15.6 Global Growth in Mobile Commerce.

Sl. no.	Name of the country	% of smartphone users that engage in m-commerce
1.	China	73%
2.	India	54%
3.	The United States	46%
4.	Japan	44%
5.	Australia	41%
6.	Mexico	39%
7.	United Kingdom	38%
8.	Saudi Arabia	31%
9.	Brazil	30%
10.	Argentina	25%

Source: Adapted from e-Marketer Goldman Sachs Google KPMG China 2014.
Retrieved on Mar 28, 2017 from http://www.pfsweb.com/blog/global-mobile-commerce-growth-infographic/ and http://www.smartinsights.com/mobile-marketing/mobile-commerce/mobile-commerce-trends/.

According to e-Marketer Goldman Sachs Google KPMG China 2014, based on Projections in UN, by 2018, 14.4% of world population (7.56 billion) will make at least one purchase using mobile phones. The development of

Technology will enhance the greater access and ease of digital payment system. Increasing penetration of smartphone, universal network connectivity, cloud computing, use of biometrics, etc., will essentially bring change and momentum in the transaction process in coming years. According to the study by BCG and Google, 2016 it is estimated that the total payments performed through digital payment instruments will be in the range of USD 500 billion by 2020, which is approximately 10 times of current levels (Ramachandran, 2016).

The Government of India envisions preparing India to become a cashless society. Starting from demonetization to the series of other stricter laws on financial issues, Government is consistent on its policy to enhance transparency by curbing on black money, fake currency, and parallel economy. This has opened the gateway to e-commerce or MEC. Popular m-commerce payments channels in India are:

- *99# Service for non-smart phone users
- Aadhaar-enabled Payment System (AEPS)
- Unified Payment Interface (UPI)
- Bharat Interface for Money (BHIM)
- Immediate Payment Service (IMPS)
- Mobile banking platform such as Buddy, V-Mobile, etc.
- E-wallet/digital wallet such as Mobikwik, Paytm, etc.
- Payment banks such as Airtel Money, etc.

15.5.1 CHALLENGES IN M-COMMERCE IN INDIA

- Issues related to security and reliability for m-commerce architecture.
- Lack of mobile network across all the parts of India.
- Lack of basic literacy/mobile literacy in India as a whole.
- Lack of any agency such as m-commerce regulator or Ombudsman to govern, monitor, superintendence the issues related to m-commerce transactions.
- Strengthening and incorporating of the provision of dynamic laws to safeguard the victims due to fraudulent practices with m-commerce transactions.
- Lack of adequate infrastructure, training to law enforcing agencies for IT-related crimes, or fraudulent practices relating to m-commerce.
- Cashless or m-commerce transactions are more costly than the traditional mode due to additional charges such as bank charges, service tax, etc., to be payable by both seller and buyer.

15.5.2 PROACTIVE ROLES OF THE STATE

India has been consistently undergoing its reform policy since many decades. Recent policies such as demonetization; digitization, etc., have attempted to increase the volume of formal economy to enhance transparency and regress the affinity toward creating and holding a parallel economy. Before demonetization took place in the country, it was evident that nearing to 86% value of bank notes were dominated by high denomination notes (erstwhile Rs. 1000 and 500 bank notes). If the State decides to raise the cashless transactions up to 20–21% from the present 2% level, the Government may reduce circulation of 20–21% value of existing high denomination bank notes. The reduction of such volume of high denomination bank notes may be supplemented by equitable value of notes from lower denominations. That will essentially help the poor and underprivileged to maintain the traditional form of transaction till they switch over to the electronic/MEC mode.

TABLE 15.7 Different Denominations, Volumes, Value, and Associated Costs for Indian Bank Notes.

Note denominations	Note preparation cost (in Rs.)[a]	Bank notes volume[b] (million pieces)	Bank notes value[b] (billion)
1	1.14	NA	NA
2 and 5	NA/0.48	11,672 (13.9%)	46 (0.3%)
10	0.96	30,304 (36.3%)	303 (2.1%)
20	1.50	4350 (5.2%)	87(0.6%)
50	1.81	3487 (4.2%)	174 (1.2%)
100	1.79	15,026 (18.0%)	1503 (10.5%)
500 (old) abandoned	2.5	13,128 (15.7%)	6564 (46.0%)
1000 (old) abandoned	3.17	5612 (6.7%)	5612 (39.3%)
500 (new)	3.09	NA	NA
2000 (new)	3.54	NA	NA
		83,579	14,289

[a] *HT*, New Delhi, Dec 20, 2016; *The Hindu*, July 2, 2015; and *TOI*, Jan 1, 2012 Report.
[b] *RBI Report on Currency Management 2014–2015*.

Source: Retrieved from http://www.hindustantimes.com/business-news/what-does-it-cost-to-print-the-new-rs-500-rs-2000-and-other-currency-notes/story-d7wlemIT1sbtibLjPzwhEM.html; http://www.thehindu.com/news/national/cost-of-printing-a-onerupee-note-is-rs-114/article7378821.ece; http://timesofindia.indiatimes.com/business/india-business/Rs-1000-note-costs-govt-Rs-3-17-to-print/articleshow/11321392.cms; and https://rbidocs.rbi.org.in/rdocs/AnnualReport/PDFs/8CURRENCYC8DA67E8EB574B8C84A75AE4D52DEBED.PDF.

The state may formulate a yardstick for measuring financial competency indicator both for individual as well as corporate to enforce compulsory participation on electronic platform for making financial transactions. The Government may fix a certain percentage of yearly transactions be made either by cheque or by electronic modes to those who qualify the competency indicator. There may be a cap of holding cash by any individual/ corporate. However, Government recently reduced upper ceiling of cash transaction up to Rs. 2 lakh from its exiting Rs. 3 lakh. It has also prohibited payments or receipts of any advance worth of Rs. 20,000 or more by cash for procuring any immovable property (Ray, 2017). This may ensure to reinforce higher accountability and transparency in the system. This would lead to create tremendous opportunities for e-commerce and more particularly to m-commerce since people of India have more mobile phones than the number of people who possess desktops or PCs. The governmental agencies have been relentlessly bringing new policies and creating awareness to motivate and promote electronic modes of transactions. But there should be incentive mechanism to motivate people to opt for electronic modes. The cashless transactions should be encouraged by rationalizing the charges, if any.

But for all of these, first, the state should concentrate on long-term investment to develop the basic services and infrastructure across the lengths and breaths of India. The Government of India should take up adequate measures to overcome the challenges so that the acceptance and adoption of m-commerce by the people will be spontaneous and the process of transformation will be smooth and sustainable.

15.6 CONCLUSION

The mobile commerce has opened up new dimensions of business. It will essentially create incredible opportunities for venturing new start-ups. Particularly for India, it will foster the people to participate in the economic activities with greater efficiency, responsibility, and transparency. With the advent of technological development, the new-generation business, trade, and industry will be value added by the mobile commerce platform. There is no doubt that the future business be it large or small will be dominated by m-commerce. The success of India Incorporation shall depend on its preparedness to move forward toward adoption of mobile-led e-commerce. Indeed, there are challenges which nobody discounts at all. All the teething problems and future challenges have to be addressed on a much wider canvas through a series of reforms and confidence building measures if the nation is committed to achieve less cash ecosystem. Strong commitments, firm political willpower, consistent policy, and the art of mastering consensus can only bring changes into the stereotype mental blocks of many of us. India will shine when the state can include and involve most of its people as an integral part of the grand success story. It is a wakeup call to the policy-makers to trade-off priorities between short-term populist model and long-term nation-building decisions. This mission can only be achieved by the active participation and commitments of all the key stakeholders. Thus the comprehensive and collaborative approach will accelerate the growth and prosperity of the nation–state framework of India.

KEYWORDS

- **mobile commerce**
- **smart phones**
- **infrastructure**
- **ICT**
- **digital movement**
- **cashless ecosystem**

REFERENCES

Avendus. *India's Mobile Internet: The Revolution has Begun*; Report, September, 2013; pp 1–126. http://www.telecomcircle.com/wp-content/uploads/2010/02/Avendus_Report-Indias_Mobile_Internet-2013.pdf (accessed Mar 10, 2017).

Banerjee, A.; Ros, A. J. Drivers of Demand Growth for Mobile Telecommunications Services: Evidence from International Panel Data, NERA (National Economic Research Associates) Economic Consulting, Marsh and McLennan Companies, 2002; pp 1–37.

Barnes, S. J. The Mobile Commerce Value Chain: Analysis and Future Development. *Int. J. Inf. Manag.* **2002,** *22* (2), 91–108.

Brown, M. Global Mobile Commerce Growth. *Infographic*, March 28, 2014 (Blog). http://www.pfsweb.com/blog/global-mobile-commerce-growth-infographic/ (accessed Mar 22, 2017).

BT Online. Here are the Top Cashless Countries in the World. *BT Online*, New Delhi, Dec 29, 2016. http://www.businesstoday.in/current/economy-politics/here-are-the-top-cashless-countries-in-the-world/story/241430.html (accessed Mar 28, 2017).

CIA World Factbook. *Population–Country Comparison–Top 100*. http://www.indexmundi.com/g/r.aspx?t=100 and https://unstats.un.org/unsd/demographic/products/vitstats/Sets/Series_A_2014.pdf (accessed Mar 22, 2017).

Coursaris, C.; Hassanein, K. Understanding M-Commerce: A Consumer-centric Model. *Q. J. Electron. Commer.* **2002,** *3*, 247–272.

Criteo. *State of Mobile Commerce Growing like a Weed, Q1*; 2015; pp 1–16. http://www.criteo.com/media/1896/criteo-state-of-mobile-commerce-q1-2015.pdf (accessed Mar 22, 2017).

Davis, F D. A Technology Acceptance Model for Empirically Testing New End Users Information Systems: Theory and Results. Doctoral Dissertation, Sloan School of Management, Massachusetts Institute of Technology, 1986.

BNY Mellon. *Global Payments 2020: Transformation and Convergence*; 2013; pp 1–49. https://www.bnymellon.com/_global-assets/pdf/business-insights/global-payments-2020-transformation-and-convergence.pdf (accessed Mar 28, 2017).

Ericsson Mobility Report on the Pulse of the Networked Society; November, 2016; pp 1–36 https://www.ericsson.com/assets/local/mobility-report/documents/2016/ericsson-mobility-report-november-2016.pdf (accessed Mar 28, 2017).

EY-FICCI Report. Speeding Ahead on the Telecom and Digital Economy Highway: Key Priorities for Realizing a "Digital Bharat;" 2015, pp 1–116. http://www.ey.com/Publication/vwLUAssets/ey-speeding-ahead-on-the-telecom-and-digital-economy-highway/$FILE/ey-speeding-ahead-on-the-telecom-and-digital-economy-highway.pdf (accessed Mar 28, 2017).

FICCI-KPMG. *The Stage is Set: FICCI-KPMG Indian Media and Entertainment Industry Report 2014*; 2014, pp 1–13. http://ficci.in/spdocument/20372/FICCI-Frames-2014-KPMG-Report-Summary.pdf (accessed Mar 28, 2017).

Fishbein, M.; Ajzen, I. Belief, Attitude, Intention and Behavior: An Introduction to Theory and Research; Addison-Wesley: Reading, MA, 1975.

Golden, S. A. R.; Regi, S. B. Mobile Commerce in Modern Business Era. *Int. J. Curr. Res. Acad. Rev.* **2013,** *1* (4), 96–102.

GSMA. *Mobile Economy Europe 2013*; 2013, pp 1–136. https://www.gsma.com/mobileeconomy/archive/GSMA_ME_Europe_2013.pdf (accessed Mar 22, 2017).

GSMA; A. T. Kearney; Wireless Intelligence. *European Mobile Industry Observatory 2011*, 2011. https://www.gsma.com/publicpolicy/wp-content/uploads/2012/04/emofullwebfinal.pdf (accessed Mar 28, 2017).

IAMAI-IMRB Mobile Internet in India Report 2014 and 2015. http://www.iamai.in/sites/default/files/research/pdf/IAMAI%20Digital%20Commerce%20Report%202014_90.pdf and https://sjeximservices.files.wordpress.com/2016/04/mobile-internet-report-2015.pdf (accessed Mar 22, 2017).

India Annual Union Budget Announcement 2014–15 and 2015–16. http://indiabudget.nic.in/budget2014-2015(i)/budget.asp. and http://indiabudget.nic.in/es2015-16/echapvol1-02.pdf (accessed Mar 22, 2017).

Indian Express. Budget 2014: Govt. Allocates Rs 500 cr for Internet Connectivity in Villages. *Indian Express*, New Delhi Ed., July 10, 2014. http://indianexpress.com/article/india/politics/budget-2014-govt-allocates-rs-500-cr-for-internet-connectivity-in-villages/ (accessed Mar 22, 2017).

International Telecommunication Union (ITU)–Telecommunication Development Bureau (Bureau de Développement des Télécommunications, BDT) Initiatives. *m-Powering*

Development Initiative; First Report of the Working Group on M-commerce, May, 2014, pp 1–27. https://www.itu.int/en/ITU-D/Initiatives/m-Powering/Documents/First%20Report%20of%20the%20Working%20Group%20on%20m-Commerce.pdf (accessed Mar 10, 2017).

Internet Live Stats; e-Marketer; KPMG in India Analysis. *India on the Go: Mobile Internet Vision 2017*; Report in KPMG-IAMAI, July, 2015, p 13.

KPMG-IAMAI. *India on the Go: Mobile Internet Vision 2017*; 2015. http://rtn.asia/wp-content/uploads/2015/07/Report.pdf (accessed Mar 22, 2017).

Kuo, Y.-F.; Yu, C.-W. 3G Telecommunication Operators' Challenges and Roles: A Perspective of Mobile Commerce Value Chain. Technovation **2006,** *26*, 1347–1356.

Kwon, O. B.; Sadeh, N. Applying Case-based Reasoning and Multi-agent Intelligent System to Context-aware Comparative Shopping. *Decis. Support Syst.* **2004,** *37* (2), 199–213.

MasterCard Advisors Measuring Progress Toward a Cashless Society; pp 1–5. http://www.mastercardadvisors.com/_assets/pdf/MasterCardAdvisors-CashlessSociety.pdf (accessed Mar 22, 2017).

Mehmood, F. Business Models and Strategies of M-Commerce: A Review. *J. Internet Bank. Commer.* **1970,** *20* (1), 1–17.

Mehmood, F. Business Models and Strategies of M-commerce: A Review. *J. Internet Bank. Commer.* **2015,** *20* (1), 1–17.

Ming, X. Y.; Qi, C. In *A Study on M-Commerce Industry Development and Market Structure in China*. International Symposium on Electronic Commerce and Security; ISECS, 2008; pp 276–280.

Muralidharan, K.; Paul, N.; Sukhtankar, S. Building State Capacity: Evidence from Biometric Smartcards in India. *Am. Econ. Rev.* **2016,** *106* (10), 2895–2929. https://www.poverty-actionlab.org/sites/default/files/publications/Building-State-Capacity_Feb2016.pdf (accessed Mar 10, 2017).

Mylonopoulos, N. A.; Doukidis, G. I. Introduction to the Special Issue: Mobile Business: Technological Pluralism, Social Assimilation, and Growth. *Int. J. Electron. Commer.* **2003,** *8* (1), 5–22.

Nielsen. *The Mobile Consumer: A Global Snapshot*; Feb, 2013, pp 1–21. http://www.nielsen.com/content/dam/corporate/uk/en/documents/Mobile-Consumer-Report-2013.pdf (accessed Mar 22, 2017).

Paavilainen, J. *Mobile Business Strategies: Understanding the Technologies and Opportunities*; Addison-Wesley Longman Publishing Co., Inc.: Boston, MA, 2002.

Parveen, A.; Habib, S.; Sarwar, S. Mobile Commerce—New Way to Business. *Int. J. Res. Dev. Manag. Rev. (IJRDMR)* 2012, 1 (1), 37–40.

PricewaterhouseCoopers International Limited (PwCIL). *Logging into Digital Banking Creating Access, Transforming Lives*; 2015, pp 1–32. https://www.pwc.in/assets/pdfs/publications/2015/logging-into-digital-banking.pdf (accessed Mar 22, 2017).

Ramachandran, A. Here Is Why India Is Poised for a Cashless (R) Evolution. IT News, *ETCIO*, Dec 1, 2016.

Ray, S. Govt. Cuts Rs 3-lakh Limit on Cash Transaction. Caps it at Rs 2 Lakh. *Hindustan Times*, New Delhi, Mar 21, 2017. http://www.hindustantimes.com/business-news/govt-cuts-rs-3-lakh-limit-on-cash-transaction-caps-it-at-rs-2-lakh/story-VKY4tgSxHXdYs-CLbN3PsaK.html (accessed Mar 22, 2017).

Reserve Bank of India. *Concept Paper on Card Acceptance Infrastructure*, 2016. https://rbidocs.rbi.org.in/rdocs/PublicationReport/Pdfs/MDRDBEDA36AB77C4C81A3951C-4679DAE68F.PDF (accessed Mar 22, 2017).

Scharl, A.; Dickinger, A.; Murphy, J. Diffusion and Success Factors of Mobile Marketing. In *Electronic Commerce Research and Applications*; 2005; Vol. 4 (2), pp 159–173.

Sharma, S.; Gutiérrez, J. A. In *Viable Business Models for M-commerce*. UK Academy for Information Systems Conference Proceedings, 2009; AIS Electronic Library (AISeL): UK, 2009; p 45. http://aisel.aisnet.org/ukais2009/45.

Tata Investment Corporation Limited. *79th Annual Report 2015–2016*; p 14. http://indiabudget.nic.in/es2015-16/echapvol1-02.pdf (accessed Mar 28, 2017).

The Boston Consulting Group (BCG); *GoogleTM. Digital Payments 2020: The Making of a $500 Billion Ecosystem in India*; July, 2016. http://image-src.bcg.com/BCG_COM/BCG-Google%20Digital%20Payments%202020-July%202016_tcm21-39245.pdf and https://www.bcg.com/en-in/d/press/25July2016-digital-payments-2020-making-500-billion-ecosystem-in-india-39417 (accessed Mar 28, 2017).

The Deloitte Global Mobile Consumer Survey, 2015. https://www2.deloitte.com/in/en/pages/technology-media-andtelecommunications/articles/deloitte-mobile-consumer-survey-2015---Deloitte-India---technolo.html (accessed Mar 28, 2017).

Tsalgatidou, A.; Pitoura, E. Business Models and Transactions in Mobile Electronic Commerce: Requirements and Properties. *Comput. Netw.* **2001,** *37*, 221–236 (Elsevier). http://www.cs.uoi.gr/~pitoura/distribution/computer-networks02.pdf (accessed Mar 28, 2017).

Yan, L. The Analysis on How China Turn to be the Global Power in M-Commerce with the Advantage of Backwardness. *Management World*, 2005, Vol. 7, pp 162–165.

Websites/WebPages (accessed Mar 22, 2017)

http://www.moneycontrol.com/news/cnbc-tv18-comments/telecom-data-wars-ideacuts-2g-3g-data-tariffs-by-90_987659.html (accessed Mar 22, 2017).

http://www.indexmundi.com/g/r.aspx?t=100&v=21andl=en (accessed Mar 22, 2017).

http://www.firstpost.com/business/economy/why-is-printing-a-rs-50-note-costlier-than-a-rs-100-note-385696.html (accessed Mar 22, 2017).

https://www.rbi.org.in/SCRIPTS/AnnualReportPublications.aspx?Id=1154 (accessed Mar 22, 2017).

https://www.mygov.in/sites/default/files/mygov_1450978729362686 1.pdf (accessed Mar 22, 2017).

http://www.tele.net.in/view-point/item/17530-the-number-of-mobile-internet-users-in-india-to-reach-314-million-by-2017?format=pdf (accessed Mar 22, 2017).

http://indiabudget.nic.in/es2015-16/echapvol1-02.pdf (accessed Mar 22, 2017).

http://www.trai.gov.in/WriteReadData/userfiles/file/NTP%202012.pdf (accessed Mar 22, 2017).

http://www.trai.gov.in/sites/default/files/AT%26T_India_AGNSI_CP_18102016.pdf (accessed Mar 22, 2017).

http://www.trai.gov.in/sites/default/files/TRAI_Annual_Report_English_16052016.pdf (accessed Mar 22, 2017).

http://progressivefix.com/wp-content/uploads/2014/10/CISCO-2014-MOBILE-VNI-for-FCC-5FEB13-copy.pdf (accessed Mar 22, 2017).

Statista. 2016. https://www.statista.com/statistics/257048/smartphone-user-penetration-in-india/ (accessed Mar 22, 2017).

http://www.pfsweb.com/blog/global-mobile-commerce-growth-infographic/ (accessed Mar 22, 2017).

http://www.smartinsights.com/mobile-marketing/mobile-commerce/mobile-commerce-trends/ (accessed Mar 22, 2017).

CHAPTER 16

M-COMMERCE AT THE BOTTOM OF THE PYRAMID: INSIGHTS FROM ASIA

FARRAH ZEBA[1], MUSARRAT SHAHEEN[2*], and RAMBALAK YADAV[3]

[1]*Department of Marketing and Strategy, IBS Hyderabad, a Constituent of ICFAI Foundation for Higher Education, India*

[2]*HR Department, IBS Hyderabad, a Constituent of ICFAI Foundation for Higher Education, India*

[3]*Department of Marketing, Institute of Management Technology, Hyderabad, India*

[*]*Corresponding author. E-mail: musarrat.shaheen@ibsindia.org; shaheen.musarrat@gmail.com*

ABSTRACT

C. K. Prahalad in his book—"*The Fortune at the Bottom of the Pyramid*," 2004—stated that the world has an abundance of poorest people and the market catering to them can be characterized as one of the fastest growing markets having untapped buying power. His suggestions were well appreciated and people even called him a "business prophet." Bill Gates, the founder of Microsoft, also commented that the book can be said to be a blueprint for earning profits and reduction of poverty. In emerging markets like India, the majority of the population resides in rural areas and falls under the section of the bottom of the pyramid (BoP). Indian market has low-income consumers who are price sensitive. Companies in the past have targeted the consumers at the BoP and have gained popularity and success. For instance, the recent success of Patanjali products targeting consumers at the BoP has proved that profits can be made even with low price and focus on volumes. After demonetization in India, consumers across India including consumers of BoP are increasingly adopting mobile wallets and online payment gateways. To ride over the challenges of cashless transactions, government has taken several initiatives such as opening bank accounts (Jan Dhan Yojana), facilitating free online transactions by waving off service taxes, rebate on the usage of debit cards, incentive schemes for small traders and rural consumers such as Digi-dhan Vyapar Yojana, as a part of Digital Business Policy, and lucky draw for the online users. Thus, the main objective of this chapter is to understand the role of mobile payment and transactions as one of the mode of m-commerce and how this is related to the consumers lying in the BoP of India. That is, in this chapter, the role of mobile and mobile-enabled services related to m-commerce and the consumers of BoP are discussed. Hurdles and benefits of m-commerce in the BoP market of India are also covered.

16.1 INTRODUCTION: BOTTOM OF PYRAMID

The focus of marketing managers is always on identifying and segmenting markets to search for consumers who have a capacity to consume, have access to the product or services produced by the companies, or who can pay a sustainable price for the product or services. It is the core activity of marketing department that has shaped the business model of all multinational corporations (MNCs) operating across the globe. In this whole process, to date, companies had relatively ignored some consumers who generally belong to rural areas or consumers who have less purchasing power. The reasons for doing so are varied. However, in the mid-1990s, a new thought and concern was raised by an eminent marketing scholar Prahalad and his colleague (Prahalad and Hart, 2002; Prahalad, 2006).

The whole process of STP (segmenting, targeting, and positioning) was revolutionized. An additional area to target for was proposed. The target area was named Bottom of Pyramid or BoP. Consumers whose annual incomes or purchasing power parity basis are less than $1500 per year were brought into the spectrum. Such type of consumers are found lying at the base of the purchasing power parity pyramid (i.e., BoP) and hence, they are popularly termed as consumers at the "Bottom of the Pyramid." The phrase "Bottom of the Pyramid" was first used by one of the former presidents of the United States—Franklin D. Roosevelt. In a radio address on April 7, 1932, he used this phrase to denote the forgotten man of the economy. In his speech, he said, "these unhappy times call for the building of plans that rest upon the forgotten, the unorganized but the indispensable units of economic power."

He requested to put "faith once more in the forgotten man at the bottom of the economic pyramid" (Roosevelt, 1932).

This concern was further expanded and elaborated in a marketing book titled "*The Fortune at the Bottom of the Pyramid.*" In this book, Professor C. K. Prahalad suggested that fortune lies at the BoP. He means to say that consumers at the BoP can be the future prospective customers for the marketers, as the world has an abundance of poorest people and the market catering to them can be characterized as one of the fastest growing markets having untapped buying power. He suggested that companies should cater to this segment to earn profits as well as contribute to the society by customizing their marketing jobs focused toward this segment. He believed that these poor people will upgrade to the middle-class level in the future due to economic growth and other related factors, hence they can be termed as the prospective customers. Further, catering to this segment will be beneficial to the companies as it is a niche area and does not have many players to compete with. Hence, he concluded that companies should not ignore this segment and should plan their strategies to enter and serve this segment. Across the world, the suggestions provided by Professor C. K. Prahalad are well acknowledged and appreciated. For instance, Bill Gates, founder of Microsoft, commented that the book can be said to be a blueprint for earning profits and reduction of poverty. Traders and manufacturers even started calling him a "business prophet" (Gunther, 2014). Since then, BoP has gained popularity and marketers started focusing on it and customers are arranged on the different level of BoP.

16.2 CONSUMER TIERS IN THE BoP

Booz & Company, a strategy and consulting firm in the United States categorized the consumers in the BoP into four tiers. At the very top of the pyramid (Tier 1) lies cosmopolitan middle- and upper-income people of developed nations and few rich and elite groups of developing nations (Fig. 16.1). The Tiers 2 and 3 (in the middle of the pyramid) include the poor consumers of developed nations and the upcoming middle classes of developing nations. They also provided estimated per capita income brackets and populations of customers lying in these sections. At the BoP lies the customer whose per capita income is less than $1500, which is considered as minimum income to sustain a decent life.

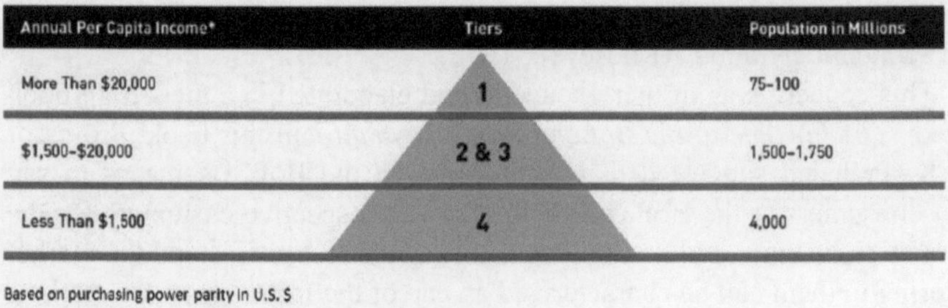

FIGURE 16.1 The world economic pyramid.
Source: Reprinted from Prahalad and Hart (2002).

The majority of the Tier 4 consumers reside in rural villages or urban slums. They are not employed and have a minimum formal education; as such, it is hard to reach them via popular distribution, credit, and communication channels. Similar to an iceberg, the massive segment of the consumers lying in this section to date has remained largely invisible to the business corporate. Different marketers have different opinions and strategies to cater to the consumers of BoP. These opinions explain the cause that, till now, why BoP has been left in isolation and why the consumers of BoP are ignored. Some of these opinions and perspectives of the marketers are discussed in the subsequent section.

16.3 MARKETERS' PERSPECTIVES ABOUT BoP CONSUMERS

The question that comes to mind is that to date why the consumers residing in the BoP have been underestimated or ignored. There could be four misperceptions from the perspectives of manufacturers and traders.

16.3.1 LOW DISPOSABLE INCOME

Manufacturers and traders believed that individual-level purchasing power is very low at the BoP, and consumers will not be able to buy the products and services which are sold at a higher price than their capacity. The reason is that the disposable income and purchasing power of BoP consumers are not at par with the high price of the MNC products and services. But, from the deeper analysis, it was found that the aggregate buying powers of the consumers lying at the BoP may not be low and possibility is that they can afford to buy goods in a group. Communities can buy goods such as computers, cellular phones, and washing machines for the common use.

Hence, the economic potential at the individual level may be low at the BoP, but *it may not be low* at the aggregate level.

16.3.2 INVESTMENT IS FOR BASICS AND NECESSITIES AND NOT FOR LUXURIES

The consumers at the BoP have low disposable income; as such, their first intention is to spend money on buying items of daily use such as necessary goods, for example, food grains, medicines, electricity, and gas stoves. These items are considered as more realistic investments to them. It provides sustenance and improves their quality of life. They believe in the comfort of the present life and not in the future life. Hence, buying a car or house may not be a viable option to them.

16.3.3 MISCONCEPTION ABOUT BoP CUSTOMERS' TECHNOLOGY KNOWHOW

GSM mobile phones are more popular in rural areas where the majority of the consumers lying in BoP are residing. Smartphones are not sought after thing and very few people know the proper usage of smartphones. Similarly, with reference to computers, very few people are buying personal computers or laptops and desire high-frequency internets facility in their houses. BoP consumers still prefer paying hourly charges to use a computer in internet cafés when needed. This could be because these people are interested in new technologies only to improve their living conditions and not to boost their status.

16.4 MARKETING POTENTIALS AT THE BoP

In addition to rising disposable income, marketers are aiming BoP segment due to various reasons described in the following sections.

16.4.1 RISE IN CONSUMERS' ASPIRATIONS

With the rise in income, a desire for product and services which are beyond the basic needs also arises. When disposable income increases, people want to improve their lifestyle and standards of living.

16.4.2 MEDIA AND INTERNET PENETRATION

Greater internet connectivity and media penetration in the rural areas have urbanized the lifestyle of the rural areas. Media and internet connectivity increases the awareness of usage and availability of goods and services among the rural consumers.

16.4.3 BRAND LOYALTY

It has been found that poor people are more loyal to their brands as they cannot make mistakes with their low disposable income. They fear to experiment with new product and services as their income is limited and restricted.

16.5 VARIED OPINIONS ABOUT THE BoP

The BoP market is not a homogeneous market, but it is made up of different segments that have different needs and requirements. Companies should understand these factors and adjust their business models to facilitate effective marketing and customer engagement. According to the statistics provided by the World Bank, International Monetary Fund (IMF), and the United Nations Conference Trade and Development (UNCTAD), doing business in the BoP segments is more challenging compared to other segments. These challenges are revolving around the economic, political, social, cultural, and financial constraints (Perdana, 2005).

Further, to resolve the social issues such as poverty and unemployment, United Nations and other related organizations have suggested several economic measures. Countries across the world are focusing on the development of their rural areas to align with these initiatives. They are developing employment opportunities to eradicate poverty. Economic policies and initiatives are developed to facilitate and motivate MNCs and other manufacturers in expanding and growing their businesses in the rural areas. These initiatives are meant to resolve the unemployment issues and increase the disposable income of the people residing in rural areas. In turn, these initiatives have created a platform for the manufacturers and traders to showcase and sell their goods and services to the consumers lying in the BoP. Some of the paradoxes which have restricted MNCs to cater to this segment are provided in Table 16.1.

TABLE 16.1 Misperception of MNCs About Consumers of Bottom of Pyramid.

1)	The poor are not our target consumers because, with our current cost structures, we cannot profitably compete for that market.
2)	The poor cannot afford and have no use for the products and services sold in developed markets.
3)	Only developed markets appreciate and will pay for new technology. The poor can use the previous generation of technology.
4)	The bottom of the pyramid is not important to the long-term viability of our business. We can leave Tier 4 to governments and nonprofits.
5)	Managers are not excited by business challenges that have a humanitarian dimension.
6)	Intellectual excitement is in developed markets. It is hard to find talented managers who want to work at the bottom of the pyramid.

Source: Adapted from Prahalad and Hart (2002).

BoP is a niche market and existing marketing strategies will not be sufficient for it. To gain access to this market, marketers have designed more innovative and product-specific strategies. Some of the strategies are discussed in Section 16.6.

16.6 MARKETING STRATEGIES FOR BoP

With the innovation in technology and e-commerce, consumers lying in the BoP are now surfacing and gaining attention from the corporate. BoP segment is more viable and has huge potential, as it is still in its infancy and has more opportunities compared to the other markets. Companies to reap benefits from the consumers of BoP should design some new and innovative ideas to make their goods and services attractive to them. They have to be more creative and work on new business models, as the consumers of this segment vary in their choice, and there exist disparity in the purchasing power. Some of the initiatives companies can employ to cater to the consumers of BoP are summarized in Section 16.6.1.

16.6.1 INNOVATION

The consumers lying in the BoP segment are cost conscious and they want to have immediate and visible benefits from the products and services, as such existing products and services are not useful to them. Companies should find a way to make a profit while offering affordable products and services.

For instance, smartphones makers such as Micormax have developed a light version smartphones which has basic features to suit the needs and requirements of the consumers of rural areas (Anwer, 2015).

16.6.2 MARGIN VERSUS VOLUME

Generally, companies focus on high gross margin. But, the purchasing power of the consumers lying in the BoP segment is very low. Companies have to redesign their strategies. They have to earn a profit on volumes. For instance, the success of Patanjali in India is based on this mantra only. The company has targeted consumers in the BoP and has proved that profits can be made even with low price and high volumes. Thus, company to earn profit from consumers lying in BoP should tighten their cost and make the supply chain lean and effective. The cost-savings strategy is the key to performance and success in the low-cost markets.

16.6.3 TARGET AGGREGATED CUSTOMERS

Since the consumers of BoP segment focus on the fulfillment of their basic needs and not on luxuries; to earn margin, companies should target aggregate customers or make their products modifiable as aggregate use. Companies can also consider the pay-per-use system. For instance, laundry system, car wash workshops, and computers in as utility center in developed countries.

Though BoP is a common feature of all developing countries, in Section 16.7, specific focus is given to the BoP sector of India, as more than 60% of Indian population still resides in the rural areas and information and communication technology (ICT) and mobile technology has grown up in the recent years. Mobile commerce has become the more sought after means in this country due to several factors which are discussed in the subsequent sections.

16.7 BoP IN INDIA

Till 2002, the idea behind BoP was not so popularized in India but with the initiative of companies such as Unilever, Cemex, and S. C. Johnson it became popular. These companies came up with innovative marketing ideas and business models to tap the business opportunities lying at the BoP. According to a report by Boston Analysis group, though India has several favorable demographic features such as young population and upcoming working group, India's population is found skewed toward the lower income group with 70% of the population having annual household earning less than $4000. They have lower disposable income and spend most of their earnings on daily items such as food, power, and fuel. Estimates say that a large part of this segment, say 78%, is composed of the rural population of India. This segment is referred to those who belong to the BoP segment.

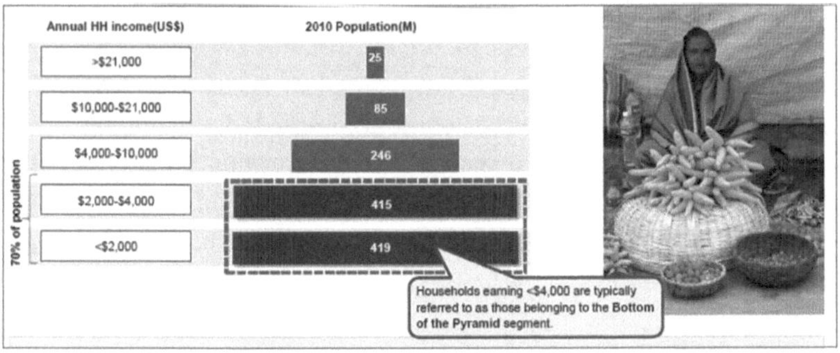

FIGURE 16.2 (See color insert.) Bottom of pyramid in India.
Source: Reprinted from the Luce (2016).

But, as per the "India's Rising Bottom of the Pyramid," report by the Boston Analytics, India's poverty rates are declining and almost half of India's GDP is contributed by private household consumptions. Due to rising income around 80% of the consumption is expected to grow and the BoP segment is moving from less than $2000 income bracket to a higher income bracket of $2000–$4000 and $4000–$10,000 (Luce, 2016) (Fig. 16.2). By 2020, an annual income of BoP segment is expected to increase to more than $21,000.

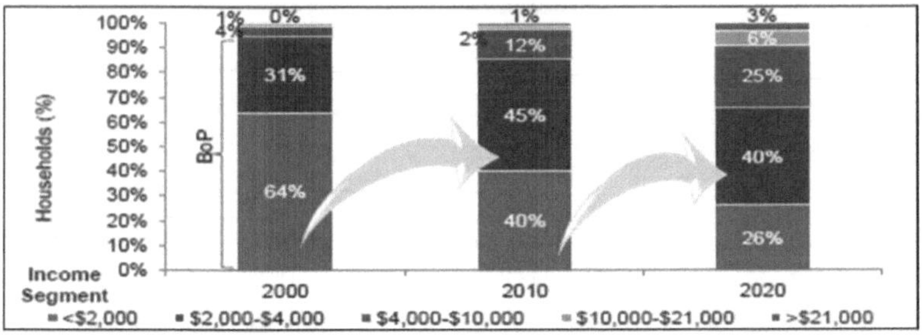

FIGURE 16.3 (See color insert.) Households by income bracket (2000–2020).
Source: Reprinted from the Luce (2016).

Average food consumption by BoP segment of India is found much higher than the middle and rich segment. It is noted that while BoP spends 59% on food, the middle and rich segment spends 47% and 36%, respectively. Statistics by the Centre for Monitoring Indian Economy (CMIE) and the National Sample Survey Office (NSSO) suggest that food and fuel together account for 70% of the total household expenditure, whereas 30% relates to other necessities (Luce, 2016) (Fig. 16.3). The statistics suggest that there has been a shift in consumption pattern, of BoP rural population, from necessities to discretionary expenses. For instance, it is noted that approximately 42% of rural households possess television set and one in every two rural households owns a mobile phone (Shukla and Bairiganjan, 2011).

Further, from the perspective of demographics, India is among the most populous countries in the world with a relatively young population. Boston Analysis group suggested that by 2020, around 64% of India's population will be the working population (Luce, 2016). Thus, India is among the emerging markets where the retail consumption will increase in the future years.

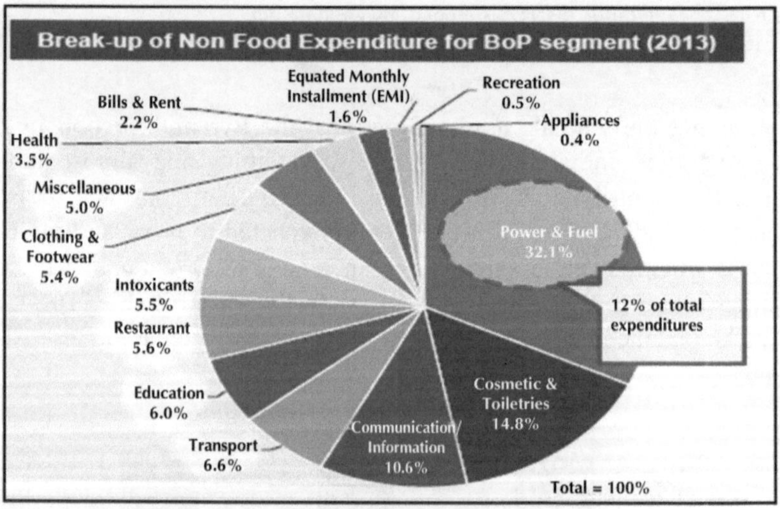

FIGURE 16.4 (See color insert.) Expenditures by the BoP segment in India.
Source: Reprinted from Luce (2016).

Discussion on m-commerce is incomplete until one throws some light on ICT. M-commerce is an extension of e-commerce. ICT and e-commerce are inseparable. They are interrelated to each other and in fact, online business is considered as the application of ICT. ICT involves usage of computers and networks to handle information. E-commerce in general and m-commerce in specific uses ICT to manage various online activities such as sharing and dissemination of information with the customers. ICT aids in all the activities of an online business. For instance, mobile banking helps in transacting on mobile for online shopping. The increase in sales of smartphone in India has led to an increase in the numbers of phone in the hands of consumers of both urban and rural areas (Fransman, 2010). A survey result has shown that the usage of smartphones in India has touched to 55% in 2013, which is considered as the fastest growth among the 30 smartphone markets in the world [International Data Corporation (IDC), 2016].

16.8 ICT AND THE BoP

ICT is an omnipresent feature in the developed as well as developing nations (Fransman, 2010). ICT has affected the life of every individual. It has gained popularity in the last two decades. However, its presence and usage among consumers of BoP are relatively at a nascent stage. ICT gave access to information and expertise which they were not able to retrieve due to the absence of an appropriate medium. ICT has a major contribution in the life of BoP consumers. More specifically, in the current scenario, when several initiatives by the companies have increased the mobile network and mobile connectivity in the rural areas (Bhatnagar and Schware, 2000). For instance, ITC Limited in India started an "*e-Choupal*" program to improve the awareness among the people of rural areas and to the farmers by providing information related to agriculture and crops by setting up kiosks and e-governance. A number of case studies have also shown a significant economic and social impact of ICT on BoP consumers in rural areas of India (Realini and Mehta, 2015) such as increased agricultural productivity and income, lower cost and fewer hassles in money transfer, educating the most remote, etc. But, the utility for BoP citizens of these facilities was very low, as they were supposed to go to "*e-Choupal*" or have their personal desktop/laptop as it may result into consuming their time and money. However, out of the several means of ICT such as television, radio, and others, mobile is more appropriate for BoP customers, as it is comparatively low cost and they can manage surfing and searching by themselves after having basic usage training. Due to ease of use of mobile phones, m-commerce has become an easy mode of business and trading for the consumers of BoP.

M-commerce is the subset of e-commerce, which includes all e-commerce transactions that can be carried out by a mobile phone (Fransman, 2010). M-commerce depends on the availability of mobile and internet connectivity on the mobile phones. Bright Simmons, Founder of mPedigree quoted that "the combination of simple technology with right resources can have an impact on people life in a very tangible ways" and m-commerce is doing the same.

The emergence of m-commerce has been very beneficial for consumers at BoP and in rural areas of all developing nations. For instance, within the second year of its launch, BBC Janala has reached over 3 million trainees via mobile phone alone in Bangladesh for providing English lessons at nominal charges. In India, Reuters Market Light (RML) provides information to the farmers about lock market price information, weather information, and news related to the agriculture via messages on mobile phones, similar to mPedigree, operating in African nations, that allows patients to confirm the legitimacy of their medicine before leaving the pharmacy just by sending a unique code and medicine information and getting an automated response in their chosen language.

16.9 M-COMMERCE IN INDIA

TRAI (Telecom Regulatory of India) has confirmed that the mobile users in India have increased to 906.62 million (2013) from 429.73 million (2008). They also predicted that the mobile users in India will be doubled and will reach a figure which will be greater than the combined population of the United States and European Nation (Peermohamed, 2016). M-commerce in India has received a concrete shape and structure with the regulation draft issued by the Reserve Bank of India (RBI) titled "Mobile Banking Transactions in India—Operative Guidelines for Banks." The document clearly specifies the guidelines for mobile payments in India. RBI made it mandatory for all the banks and mobile operators to have an alliance with each and use Smart Messaging Services (SMS) as a medium for banking transactions. It means SMS will be used as a medium of communication between the bank and their customers. To adhere to the guidelines of RBI, banks in India initiated online applications and Wireless Application Protocol (WAP) systems. Every bank provided an online/internet banking facility to their customers which was password protected. Consumers can transfer, purchase, and trade through these internet banking accounts which were linked to their savings or current accounts. Please refer to Figure 16.5 for mobile phone users, smartphone users, and internet users as on May 2017.

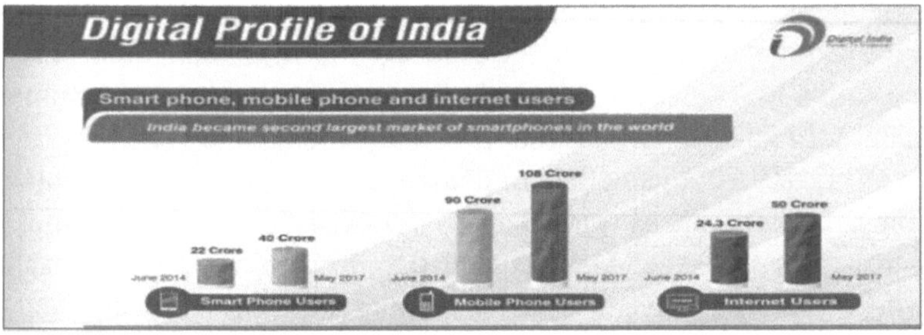

FIGURE 16.5 Digital profile of India.
Source: Reprinted from www.digitalindia.gov.in.

A mobile readiness survey by the MasterCard Mobile Payments Readiness Index (MPRI) suggests that mobile payments can be categorized into three types—person to person (P2P), mobile e-commerce (m-commerce), and mobile payments at the point of sale (POS). They calculated the mobile readiness scores of around 34 countries. Singapore topped the list with a mobile readiness score of 45.6 followed by China and the United States with scores of 42 and 41.5, respectively. India ranked 21st with a score of 31.5 on a scale of 100. MPRI stated that only 14% of mobile users in India are familiar with P2P and m-commerce transactions and approximately 10% are familiar with POS transactions. Thus, the statistics confirmed that m-commerce is still growing in India. RBI data validate the same. Around 3.7 crore mobile transactions took place between February 2012 to November 2012, which was around 1.7 times more in volumes in a span of 10 months. But, these statistics pertain to the urban population, and the rural and the deprived customers who do not enjoy the benefit of convenient banking are not included. These customers neither have bank accounts nor have connectivity on their mobile phones. To improve the situation and make the mobile transactions successful, in 2012, deputy governor of RBI, H. R. Khan raised a concern and called upon the banks and mobile service providers to strategize together and make the service workable and adaptable (Peermohamed, 2016).

One of the major hindrances behind low mobile usage was the connectivity and fewer smartphone users. Mobile users in India are not low, but only about 250 million users have smartphones (Luce, 2016). App-based services by several companies become a blessing in disguise for India. These applications have in-built technology which provides uninterrupted service on 2G networks also. For instance, the cab service providers Ola have designed their app in such a way that it works even in patch networks. Companies such as Reliance are investing crores in building 4G infrastructures across the country.

M-commerce gained momentum with the demonetization event in India. According to a spokesperson from the Confederation of Indian Industry (CII), demonetization as a "move was exactly the fillip the fledgling m-commerce ecosystem in India was looking for" [Peermohamed, 2016; The Press Trust of India (PTI), 2016]. As an effect of demonetization, usage of smartphones for financial transactions has grown at an explosive rate in India. Internet use on mobile has increased to 371 million in 2016, which is expected to reach 50 million by 2019. Online business and shopping are expected to get doubled by 2020. It is reported that "the current online shoppers are in the range of 80–100 million and expected to increase to 175–220 million by 2020." It is found that more than 60% of online customers prefer their mobile devices to do shopping. Customers were found using their mobile phones to buy grocery, book cabs and movie tickets and even food deliveries, and payment of utility bills. Almost everything can be bought and charges are rendered via simple mobile apps. Mobile wallets such as Paytm, MobiKwik, FreeCharge, BookMyShow, and others have become more popular after demonetization, as mobile users were hesitant to store or disclose their bank account or debit/credit cards details due to the fear of theft of their details.

Several initiatives are taken by Indian Government to make the economy cashless. Please refer Table 16.2 for some of the initiatives taken by Indian government toward digital India (Peermohamed, 2016; PTI, 2016).

TABLE 16.2 Indian Government Initiatives for Digital India.

1)	Incentives and discounts of 0.75% were given on the fuel purchase through digital means.
2)	Service tax was waved off on digital transactions up to 2000 rupees.
3)	Indian railways provided 0.5% discounts to customers for monthly or seasonal booking of tickets from 1st Jan 2017, if the payments are remitted through an online application.
4)	A discount of 5% was provided to the customers if charges for 'paid services' of Indian railways such as catering, accommodation, and retiring rooms digital means are rendered through digital means.
5)	Discount of 8%–10% was given to customers on the purchase of new life and general insurance policies through customer online applications.
6)	At toll plazas on national highways, a discount of 10% was available for mobile/digital payments between the years 2016–2017.
7)	The government makes it sure that the transaction fees associated with digital payments is not passed on to the customers and all such expenses shall be borne by them.
8)	The government also makes it sure that merchants or traders should not bear more than 100 rupees as monthly rental for POS terminals or Micro ATMS usages.

POS, point of sale.

Source: Package for Promotion of Digital and Cashless Economy (adapted from *The Press Information Bureau Report*, 2016).

16.10 M-COMMERCE AND BoP IN INDIA

Indian government's initiative behind "Digital India" has a key role to play in the growth of m-commerce in India. The vision of Digital India is to improve digital infrastructure as a utility to every citizen, to improve e-governance and services, and digital empowerment of citizens. Under digital infrastructure, some of the objectives are to:

- Develop infrastructure for high-speed internet as core utility to the citizens,
- Make mobile phones and bank accounts easily available to enable citizens to participate in digital and financial space, and
- Provide safe and secure cyberspace.

Under e-governance and services, some of the objectives are to:

- Connect and integrate all the departments of government,
- Provide a real-time facility for online and mobile services, and
- Make financial transactions electronic and cashless.

Under digital empowerment, making digital literacy universal and making all resources digitally accessible are some of the goals. Digital India has nine pillars or milestones—broadband highways, universal access to phones (UAP-enabled services), electronics manufacturing, online early harvest program, information technology-enabled jobs, access to information for all, Public Internet Access Program, e-Governance, and e-Kranti (PTI, 2016).

After demonetization, Digital India initiatives led to increase in payment volumes and innovations to further make digital payments successful. To make mobile as a financial transaction tool and empower citizens with mobile payments, proximity payments such as tap and pay, NFC, and sound wave gestures have progressed. Some of the new modes of payments such as small merchants' QR codes, smart payments at the toll and other government utilities, UPI, Aadhaar-linked pay, and seamless and cashless checkout have emerged as an outcome of Digital India initiative.

Making the citizens comfortable with the initiatives behind Digital India was easy in urban areas and elite classes. But, for consumers residing in rural India and/or below the poverty line, progressing and penetrating these initiatives were a major challenge and hurdle for the Indian government. To expand digital payments in rural areas, the government with the help of NABARD (National Bank for Agriculture and Rural Development, India) deployed two POS devices each in 1 lakh villages having a population less than 10,000. These POS machines are installed at primary cooperative societies, agricultural input dealers, and milk societies to propagate agri-related transactions with the help of digital means such as POS and mobile phones. The aim is to support farmers of these villages and provide them necessities and amenities to transact in cashless mode. Another initiative is that with the help of NABARD and other rural and cooperative banks, government issued "RuPay Kisan Cards" to around 43.2 million farmers and residents of rural areas. The motive was to make digital transaction feasible for them at the POS machines and micro ATMs.

Apart from these initiatives to promote and motivate, traders and consumers embrace digital payments, the government announced several prizes and discounts schemes such as Digi-Dhan Vyapar Yojana and Lucky Grahak Yojana. Under the scheme, cash awards were to be given to consumers and merchants who employ digital payment instruments such as mobile devices and computers for their personal consumptions and expenditures. To include common and poor men under these schemes, all the transactions between INR 50 and INR 3000 were made eligible for prizes which range from INR 1000 to INR 10 million. Further, since every poor man was not having smartphones, Unstructured Supplementary Service Data services (USSD) were started to support mobile transactions even on GSM phones. People in the rural area used Aadhaar Enabled Payment Systems (AEPS).

Mobile service providers also took part actively in these initiatives and supported the government to facilitate mobile transactions. Airtel India, one of the biggest mobile service providers, started *Airtel Money* app which allows mobile users to recharge their prepaid accounts and render their postpaid bills. Users can also shop online by using the e-cash uploaded in the digital wallet. To make the transaction secure, a secret four-digit mPin was used by this app. Reliance, another mobile service provider initiated *JioMoney* app, which can be used as digital wallet to do online shopping and transactions. Users were offered great discounts and offers to motivate them to use the wallet. Users can also bookmark the retailers from whom they have done their purchases. This made their online transactions quicker and easier. Even commercial banks in India also played a major role and offered personalized mobile wallets through their dedicated apps to facilitate mobile payments and transactions. For instance, *HDFC PayZapp* is an online app of HDFC Bank. It has simplified online payments in India by providing online e-commerce platform to its nearly 12 million customers (Adhikari, 2016). Users can buy, compare, and cancel their flight and hotel tickets and can pay utility bills. The app provides a link with the debit/credit or account. It also stores login details, so that one can transact hassle free without remembering the details. Similar to this, ICICI Bank started *ICICI Pockets* app for its customers. This app was powered with VISA facility which facilitates transactions on any website and also allows transfer of money to the e-mail addresses and WhatsApp contacts. Even RBI provided one mobile app called *MoneyOnMobile*. It allows users to purchase goods and services from the merchants who are with the app. The unique feature of the app is that the app functions in multiple languages. It leads to penetration of the reach of the app even in remote areas of the country and covered millions of people.

Apart from India, presence and mode of business through mobile is quite evident in other countries also. In section 16.11, those countries where m-commerce is used aggressively as a medium of transaction and business especially by people with lower purchasing power or by those who are residing in the rural areas of these countries are discussed.

16.11 M-COMMERCE AND THE BOP IN OTHER ASIAN COUNTRIES

M-commerce or mobile commerce is the platform where buying and selling of goods and services are done through wireless mobile phones or tablets. It is considered as next-generation e-commerce which enables consumers to connect to the companies' websites and other online portals for their purchases without getting plugged to Local Area Network (LAN). The technology which works behind m-commerce is WAP, where a web-ready micro-browser is used, in the smartphones, to access websites or portals. Through Bluetooth technology and internet connectivity, the smartphones can easily get connected to FAX, e-mail, and the bank servers to complete the processes of online transactions. These services are considered as mobile version 2.0, where mobile phones are used beyond its basic usage of voice application and services such as to send and receive text messages. In mobile 2.0, mobile handsets are used for a variety of other services such as payments, information access, and retrieval.

Middle and rich segments are among the majority of the m-commerce services as they are among the major users of the smartphone. They reside in urban areas and have access to high broadband services and high internet connectivity. But, how feasible and comfortable m-commerce activities are for the consumers lying at the BoP. Consumers lying at the BoP are considered as potential customers due to varied reasons discussed above. Government and the manufacturers have started several initiatives to reach and cater to the needs of the consumers at the BoP. High-speed internet and connectivity are the first initiations. Similar to the Digital India campaign of Indian government, where more than 42,000 villages are targeted to provide connectivity facility by 2018, other developing countries have also started several projects to reach and cater to the consumers of the BoP such as "Easy Paisa" (in Pakistan), "M-Pesa" (in Kenya), "Buzz City" (in Thailand), "Cell Bazaar" (in Bangladesh), and "m-payments" (in Sri Lanka) (Soma, 2010).

16.11.1 PAKISTAN: EASY PAISA

Under "Easy Paisa," Pakistan Telecom Authority (PTA) in collaboration with Telenor mobile service provider and State Bank of Pakistan penetrates to rural areas of Pakistan and provides access to the consumers in the rural areas to use their mobile phones to pay their utility bills and transfer money through banks. Telenor acquired a microfinance institution called Tamer Bank to offer such services of bill payment and money transfer in rural areas of Pakistan. The Chief Strategy Officer of Telenor reported that "at present, we have 5000 bill payments a day." He further stated that "we have successfully facilitated transactions worth $25 million—money transfer of Rs. 1 billion, 6 lakh utility bill payments in 6 months of operation. There are over 3500 certified merchants available for financial transactions." Apart from this, Telenor Pakistan has extended several other facilities to the consumers of BoP in Pakistan such as Tele Doctor, Tele Kisan, and Tele-Lawyer (Soma, 2010).

16.11.2 KENYA: M-PESA

In Kenya, with 75% of its population residing in rural areas, ICT connects around 19 million people, and mobile penetration is near to 49%. "M-PESA" is the success story of m-commerce in Kenya. Its success is measured in terms of its popularity and usage. Around 40% of all adults use "M-Pesa." This initiative was launched by Vodafone and the Department for International Development (DFID) as a pilot project to facilitate microfinance loans. Later, on the local telecom service provider Safari.com in alliance with Vodafone extended this pilot project to include other services such as P2P money transfer, salary payments, bill payments, and ATM withdrawals. "M-Pesa" has been reported to have customers list in lakhs (Soma, 2010).

16.11.3 THAILAND: BUZZ CITY

"Buzz City" is the prepackaged software provider in Thailand. It has revolutionized the m-commerce business in Thailand. It manages online advertising business in Thailand. It is reported that Thai is among the fastest developing digital economies of the world. Around 59% of mobile users shop online and use their phone to take purchase decision (Soma, 2010).

16.11.4 BANGLADESH: CELL BAZAAR

"Cell Bazaar" is a virtual marketplace started by Kamal Quadir in Bangladesh. The motive behind the start-up of this virtual marketplace was to facilitate and help the poor people who were living in the rural areas of Bangladesh and have limited access to conventional means of market and communication. These rural people include local merchants, fishermen, and traders who were residing in the remote areas of Bangladesh. Cell Bazaar business model was very user friendly. It allows trading through SMS via the mobile phones. Because of the ease of use, Cell Bazaar gained popularity among consumers and businessman of the BoP of Bangladesh. Through this portal, traders and farmers were able to fetch a fair price for their produce, have a shorter period for trading, and get least exploited by the middlemen (Soma, 2010).

16.11.5 SRI LANKA: M-PAYMENT

m-Payment in Sri Lanka was initiated in the name of "eZ Cash" and "mCash." These are mobile money accounts which are approved and attached to all major banks of Sri Lanka. A person can transact even if he does not have a bank account. The services enable one to transact using their mobile service providers such as Dialog, Etisalat, and Hutch mobile phone anytime and anywhere. These mobile money accounts facilitate cash account in the users' mobile phone which can be topped up and used for a range of cash transactions directly from the mobile phones. Through these mobile money accounts a person can also deposit, transfer, and withdraw funds to purchase goods and services and pay his utility bills (Soma, 2010).

16.12 THE WAY AHEAD

The potential of mobile commerce can be projected through its growth in different parts of the world. In the United States itself, transaction through mobile commerce has gone up from $0.6 billion in 2010 to $11.1 billion in 2015 (Hałabuda, 2015). Consumers of online market place are now given a new term "digital spenders," as shopping through the device has become an integral part of their shopping habits and a matter of daily affairs. The role of m-commerce is also evident from the fact that Google Inc. has planned to penalize the nonmobile-friendly sites (Anonymous, 2012; Hałabuda, 2015). M-commerce's impact and penetration are also visible in a revolution of mobile in India; Myntra, an online merchant has shut down its website to sell exclusively through mobile apps. CEO of Flipkart, Sachin Bansal, stated that if the experiment with Myntra works well, then they will shut down the Flipkart's website also and will sell through dedicated mobile apps. It means that India's largest e-commerce company is planning to go mobile. Another instance is the initiative taken by cab service provider—Uber. Through "Urberfication," they are making their consumers avail offers and book cab online (Hałabuda, 2015).

Similarly, the BoP has grown up across the world. It has gained a renewed attention in the contemporary marketplace. Companies are designing and remapping their strategies to reach to the consumers of BoP. It is well understood that traditional marketing strategy will not be adequate to cater to the consumers of BoP. The proponent of BoP, C. K. Prahalad, himself agreed to it and suggested 4As instead of 4Ps to capitalize on this opportunity (Anonymous, 2012). According to the proponent of BoP, marketers can cater to BoP sector by creating *awareness* among the BoP consumers about the usage and benefits of the product and services. That is, companies can create awareness by making the BoP consumers comfortable with the usage and applicability of the smartphones and its various applications. The second suggestion was to make the product and services easily *accessible* to the BoP consumers. M-commerce will be useful medium to accomplish this. Booking, exchange, and service requests can be easily done through mobile phones, by visiting to the companies' websites or their dedicated apps and using net banking or e-wallet to complete the transaction. The third suggestion was that the product and services should be offered at an *affordable* price. This is because the consumers at the BoP are very price sensitive. Product and services can be sold at a much cheaper price than the retail outlets through m-commerce, as m-commerce reduces the cost associated with land, labor, and inventory management of a retail outlet. The fourth suggestion was that the product and services should be made *available* to them. Through mobile apps and access to the websites of the companies, consumers can browse through a number of varieties and can compare sizes, color, and price before the final purchase. But, the retail outlets due to space restriction and a cap on inventory cannot have more varieties. Thus, marketers should take this opportunity of deploying m-commerce as the best medium to reach to the consumers of BoP. To cater to the BoP consumers, companies can design their marketing strategies around these factors.

M-commerce is pegged to add an advantage to the companies by aiding in reaching the consumers at BoP. Through mobile phones, consumers can be easily made aware of the various benefits and usages of the product services. Price ranges and discounts can also be flashed on the dedicated apps of the companies to attract the consumers of BoP. Consumers can have easy access to the product and services of the companies through their mobile phones. With a click, they can surf through the product and services available. Also, the after sales services and customer feedback can be availed on the mobile phones, which help in building a long-term relationship with the consumers of BoP. Thus, m-commerce is one of the medium by which companies can incorporate the 4As to meet and cater to the needs of the consumers of BoP (Anonymous, 2012).

Marketers can develop dedicated company's app to provide customized online shopping experiences to the customers (Balasubramanian et al., 2002). To complete the transaction through app, users have to first register themselves. In this way, marketers can have the contact details and access to their customers. It will help them in informing their customers about the promotional activities, seasonal discounts, and festive offers. To retain their customers' interest and engagement with these apps, companies can provide them customized services by suggesting them related product according to their choices and last searches. Also, companies can manage their revenue and provide huge discounts to attract cost-sensitive consumers through mobile commerce, as the cost associated with land, labor, and inventory management in a retail store can be eliminated. Further, lite version of apps also provides an added advantage to the companies, as consumers can browse through these apps even with low internet speeds. The growth of m-commerce is omnipresent as suggested by the Goldman Sachs in a forecast that m-commerce will cross the value of e-commerce, that is, $626 billion by the year 2018 (Hałabuda, 2015).

KEYWORDS

- **m-commerce**
- **bottom of pyramid**
- **emerging markets of Asia**
- **rural consumers**
- **information and computer technology**

REFERENCES

Adhikari, A. *Been There, Done That,* 2016. https://goo.gl/JiZYVH (accessed May 20, 2017).

Anonymous. *4As and 5ds of Marketing for BoP Population,* 2012. http://www.thebopstrategy.com/ (accessed May 20, 2017).

Anwer, J. *7 Reasons Why Micromax has Beaten Samsung in India,* 2015. http://indiatoday.intoday.in/technology/story/micromax-samsung-india-smartphones/1/416878.html (accessed Aug 27, 2017).

Balasubramanian, S.; Peterson, R. A.; Jarvenpaa, S. L. Exploring the Implications of M-Commerce for Markets and Marketing. *J. Acad. Mark. Sci.* **2002,** *30* (4), 348.

Bhatnagar, S.; Schware, R. *Information and Communication Technology in Rural Development. Case Studies from India*; World Bank Institute: Washington, DC, 2000.

Fransman, M. *The New ICT Ecosystem: Implications for Policy and Regulation*; Cambridge University Press: Cambridge, 2010.

Gunther, M. *The Base of the Pyramid: Will Selling to the Poor Pay Off?* 2014. https://goo.gl/d4nhNL (accessed May 20, 2017).

Hałabuda, P. *Introduction to M-Commerce: Benefits of M-Commerce,* 2015. http://whallalabs.com/introduction-to-m-commerce/ (accessed May 5, 2017).

International Data Corporation (IDC). *Indian Smartphone Market Crosses 30 Million Unit Shipments Milestone for the First Time Ever In CY Q3 2016: IDC India*, 2016. https://www.idc.com/getdoc.jsp?containerId=prAP41922916 (accessed Apr 5, 2017).

Luce, K. *The Rising Bottom of the Pyramid in India, A Perspective from Boston Analytics.* 2016. https://www.slideshare.net/KimberleeLuce/the-rising-bottom-of-the-pyramid-in-india (accessed Aug 27, 2017).

Peermohamed, A. *Fortune at the Bottom of Global Pyramid for Indian Internet Firms,* 2016. https://goo.gl/1N4zfL (accessed May 25, 2017).

Perdana, A. A. *Risk Management for the Poor and Vulnerable*; Economics Working Paper Series; Centre for Strategic and International Studies: Jakarta, 2005. http://saber.eaber.org/node/22005 (accessed May 25, 2017).

Prahalad, C. K. *The Fortune at the Bottom of the Pyramid*; Pearson Education: India, 2006.

Prahalad, C. K.; Hart, S. L. The Fortune at the Bottom of the Pyramid. *Strateg. Bus.* **2002,** *26,* 54–67.

Realini, C.; Mehta, K. *Financial Inclusion at the Bottom of the Pyramid*; Friesen Press: Victoria, BC, Canada, 2015.

Roosevelt, F. D. *The 'Forgotten Man' Speech*; Radio Address from Albany, New York, 1932. https://goo.gl/h6UmRK (accessed May 20, 2017).

Shukla, S.; Bairiganjan, S. *The Base of Pyramid Distribution Challenge*, Centre for Development Finance, Institute for Financial and Management Research, 2011. https://goo.gl/BAjohc (accessed May 30, 2017).

Soma. *Connect to the Bottom of the Pyramid* (News Article), 2010. http://www.thehindubusinessline.com/todays-paper/tp-eworld/connect-to-the-bottom-of-the-pyramid/article1013219.ece (accessed Aug 27, 2017).

Tarafdar, M.; Singh, R. In *A Market Separations Perspective to Analyze the Role of ICT in Development at the Bottom of the Pyramid*, The Proceedings SIG GlobDev Fourth Annual Workshop, Shanghai, China, Dec 03, 2011; Vol. 3. https://aisel.aisnet.org/globdev2011/19.

The Press Information Bureau. *Package for Promotion of Digital and Cashless Economy*, 2016. https://goo.gl/hkxA31 (accessed May 25, 2017).

The Press Trust of India (PTI). *Demonetisation to Help M-Commerce in India*; Report, 2016. http://indianexpress.com/article/technology/tech-news-technology/demonetisation-to-help-m-commerce-in-india-report-4408203/ (accessed May 25, 2017).

CHAPTER 17

VIGOR OF M-WORD OF MOUTH IN THE CONSUMER EPOCH

YOG MISHRA[1*], ANURAG SINGH[1], and PUNITA DUHAN[2]

[1]*Institute of Management Studies, Banaras Hindu University, Varanasi, Uttar Pradesh, India*

[2]*Department of Training and Technical Education, Meera Bai Institute of Technology, Government of Delhi, Delhi, India*

[*]*Corresponding author. E-mail: yog.mishraa@gmail.com*

ABSTRACT

As m-commerce is increasing its penetration in Indian households as a result of expansion of digital platforms, marketers are also looking for new ways to meet customer expectations and in order to do the same; they are inventing new digital tools and techniques to gain the attention of the consumer.

Mobile word of mouth, a subset of electronic word of mouth, has gained much as digital tool because now more and more consumer are accessing internet via mobile devices. Mobile word of mouth has given empowerment to both consumer and marketer. Now, consumer can update real-time experience of a product or service as soon as he/she comes in contact with the service provider. On the other hand, Mobile word of mouth has become an important tool in the hands of marketer who wants to shape the consumer purchase decision via their Web Push Message and short message service (SMS). Accordingly, customer also exhibits altered purchase behavior due to Mobile word of mouth vis-à-vis traditional word of mouth. In this study, researchers have attempted to document this change in consumer behavior. Alongside, researchers have tried to explore the determinants of mobile word of mouth such as brand commitment, ease of use, individual connectivity, motivation behind mobile referral campaign, and information source.

17.1 INTRODUCTION

In current scenario, a new generation of users is emerging, which is tech savvy and wants real-time data. Similarly, there have been significant changes in the digital services which are expanding their presence across India as a result of improvement and rapid expansion of telecommunication infrastructure and services (Argade and Chavan, 2015). As per TRAI report (2017); there has been a significant increase in number of internet users in India that has crossed 350 million. Out of this internet using population, it is essential to notice that 93.91% of the total subscribers are using these services on their mobile handsets. Internet subscription is growing at a rate of 9.72% and smartphone users are expected to grow at a rate of 16% in next 5 years in India, which makes India the fastest growing market in world in terms of internet and mobile penetration.

Mobile devices enable ubiquitous access. Due to its smaller size and portability, consumer finds it comfortable in using and as a result, consumer shows different types of behavior. However, the observed consumer behavior may also differ because of factors through which a user receives mobile word of mouth (m-WoM) communication. Preference can be due to their personal likings and characteristics (e.g., habit of only using mobile devices or desktops), or contextual factors that can influence device accessibility (e.g., social appropriateness or travel time). Despite the fast growth of mobile industry, it is notable that there is a dearth of work, which has attempted to explore differences between mobile and non mobile users' behavior. Some of the researches done in this context have been focused more on contextual factors (Goh et al., 2009; Shankar and Balasubramanian, 2009; Sultan et al., 2009; Shankar et al., 2010; Ghose and Han 2011; Noulas et al., 2011; Ghose et al., 2012; Molitor et al., 2016) and preferential factors have been completely ignored.

Accordingly, present study emphasizes on various preferential determinants of m-WoM such as brand commitment, ease of use, interpersonal connectivity, motivation behind mobile referral campaign, and information source. This study further highlights the change in user's perceived experience and behavior in m-WoM vis-à-vis traditional WoM campaign.

17.2 METHODOLOGY

Research methodology is the method that motivates a researcher in generating solutions to research questions and to assess this, available data, and literature is thoroughly studied. Kind of issue leads to the kind of study which is assessed after compiling the data or analyzing the literature. Success of our study depends on the accuracy of the study and systematic application of the research method. Thus, a researcher can get the desired result by choosing the appropriate method of research in a systematic way.

This chapter attempts to throw light on the conceptual issues related to m-WoM in relation to consumer behavior on m-commerce platform. Thus, data was collected through various secondary data sources such as research journals, government data, marketing reports, and books. Finally, to assemble the issues considered in objectives and to reach on the findings, content analysis approach was followed.

17.3 M-COMMERCE AT A GLANCE

M-commerce is broadening the marketplace for both, marketer and the consumer, and thus making a paradigm shift between companies and the stakeholders (Keen et al., 2001). And this shift is because of the fact that in m-commerce customer does not arrive physically at the marketplace but seller tries to reach the customer via m-commerce applications (Shankar et al., 2010).

M-commerce refers to wireless transaction with the usage of mobile devices (Mishra, 1970). According to Chaffey (2007), "Electronic transactions and communications conducted using mobile devices such as laptops, PDAs (Personal Digital Assistant), and mobile phones, and typically with a wireless connection" is known as m-commerce transaction. As per Tsalgatidou and Pitoura (2001) "Mobile electronic commerce (MEC) operates partially in a different environment than Internet e-commerce due to the special characteristics and constraints of mobile terminals and wireless networks." M-commerce requires wireless telecommunication network setting to carry out commercial transactions by using mobile technology (Moczarny, 2011). M-commerce interface is diverse in terms of usage, users, and value chain (Feng et al., 2006). M-commerce adaptability is high as compared to e-commerce as m-commerce is used via mobile handset and thus does not require costly and technical apparatus every time it is connected to internet (Pavlou, 2002; Jarvenpaa et al., 2003). M-commerce customer does not arrive physically at the marketplace rather seller tries to reach the customer via m-commerce applications (Shankar et. al., 2010).

Content delivery mode and Transaction Mode (Yang et al., 2008) are two ways through which companies reach to their customer. Content delivery mode takes responsibility of delivering the content or notification about a product or services to the customer and transaction mode involves business transaction which looks after monetary exchange between the customer and the buyer (Lee et al., 2008). Many researchers have explored the factors responsible for using the m-commerce devices (Lim, 2008) and have pointed the factors such as perceived ease of use on intention which has a direct or indirect effect on perceived usefulness. Early adopters actively engage in information seeking to learn more about the benefits of using new technology (Amin et al., 2009).

17.4 JOURNEY FROM E-WORD OF MOUTH TO M-WoM

E-word of mouth (e-WoM), in recent years, has gained significance because of increased use of social media. Social media has effect on consumer's purchase intentions because now customer spends more and more time on digital devices such as smartphone (Tsimonis and Dimitriadis, 2014). Companies are rapidly aligning themselves with e-WoM. This not only gives them the advantage of fetching real-time data but also increases their reach to large number of consumers without spending higher expenses (Litvin, 2008).

In e-WoM, customers write about their thoughts, opinions, and feelings about products and services. This happens either online by directly emailing the organization concerned or by writing on blogs (Schindler and Bickart, 2005). Mobile-based applications have further increased this penetration as more and more customers are accessing different websites on web-based application platforms. This has been further fuelled by increase in the speed of network such 3G and 4G.

M-WoM solves a major problem for marketers, that is, to draw attention of a customer toward their brand (Barwise and Strong, 2002; Pousttchi and Wiedemann, 2006). Three strategies involved in mobile viral marketing are push, pull, and viral strategies. In push, marketers try to request the consumers to distribute or to communicate the material. In pull, case is exactly opposite and here user request content from marketers. In viral, information exchange happens among the consumers. The latter strategy is known as m-WoM (Wiedemann, 2007).

M-WoM can be summarized as a process of distribution or communication. In this, the communication depends upon user to spread the mobile viral content via mobile communication techniques and various digital devices to other user in their social group and to request these users to further forward the content (Wiedemann, 2007).

The swift increase in smartphone ownership has opened up a new pitch for WoM communication, that is, mobile world of mouth. Like the earlier fixed digital platform, that is, e-WoM—m-WoM can be understood as a message and delivery concept. It can be concluded that it is a type of marketing communication where a customer, who possesses an advertising message, passes it to another customer and it goes on "like a rampant flu virus" (Montgomery, 2001).

As given in earlier studies (Jelassi and Enders, 2004), m-WoM helps the marketer to increase the number of recipients significantly, enhancing the impact of marketing communication at a low expense, and subsequently reducing the distribution expenses. Existing studies have reported that personal messages have more credibility among the user as compared to messages coming directly from the self-interested marketer. In addition, this also increases frequency of participation among the user if he gets the message form a familiar source as initial contact (Marini and Wiedemann, 2006). Available literature focuses more on worth of m-WoM from the marketers' viewpoint and only few studies are available which encompass motivations, attitudes, and behaviors of consumers involved in this marketing communication instrument (Chen et al., 2008; Wiedemann et al., 2008).

17.5 DETERMINANTS OF M-WoM

M-WoM depends on various determinants which shape user's experience toward a product or service. It is essential for the marketers to understand these determinants; otherwise they cannot reach their target audience. Few determinants of m-WoM are as follows:

17.5.1 BRAND COMMITMENT

Brand commitment plays an important role as it "involves an exchange where belief on ongoing relationship is paramount and efforts should be maximized to maintain the same" in the relationship–marketing scenario (Morgan and Hunt, 1994). Earlier researchers describe commitment as an attitudinal construct that can be defined as "a continuous pledge of hidden or clear relationship between exchange partners" (Dwyer et al., 1987). Previous literature on organizational behavior principally regards commitment as the most effective and a persistent component, but marketing scholars have described this commitment as affective in nature. In the broad understanding of marketing relationship, affective commitment encompasses shared values, trust, benevolence, and relationalism as important attributes (Fullerton, 2005; Morgan and Hunt, 1994; Verhoef et.al., 2002). In spite of the type of service used by the consumer, marketers usually accept it as affective commitment when consumer prefers a service provider over other (Fullerton, 2005). In addition to this, a satisfied consumer is likely to recommend that service to other consumers on the parameter of service provided (Anderson, 1998).

17.5.2 RELATIONSHIP WITH MOBILE DEVICES

Earlier research suggests that mobile devices play an important role in day-to-day activities of relatively young users (De Kerckhove, 2002) and it is replacing traditional digital devices as the primary source of information. It has been also found that young users response to the mobile messages in a better way. One survey suggests that as compared to 68% of the total population, a spike of almost 92% was witnessed when it was measured on the parameter of response to a message on mobile devices for young users (Ito and Okabe, 2005). Mobile gives them an advantage of accessing information wherever and whenever they want, through the use of search function. In mobile internet, search can "automatically identify user's geographic location, with mobile search applications delivering the desired information" and this gives a relatively significant search result but this facility is not automatically available in personal computer (PC)-based search. For relatively young users, mobile phones help them stay linked, up-to-date, and entertained. As this relationship between user and mobile device is described as rational in nature, marketers describe it as brand relationship. This relationship acts as a guide in their assessment of the benefits from service provider in everyday communication for young user. Through a series of interpersonal interactions, benefits are delivered to the user is such a way that it not only gives them the entertainment quotient but also shows concern for users' needs (Aggarwal, 2004).

But in the case of users who are relatively older, they find it difficult to connect to this technological change as they find it far more complicated and inconvenient to adapt to these rapid changes and so their presence on this digital platform is low whether it is internet on PCs or mobile devices.

17.5.3 INTERPERSONAL CONNECTIVITY

Interpersonal connectivity can be explained as "social remuneration that an individual receives by establishing and maintaining contact with others, such as social support, companionship, and familiarity." It acts as antecedents for consumers' participation in a virtual group community (Dholakia et al., 2004).

WoM, of any kind, requires a necessary component that it happens within a relationship (Anderson, 1998), and so it can be asserted that WoM network "engages people in word of mouth in order to create a social network that enhances relationship between the people of that group" (Bristor, 1990), and so it acts as a foundation for interpersonal influence (Grewal et al., 2003). In the case of m-WoM, it has been seen that Interpersonal connectivity has a straight and affirmative impact on the motives to engage in the mobile-based referral campaign. Fundamentally, a m-WoM system consists of a strongly established, deep rooted, and everyday-based circle of people, such as relatives, associates, and schoolmates. While circulating useful and interesting information among the potential user, a user must keep in mind that it should ultimately serve the interest of the group.

However, as compared to e-WOM network, the concept of "group" is more tightly packed in m-WoM network such as website bulletin boards, web-based chat rooms, opinion platforms, etc., because m-WoM network consists of known users but in e-WOM, users are unknown to each other.

17.5.4 MOTIVATION BEHIND MOBILE REFERRAL CAMPAIGN

Monetary incentives are the basic motive behind any kind of sales promotion. But other than this, there are other benefits which are monetary and nonmonetary in nature. Hedonic and utilitarian are two broad classifications of benefits in sales promotion.

Hedonic benefits can be described as intrinsic, noninstrumental, experiential, and affective in nature, and are often "without noticing its practical purposes, it is just appreciated for their own sake." But in the case of utilitarian benefits, customer value is given paramount importance and it is more cognitive and instrumental in nature (Chandon et al., 2000). Thus, mobile-based referral campaign should be based after consideration of these two types of benefits. First, entertainment can be an important value while forming a favorable attitude toward the campaign as it has been seen that customer participation in a virtual community depends on the value derived from the fun and leisure obtained through playing, or otherwise interacting with friends (Dholakia et al., 2004). Second, as human beings are a social animal and they find social interaction as an important value in their life, they may find this interaction to be important to perceive the benefit of the mobile-based referral campaign to be personally and socially meaningful. In particular, when the dispersal of campaign information is planned at a personal level within a known group, the reason behind may serve a purposive value.

17.5.5 SOURCE OF CAMPAIGN INFORMATION

Consumer involvement in any referral campaign or any other form of m-WoM depends on the source from which the information was received. In common practice, companies use similar content-based mobile websites that are available for PC-based websites by just adjusting the screen size. E-mail magazines and site links are the most popular message vector used by the marketer for user who wants to seek information associated with the topic. This commonly used message vector is commercial or nonpersonal sources of information. But in m-WoM, a different type of approach is more frequent, that is, pass-along e-mails from their peers.

It has been witnessed that user forwards those messages which they receive from their peer group and are very interesting and engaging in nature. Many service providers even strengthened their security system to filter these mails so that they are not classified as spam. Personal chit-chat of the known peer group further validates this information as it produces stronger motives, attention, and intentions. The last step includes pass-along e-mail. It includes indirect but very intimate form of information exchange between closed and known group (i.e., friends). In contrast, mobile websites and e-mail magazines are nonpersonal information sources.

Literature review evidences that the issues such as determinants of m-WoM and there interdependence on each other and its effect on m-WoM and in turn, effects of m-WoM on consumer behavior were disorganized.

According to earlier literature, it is found that m-WoM, as a subset of e-WoM, is increasing its presence among the consumers because of improvement in connectivity and slashed prices of smartphones. Though, there have been several studies on determinants of m-WoM, little work has been done to understand brand commitment or relationship with the medium in a single framework of electronic referral. Ambiguity has been found in studies assessing the process of traditional WoM and m-WoM.

17.6 OBJECTIVES

1. To assess the effects of determinants of m-WoM.
2. To explore the process of m-WoM on consumer behavior.

17.7 ASSESS THE EFFECT OF DETERMINANTS OF M-WoM

Present study encompasses assurance to the promoted brand, eases of use with the mobile device, and interpersonal connectivity, and considers them the antecedents of motives to participate in the mobile referral campaign. In earlier literature, interpersonal connectivity plays the most significant role in establishing a m-WoM network. Earlier research suggests that there are two values, that is, entertainment and purposive, which are the reasons behind user participation in mobile promotion. These values run parallel to the hedonic and utilitarian values. Thus, after going through the existing literature, present study proposes proposition which is a mixture of hedonic–utilitarian value up to some extent. Literature reviews suggest that the entertainment value has a greater impact on attitude as compared to purposive value (Chandon et al., 2000). But still purposive value appears to gain a solid ground, suggesting that the stimulation of a sense of "accomplishing some predetermined instrumental purpose" was as important for both, that is, m-WoM and PC-based virtual network (Dholakia et al., 2004).

Brand commitment, ease of use, and interpersonal connectivity directly and positively affect the entertainment value of m-WoM. In addition to this, it was also found that the effect of interpersonal connectivity was stronger but in the case of brand commitment, the effect was modest in nature (Okazaki, 2008). Though the determinant brand commitment also shows positive effect on purposive value, this effect is moderate in nature as compared to entertainment value. So, it can be asserted that information, which is fun and entertaining in nature, will have stronger impact on the user.

Now, as learnt from previous research, users who learn about pass-along e-mail go through two paths toward the m-WoM campaign: (1) from interpersonal connectivity to purposive value and (2) from purposive value to attitude. It was also found that user, who learnt of the campaign via pass-along e-mails, accepted peers recommendations even when they were not much interested in the subject. So users are likely to use m-WoM to satisfy both entertainment and predetermined instrumental purposes because of strong social ties.

17.8 PROCESS OF M-WoM ON CONSUMER BEHAVIOR

In traditional WoM communication, communication is interpersonal in nature and it depends upon experiences of only one individual, and it is evident from the previous literature that experience of an individual depends on many interpersonal and extraneous variables which affects the entire thought process of an individual.

But the model of m-WoM relies on consumers who are responsible for transmitting mobile viral content via various available mobile communication techniques and devices to other potential customers in their social group and to request those users to forward the content (Wiedemann, 2007). Such content are labeled as mobile viral content and it adheres predominantly to mobile services or ads. *Mobile vector* is responsible for carrying this content. In biology, a vector is an organism that acts as a transporter which itself does not cause disease but acts as a transporter that carry pathogens from one host to another. Replicating this phenomenon into m-WoM, a mobile vector transports the content such as a text or a link using the Wireless Application Protocol (WAP). Sometimes, pure text messages such as short message service (SMS) are used for this purpose and in other cases, WAP push messages, a specially formatted type of message that directly connects the browser, are used. Individuals who transmit this content are labeled as *communicators*, whereas individuals who are at the receiving end are termed as *recipients* (Fig. 17.1) (Pousttchi and Wiedemann, 2006).

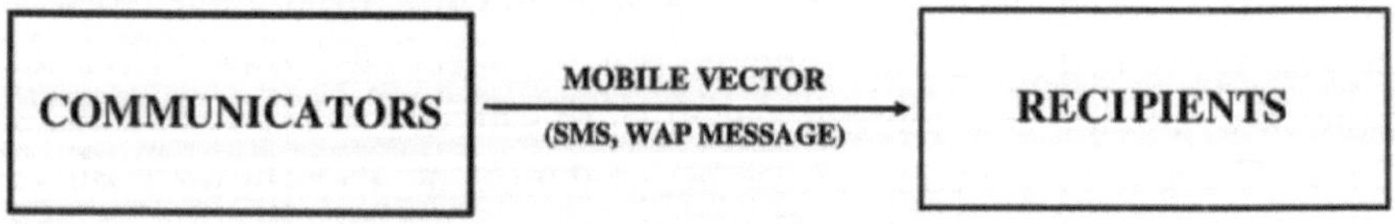

FIGURE 17.1 Mobile word of mouth process.
Source: Adapted from Pousttchi and Wiedemann (2006).

Earlier studies demonstrated that temporal distance is associated with greater abstraction in thoughts and ideas due to the use of high-level construal (Liberman et al., 2007), and so it can be said that mobile reviews should take place with less delay. That is, if two reviewing scenarios are identical in that they share the same author and pertain to the same restaurant, except that one review is posted on the spot while the other is posted following some delay, the above theory suggests that we would expect to observe differences in review content. The review posted on the spot (i.e., low temporal distance) contains detailed information, such as the taste, the ambiance, and the response of the service staff (concrete information), and exhibits deep positive and negative emotions. In contrast, the review posted after some delay (i.e., high temporal distance) would contain less information, as fine-grained details fade from memory with time.

17.9 CONCLUSION

With the onset of 21st century, there have been rapid changes in digital platform and most of this digital revolution rotates around mobile internet. Changes in devices such as smartphone and improvement in connectivity such as 3G, 4G, and other wireless services has further fuelled this revolution and it is shaping a virtual world around the user.

A user, on the basis of effect of individual determinants of m-WoM such as brand commitment, ease of use, interpersonal connectivity, motivation behind mobile referral campaign, and information source, accesses this information that he/she receives from a communicator and then he/she interprets that information based on multiple cues.

This interpretation is further reflected in his/her consumer behavior where the user makes the purchase decision depending upon his/her understanding of the information which reaches to the user in the form of WAP Push Messages or SMS. Marketers, sometimes, acts as communicators in this model and try to effect the consumer decision making in favor of a particular product or a service through m-WoM.

Marketers should carefully design the content that is timely, up-to-date, and personalized according to individual user identities. Other than precision, it should also be taken care that content should be amusing and enjoyable but communicators should take care that messages should not be over attempted otherwise it will upset the balance and reduce the effectiveness of the content.

KEYWORDS

- **m-commerce**
- **mobile word of mouth**
- **e-word of mouth**
- **consumer behavior**
- **telecommunication**

REFERENCES

Aggarwal, P. The Effects of Brand Relationship Norms on Consumer Attitudes and Behavior. *J. Consum. Res.* **2004,** *31* (1), 87–101.

Amin, A.; Zhang, J.; Cramer, H.; Hardman, L.; Evers, V. In *The Effects of Source Credibility Ratings in a Cultural Heritage Information Aggregator*. Proceedings of the 3rd Workshop on Information Credibility on the Web, Apr 20, 2009; Association for Computing Machinery: New York; pp 35–42.

Anderson, E. W. Customer Satisfaction and Word of Mouth. *J. Serv. Res.* **1998,** *1* (1), 5–17.

Argade, D.; Chavan, H. Improve Accuracy of Prediction of User's Future M-commerce Behaviour. *Procedia Comput. Sci.* **2015,** *49*, 111–117.

Barwise, P.; Strong, C. Permission-based Mobile Advertising. *J. Interact. Mark.* **2002,** *16* (1), 14–24.

Bristor, J. M. Enhanced Explanations of Word of Mouth Communications: The Power of Relationships. *Res. Consum. Behav.* **1990,** *4* (1), 51–83.

Chaffey, D. *E-business and E-commerce Management: Strategy, Implementation and Practice*; Pearson Education: London, 2007.

Chandon, P.; Wansink, B.; Laurent, G. A Benefit Congruency Framework of Sales Promotion Effectiveness. *J. Mark.* **2000,** *64* (4), 65–81.

Chen, W. K.; Huang, H. C.; Chou, S. C. T. In *Understanding Consumer Recommendation Behavior in a Mobile Phone Service Context.* ECIS, 2008; pp 1022–1033.

de Kerckhove, A. Building Brand Dialogue with Mobile Marketing. *Young Consum.* **2002,** *3* (4), 37–42.

Dholakia, U. M.; Bagozzi, R. P.; Pearo, L. K. A Social Influence Model of Consumer Participation in Network-and Small-group-based Virtual Communities. *Int. J. Res. Mark.* **2004,** *21* (3), 241–263.

Dwyer, F. R.; Schurr, P. H.; Oh, S. Developing Buyer-seller Relationships. *J. Mark.* **1987,** 11–27.

Feng, H.; Hoegler, T.; Stucky, W. In *Exploring the Critical Success Factors for Mobile Commerce*. International Conference on Mobile Business 2006 (ICMB'06), Jun 26, 2006; IEEE, 2006; pp 40-40.

Fullerton, G. How Commitment Both Enables and Undermines Marketing Relationships. *Eur. J. Mark.* **2005,** *39* (11/12), 1372–1388.

Ghose, A.; Han, S. P. An Empirical Analysis of User Content Generation and Usage Behavior on the Mobile Internet. *Manag. Sci.* **2011,** *57* (9), 1671–1691.

Ghose, A.; Goldfarb, A.; Han, S. P. How Is the Mobile Internet Different? Search Costs and Local Activities. *Inf. Syst. Res.* **2012,** *24* (3), 613–631.

Goh, K. Y.; Chu, H.; Soh, W. *Mobile Advertising: An Empirical Study of Advertising Response and Search Behavior*, ICIS 2009 Proceedings, 2009, p 150.

Grewal, R.; Cline, T. W.; Davies, A. Early-entrant Advantage, Word-of-mouth Communication, Brand Similarity, and the Consumer Decision-making Process. *J. Consum. Psychol.* **2003,** *13* (3), 187–197.

Ito, M.; Okabe, D. Technosocial Situations: Emergent Structurings of Mobile Email Use. In *Personal, Portable, Pedestrian: Mobile Phones in Japanese Life*; The MIT Press, 2005; Vol. 20 (6), pp 257–273.

Jarvenpaa, S. L.; Lang, K. R.; Takeda, Y.; Tuunainen, V. K. Mobile Commerce at Crossroads. *Commun. ACM* **2003,** *46* (12), 41–44.

Jelassi, T.; Enders, A. *Leveraging Wireless Technology for Mobile Advertising*, ECIS 2004 Proceedings, 2004, p 50.

Keen, P. G.; Mackintosh, R.; Foreword By-Heikkonen, M. In *The Freedom Economy: Gaining the M-Commerce Edge in the Era of the Wireless Internet*; McGraw-Hill Professional, 2001; p 41.

Lee, J.; Park, D. H.; Han, I. The Effect of Negative Online Consumer Reviews on Product Attitude: An Information Processing View. *Electron. Commer. Res. Appl.* **2008,** *7* (3), 341–352.

Liberman, N.; Trope, Y.; Wakslak, C. Construal Level Theory and Consumer Behavior. *J. Consum. Psychol.* **2007,** *17* (2), 113–117.

Lim, A. S. Inter-consortia Battles in Mobile Payments Standardisation. *Electron. Commer. Res. Appl.* **2008,** *7* (2), 202–213.

Litvin, S. W.; Goldsmith, R. E.; Pan, B. Electronic Word-of-mouth in Hospitality and Tourism Management. *Tour. Manag.* **2008,** *29* (3), 458–468.

Marini, S.; Wiedemann, D. G. Entwicklungen im Bereich Mobile Advertising aus Dersicht von Experten. In *Ergebnisse der Expertenbefragung MM 1. Studienpapiere der Arbeitsgruppe Mobile Commerce*; Pousttchi, K. Ed.; Wi-mobile Research Groups, University of Augsburg: Augsburg, Germany, 2006; pp 1–49.

Mishra, S. Adoption of M-commerce in India: Applying Theory of Planned Behaviour Model. *J. Internet Bank. Commer.* **1970,** *19* (1), 1–17.

Mizuko, I. T. O. Personal Portable Pedestrian: Lessons from Japanese Mobile Phone Use. *Asia-Pacif. J. Jpn. Focus Vol.* **2005,** *3* (5)

Moczarny, I. M. *Dual-method Usability Evaluation of E-commerce Websites: In Quest of Better User Experience* (Doctoral Dissertation), 2011.

Morgan, R. M., Hunt, S. D. The Commitment-trust Theory of Relationship Marketing. *J. Mark.* **1994,** *1*, 20–38.

Molitor, D.; Reichhart, P.; Spann, M.; Ghose, A. Measuring the Effectiveness of Location-based Advertising: A Randomized Field Experiment. 2017. https://ssrn.com/abstract=2645281; http://dx.doi.org/10.2139/ssrn.2645281.

Montgomery, A. L. Applying Quantitative Marketing Techniques to the Internet. *Interfaces* **2001,** *31* (2), 90–108.

Moorman, C.; Zaltman, G.; Deshpande, R. Relationships Between Providers and Users of Market Research: The Dynamics of Trust Within and Between Organizations. *J. Mark. Res.* **1992,** *29* (3), 314.

Noulas, A.; Scellato, S.; Mascolo, C.; Pontil, M. An Empirical Study of Geographic User Activity Patterns in Foursquare. *ICWSM* **2011,** *11*, 70–573.

Okazaki, S. Determinant Factors of Mobile-based Word-of-mouth Campaign Referral Among Japanese Adolescents. *Psychol. Mark.* **2008,** *25* (8), 714–731.

Pavlou, P. A. In *What Drives Electronic Commerce? A Theory of Planned Behavior Perspective*. Academy of Management Proceedings, Aug 1, 2002; Academy of Management: Briarcliff Manor, NY, 2002; Vol. 2002 (1), pp A1–A6.

Pousttchi, K.; Wiedemann, D. G. In *A Contribution to Theory Building for Mobile Marketing: Categorizing Mobile Marketing Campaigns Through Case Study Research*. International Conference on Mobile Business 2006 (ICMB'06), June, 2006; IEEE, 2006; pp 1–1.

Pousttchi, K.; Wiedemann, D. G. In *Success Factors in Mobile Viral Marketing: A Multi-case Study Approach, Management of Mobile Business, 2007,* ICMB 2007. International Conference on July, 2007, IEEE: 2007; pp 34–34.

Schindler, R. M., Bickart, B. Published Word of Mouth: Referable, Consumer-generated Information on the Internet. In *Online Consumer Psychology: Understanding and Influencing Consumer Behavior in the Virtual World*; 2005; Vol. 32, pp 35–61.

Shankar, V.; Balasubramanian, S. Mobile Marketing: A Synthesis and Prognosis. *J. Interact. Mark.* **2009,** *23* (2), 118–129.

Shankar, V.; Venkatesh, A.; Hofacker, C.; Naik, P. Mobile Marketing in the Retailing Environment: Current Insights and Future Research Avenues. *J. Interact. Mark.* **2010,** *24* (2), 111–120.

Sultan, F.; Rohm, A. J.; Gao, T. T. Factors Influencing Consumer Acceptance of Mobile Marketing: A Two-country Study of Youth Markets. *J. Interact. Mark.* **2009,** *23* (4), 308–320.

Telecom Regularity Authority of India (TRAI). *Yearly Performance Indicators of Indian Telecom Sector–2016*; New Delhi, 2017; p 31.

Tsalgatidou, A.; Pitoura, E. Business Models and Transactions in Mobile Electronic Commerce: Requirements and Properties. *Comput Netw.* **2001,** *37* (2), 221–236.

Tsimonis, G.; Dimitriadis, S. Brand Strategies in Social Media. *Mark. Intell. Plan.* **2014,** *32* (3), 328–344.

Verhoef, P. C.; Franses, P. H.; Hoekstra, J. C. The Effect of Relational Constructs on Customer Referrals and Number of Services Purchased from a Multiservice Provider: Does Age of Relationship Matter? *J. Acad. Mark. Sci.* **2002,** *30* (3), 202.

Wiedemann, D. G. Exploring the Concept of Mobile Viral Marketing Through Case Study Research. In *MMS;* 2007; pp 49–60.

Wiedemann, D. G.; Haunstetter, T.; Pousttchi, K. In *Analyzing the Basic Elements of Mobile Viral Marketing-an Empirical Study.*7th International Conference on Mobile Business (ICMB'08), July, 2008; IEEE, 2008; pp 75–85.

Yang, Q.; Huang, L.; Xu, Y. Role of Trust Transfer in E-commerce Acceptance. *Tsinghua Sci. Technol.* **2008,** *13* (3), 279–286.

INDEX

E

F

G

H

I

K

M